WELDING
fundamentals

Welding Fundamentals

Mike Gellerman

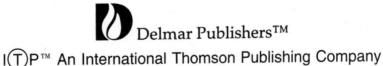

Delmar Publishers™

I(T)P™ An International Thomson Publishing Company

Albany • Bonn • Boston • Cincinnati • Detroit • London • Madrid • Melbourne
Mexico City • New York • Pacific Grove • Paris • San Francisco • Singapore • Tokyo
Toronto • Washington

NOTICE TO THE READER

Cover photos: Courtesy of the American Welding Society.

Delmar Staff

Senior Administrative Editor: Vernon Anthony
Project Editor: Patricia Konczeski
Production Coordinator: Dianne Jensis
Art/Design Coordinator: Cheri Plasse

COPYRIGHT © 1995
By Delmar Publishers
a division of International Thomson Publishing Inc.
The ITP logo is a trademark under license

Printed in the United States of America

For more information, contact:

Delmar Publishers
3 Columbia Circle, Box 15015
Albany, New York 12212-5015

International Thomson Publishing Europe
Berkshire House 168 - 173
High Holborn
London WC1V7AA
England

Thomas Nelson Australia
102 Dodds Street
South Melbourne, 3205
Victoria, Australia

Nelson Canada
1120 Birchmount Road
Scarborough, Ontario
Canada M1K 5G4

International Thomson Editores
Campos Eliseos 385, Piso 7
Col Polanco
11560 Mexico D F Mexico

International Thomson Publishing GmbH
Königswinterer Strasse 418
53227 Bonn
Germany

International Thomson Publishing Asia
221 Henderson Road
#05 - 10 Henderson Building
Singapore 0315

International Thomson Publishing - Japan
Hirakawacho Kyowa Building, 3F
2-2-1 Hirakawacho
Chiyoda-ku, Tokyo 102
Japan

5 6 7 8 9 10 XXX 01 00 99 98

Library of Congress Cataloging-in-Publication Data

Gellerman, Mike.
    Welding fundamentals / Mike Gellerman.
        p.   cm.
    Includes index.
    ISBN 0–8273–5937–3
    1. Welding.   I. Title.
TS227.G43   1994
671.5'2--dc20

                                    94–35393
                                       CIP
                                        AC

Dedication

To welding students and students of welding

Acknowledgments

I would like to thank the following people for their contributions in welding:

John W. King, Green Bay, Wisconsin
Dan Maynard, Indianhead Technical College, Superior, Wisconsin
Larry Jeffus, Garland, Texas

I would also like to thank the following people:

Reynaldo L. Martinez, Jr., Ph.D., Oklahoma State University, Stillwater, Oklahoma, for assistance in the development of unit questions
Judith Mara Riotto, whose copyediting pulled the text together
Susan Geraghty, for scrutiny in the production of the text

Vern Anthony, Senior Administrative Editor at Delmar Publishers Inc., who had foresight for this project

I would also like to thank the staff at Delmar for a quality effort in all aspects involving this book.

Acknowledgments are given to the authors for the use of art from the following books published by Delmar Publishers:

Griffin, Ivan H., Roden, Edward M., Jeffus, Larry, and Johnson, Harold V. 1984. *Welding Processes,* 3rd ed.
Jeffus, Larry. 1980. *Safety for Welders.*
Jeffus, Larry. 1993. *Welding: Principles and Applications,* 3rd ed.
Scharff, Robert A. 1992. *Motor Auto Body Repair,* 2d ed.

Contents

Preface

TO THE STUDENT

I appreciate your seizing the moment to read this Preface. I hope it will be worth your time to understand some of the ideas behind this book.

My goal has been to write a user-friendly textbook on welding. I have tried to take some of the complexity out of the typical welding instructional material and make the subject more easily understood. For example, each unit introduces **new words** that appear in bold type with their explanation in the glossary at the end of the text. A text that frustrates learning can put a damper on the enthusiasm you bring to the first day of class. Every step in the learning process should boost your interest in welding and encourage you to work hard to accomplish even more. I want this book to help fire your enthusiasm for welding throughout the entire program.

I enjoy almost everything that has to do with welding, and I believe you will perform up to your potential if you enjoy welding, too. Anytime you can take pieces of functionless metal and give them purpose, you can have pride in an accomplishment that, in a small way, makes this world a better place to live.

I hope you will take an active part in your own learning. This is one reason for giving you an opportunity in each unit to pose some of your own questions. There is nothing to stop you from going even further and developing additional exercises, too. Finally, I have tried to provide enough information in each exercise so that you can work independently without having to go to your instructor every time something comes up.

Thank you for your attention, good luck, and successful welding.

TO THE INSTRUCTOR

I have designed this book to measure the performance of the tasks laid out in the exercises. I am not presenting anything new in this regard. Welding has been taught by measuring performance for decades, and only recently have some of the other areas of education begun to catch up. What may be new to you, however, is that I have designed this book to include only those exercises necessary to learn the required welding skills. Thus, I have concentrated on 88 exercises that I believe will accomplish the goal of developing skilled welders who can go out to pass the welding test required to get a job. Obviously, you might differ with me on which exercises you feel are important, and that is fine. I have left room on the Progress Sheet for you to include additional exercises. (You will find the Progress Sheet in the Instructor's Guide.)

The exercises are designed to test the student's work at each step. In my own experience I have found that students like to know how they stand in relation to others, and they like to monitor their own progress. In addition, testing every single weld shows students firsthand how well they are doing. A weld that fails the test by not meeting the standards established in the course makes the student more readily open to suggestions regarding what can be done better the next time. There is little argument from a student who has observed a defective weld fail.

I have written this book with the American Welding Society in mind and have tried to be consistent in the use of their terminology. I think it is important to follow their standards so that everyone in the welding community is speaking the same language. I found the writing of this book much easier by referring to some of the standards that the AWS has established. I used their *Structural Welding Code* to develop some of the exercises because it is one of the codes that companies use for testing welders.

The breakdown of steel required for one student to complete all of the exercises one time, without recycling any material, is laid out in the Instructor's Guide. The total weight of the steel for one student is approximately 225 pounds. Of course, your actual needs might vary. For example, not all exercises might be completed, you might provide substitute exercises, and some exercises might be carried out several times before students successfully pass a given test. The In-structor's Guide also offers some suggestions that might be helpful in presenting the information in this text.

Finally, I would like to ask your help in making this book even better. If you have any suggestions for improvement and are the first to write, I will try to acknowledge your contribution the next time around.

UNIT 1

INTRODUCTION

"After you pick up skill, welding gives a tremendous feeling of power and control over the metal."
—Robert M. Pirsig, *Zen and the Art of Motorcycle Maintenance*

GOAL

- Develop an appreciation for welding through a basic understanding of metals

QUESTIONS

- What are some of the physical properties of metal?
- What are some of the ingredients in metals?
- What effect does heat have on metal?
- What is tempering?

THE FUTURE OF WELDING

These are exciting times in **welding** as technology develops and pushes into the future. The National Aeronautics and Space Administration (NASA) has become a symbol of the future and the advancement of technology, and welding has played an important role in the U.S. space program (Figure 1–1). The main deck of the launch pad is made of 1-inch **steel** plate, which is subjected to 7,000° F temperatures at liftoff. To protect the launch pad, another 1-inch steel plate is positioned on top of the main deck. This false deck **cracks** under **stress** after every few flights. Some cracks have measured approximately 20 feet. The **shielded metal arc welding** process, with E7018 **electrodes,** is used to repair the cracks on the A36 steel plate. (A36 is a popular type of steel used in construction. Shielded metal arc welding and E7018 electrodes are covered in Unit 4.)

Space flights have helped to make us aware of how fragile this planet is in the scheme of the universe. The health of the earth will be everyone's concern as technology moves us into the twenty-

FIGURE 1–1 Space Shuttle *Atlantis. (Courtesy of NASA, Kennedy Space Center, Florida.)*

first century. The recycling of materials and supplies used in welding will become very important. Recycling is already quickly becoming a part of everyday life as household plastic, aluminum, glass, and paper products are sorted to be reused (Figure 1–2). New developments in welding will continue to

1

Anatomy of Recyclables

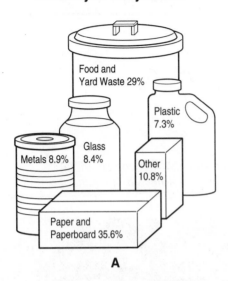

FIGURE 1–2   Recycling: (A) A breakdown of recyclable materials. *(Courtesy of Wheelabrator Frye Technologies, Inc.)* (B) How different materials can be reused. *(Courtesy of Wheelabrator Frye Technologies, Inc.)*

greatly improve the health and safety of the workplace environment (Figure 1–3). Every living creature will benefit in the end from our appreciation of the earth, as will the rocks, the air, and the water.

Job Outlook

The importance of welding will not diminish in the technological world of tomorrow. The skill of welding will still require training and conscientious practice. Although **robots** will replace monotonous jobs on production lines and jobs in specialized or hazardous environments, the skilled **welder** will still be in demand. Skilled welding is much more than fusing two pieces of metal together. Among other things, skilled welders can recognize **quality welds** and can make them. Skilled welders also understand the effects of welding on the **base material.**

There is a world of opportunity waiting for those who work hard to learn welding. Skilled welders will find jobs in transportation, for example in the production of trucks, trains, ships, and airplanes; in the building of **steel** structures, such as skyscrapers and bridges; in the construction of earthmoving equipment, such as bulldozers, shovels, and graders; and in agriculture and logging, manufacturing tractors and specialized equipment like skidders and combines. This list does not include

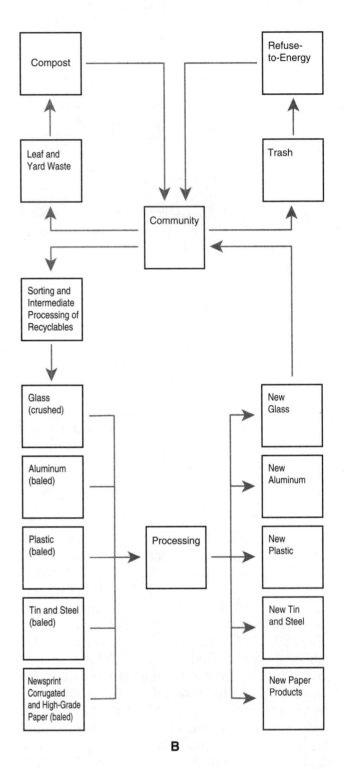

the welded manufactured products used in the home that make living so much easier.

And if manufacturing is not enough, so many products can be repaired by welding. Skilled welders are needed to maintain and rebuild equipment to save the energy costs that would be needed to replace this equipment (Figure 1–4). Auto body shops perform repair welding, and marinas retrofit

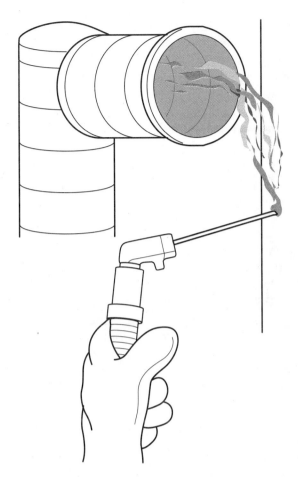

FIGURE 1–3 Exhaust system.

boats. Maintenance welding of equipment plays an important role in industries like mining, manufacturing, construction, agriculture, and transportation (Figure 1–5). Equipment that is repaired does not have to be replaced.

Those who learn this skill of welding will gain self-confidence. The knowledge and ability that come from learning a skill like welding will also make them contributing members to the growth of society. Technology needs people who want to be a part of building our industries as we move into the next century. Welding will contribute to that growth.

THE DEVELOPMENT OF WELDING

Welding is not new. Depending on the definition, welding can be traced back 5,300 years with the discovery of the remains of the "Iceman" in the Alps. His ax was made of formed **copper,** still razor sharp (Figure 1–6.) The Middle Ages brought

FIGURE 1–4 Welder at work. *(Courtesy of R.H. Blake, Inc., Cleveland, Ohio.)*

FIGURE 1–5 Equipment. *(Courtesy of Amoco Corporation and Caterpillar Inc.)*

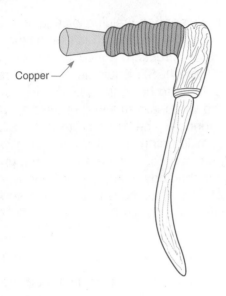

Copper

FIGURE 1–6 Ancient copper ax.

blacksmithing to a level of artisanry never surpassed in the manufacture of armor and weaponry (Figure 1–7). Modern welding really began in the

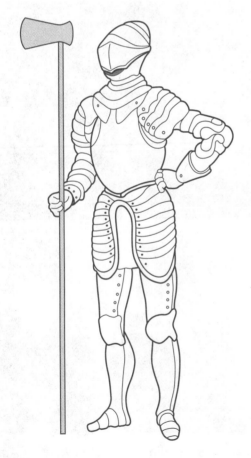

FIGURE 1–7 Suit of armor.

nineteenth century during the Industrial Revolution, when the development of structural iron was combined with the development of electric power. Construction using iron and steel created a need for new methods of joining these new metals. Gas and electric welding were the result.

Welding developed and grew in the twentieth century. In the early 1900s, it was discovered that bare electrodes could be improved by the addition of a **covering** to shield the **weld pool** from air contamination. **Arc welding** became more important for manufacturing in the 1920s, when an **extrusion process** made covered electrodes less expensive. Up until this breakthrough, each electrode had to be hand-wrapped.

Welding technology continued to develop into the 1930s, when **submerged arc welding** and other arc welding processes came on the scene using a continuous **wire** fed from spools. Continuous wire increased the manufacturing output because more welding could be completed without stopping. The addition of a gas shielding to protect the weld pool made welding possible without additional time spent to remove the **slag.** In the 1980s, the popularity of continuous wire welding processes grew. This pushed the volume of sales for spool wire ahead of the covered electrodes used in the shielded metal arc welding process.

The 1990s have brought computer technology to welding, producing lightweight **welding machines,** that can be easily converted to one of several welding processes under the various electric **input** voltages (208, 230, 460) using **single-phase power** or **three-phase power.** Developments in robotics continue in manufacturing, where quality welding is required for cost-saving, tedious, and dangerous work.

From applications that put us into space, today's welding has application in the laboratory. In antiseptic environments, technicians in white smocks perform **laser beam welding** and **electron beam welding** with equipment costing many thousands of dollars. Welding not only enhances our lives practically, but emotionally, too, through the creative work of artists (Figure 1–8). Advancements in space-age technology hold promise for yet unknown methods of welding.

THE WELDED JOINT

Welding begins with the simple idea of joining two pieces of metal together in a **joint.** Five basic

FIGURE 1–8 The art of welding applied at Walt Disney World's Swiss Family Robinson Treehouse. *(Courtesy of Wendy Jeffus, in Jeffus, 1988.)*

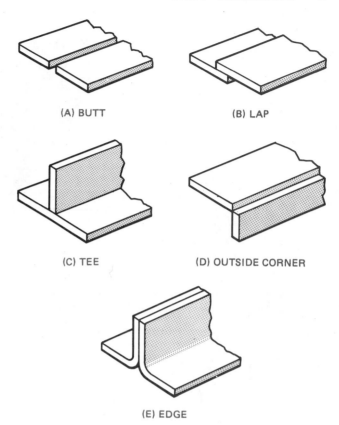

FIGURE 1–9 The five basic joints used in welding: (A) Butt joint. (B) Lap joint. (C) Tee joint. (D) Corner joint. (E) Edge joint. *(Courtesy of Jeffus, 1993.)*

joints are used in welding. Whenever two pieces of metal are joined together, they fit into one of these five arrangements. These five basic joints are the foundation on which every welded project is built: the **butt joint,** the **tee joint,** the **lap joint,** the **corner joint,** and the **edge joint** (Figure 1–9).

The selection of the **joint design** to be used in welding application is very important. A **welding engineer** is usually responsible for the joint design, though this responsibility could belong to the welder. Careful consideration must be given to the selection of the joint.

A butt joint may be selected for lengthening a steel beam. A tee joint may be used for attaching a support girder at a 90° angle to a beam. A lap joint would be the result of welding one **pipe** that is telescoping from inside another pipe. A corner joint may be used when welding up a metal container, and the edge joint is commonly used on thin metal where **melt-thru** and **distortion** can cause problems.

Preparing metal for welding may amount to nothing more than the use of a grinder to remove **rust** or **mill scale** from the metal. At other times, the thickness of the metal could require **beveling** the pieces of the joint to ensure complete **root penetration.** Beveling removes metal along the end of the pieces to be joined (Figure 1–10). The grinder and the **oxyacetylene cutting** torch are

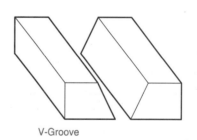

V-Groove

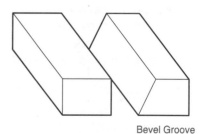

Bevel Groove

FIGURE 1–10 Beveled joints.

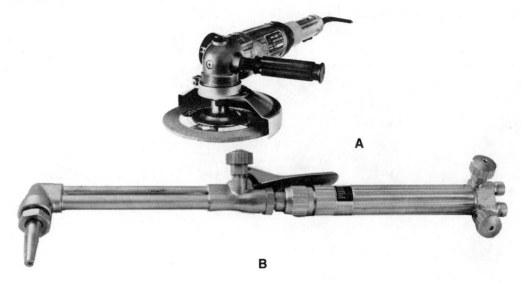

FIGURE 1–11 Two tools for beveling plate: (A) Grinder. *(Courtesy of Scharff, 1992.)* (B) Cutting torch. *(Courtesy of Griffin, Roden, Jeffus, and Briggs, 1984.)*

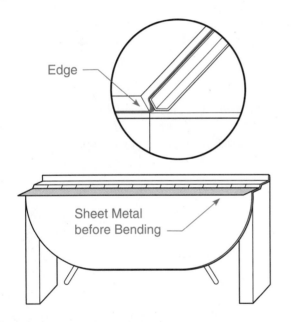

FIGURE 1–12 Box and pan brake used to put a lip on sheet metal.

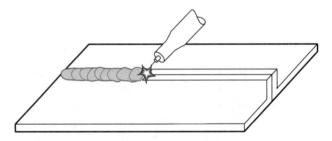

FIGURE 1–13 Edge joint.

two methods among many used for beveling plate (Figure 1–11).

The edge joint can be prepared by bending a lip on a piece of **sheet metal,** using mechanical force like that provided by a box and pan brake (Figure 1–12). Putting an edge on thin metal can help to minimize distortion during welding (Figure 1–13).

THE PRINT

Before beginning the welding exercises, students are given instructions. The instructor may give general instructions and use the welding exercise to supply more detailed instructions. Hand-drawn sketches are sometimes used to manufacture a part or object, but **blueprints** are the formal method generally used to communicate the detailed instructions of how a part is to be fabricated and welded together. The information given on a blueprint includes the type of joints to be welded, the **welding symbols,** and much more. The welding symbol can be very important, explaining how each joint is to be welded (Figure 1–14). The completed part is the result of having followed the instructions (Figure 1–15).

WELDING PROCESSES

The **American Welding Society** (AWS) is an organization dedicated to the advancement of welding. According to the American Welding Society, welding is defined as any **process** by which

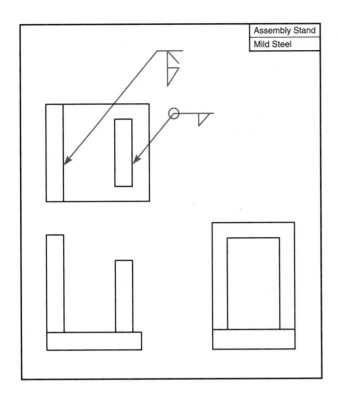

FIGURE 1–14   Print with a welding symbol.

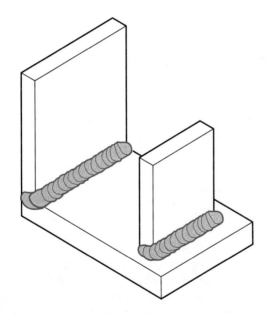

FIGURE 1–15   Isometric drawing of completed part.

materials are joined together by heat. This organization has a major role in setting standards that are followed throughout the welding industry. The American Welding Society lists more than 90 major welding and related processes (Table 1–1). A process is a method used to join or cut metal. For

example, **oxyacetylene welding** is one welding process; **air carbon arc cutting** is another related process used to prepare metal for welding. This book covers in detail some of the popular cutting, **semiautomatic,** and **manual** welding processes that require the skill of people who can make quality welds.

METAL

Metal is the material joined by welding. Metal can have numerous structural shapes with unlimited possibilities. Metal can be light or heavy, hard or soft, to name four of its many characteristics. Metal has probably played a greater role in the development of civilization than any other material.

Metal Production

Metal production begins with **iron ore** and **bauxite ore,** two materials that are mined and shipped to mills, where they are refined (Figures 1–16 and 1–17). Iron ore is used in making steel, and bauxite ore is used in making **aluminum.** To give a piece of metal the label *steel* or *aluminum* is to begin a general form of communication. Unfortunately, labels tell us very little about these metals because not all steel or aluminum is the same. The **types** of steel and aluminum produced depend on the needs of the customer. There are hundreds of types of steel and aluminum on the market.

The many types of metal are the result of requirements determined by the needs of the customer. **Strength, corrosion resistance, machinability, hardness,** and **weldability** are only five of the **physical properties** determined by the manufacture of steel and aluminum. For example, a commercially pure aluminum is highly weldable but does not have the strength required for a bicycle frame. **Magnesium** is one ingredient added to aluminum for increased strength. The addition of sufficient amounts of magnesium and **silicon** changes both the grouping into which the aluminum fits and its physical properties. For example, 6061 aluminum (mixed with 1% magnesium and 0.6% silicon) is strong enough to be used in boats, furniture, and bicycle frames.

The addition of ingredients to a batch of steel or aluminum determines its physical properties (Figure 1–18). Silicon, **zinc,** and **manganese** are three elements added as ingredients to make a

TABLE 1–1:   PROCESS WHEEL: 93 WELDING AND ALLIED PROCESSES

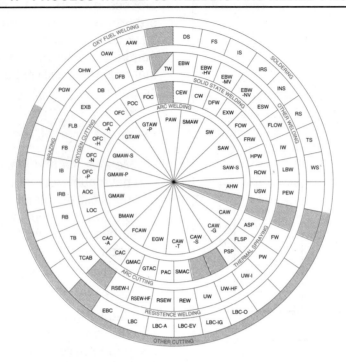

Arc welding (AW)		Plasma arc welding	PAW
Atomic hydrogen welding	AHW	Shielded metal arc welding	SMAW
Bare metal arc welding	BMAW	Stud arc welding	SW
Carbon arc welding	CAW	Submerged arc welding	SAW
Carbon arc welding—gas	CAW-G	Submerged arc welding—series	SAW-S
Carbon arc welding—shielded	CAW-S	**Solid state welding (SSW)**	
Carbon arc welding—twin	CAW-T	Coextrusion welding	CEW
Electrogas welding	EGW	Cold welding	CW
Flux cored arc welding	FCAW	Diffusion welding	DFW
Gas metal arc welding	GMAW	Explosion welding	EXW
Gas metal arc welding—pulsed arc	GMAW-P	Forge welding	FOW
Gas metal arc welding—short circuiting arc	GMAW-S	Friction welding	FRW
Gas tungsten arc welding	GTAW	Hot-pressure welding	HPW
Gas tungsten arc welding—pulsed arc	GTAW-P	Roll welding	ROW

Cont. on next page

FIGURE 1–16   Iron ore mine.

required type of steel or aluminum. Steel and aluminum are only two metals to consider, though. **Cast iron,** copper, **brass,** and **stainless steel** are four more. Changing the ingredients in the recipe changes the physical properties of the metal (Figure 1–19). As a result, some metals are easily welded, while other metals are not as easily welded. In addition, some types of metal cannot be practically welded at all.

Choosing a Metal

The selection of metal for a particular application is partly based on the physical properties that

TABLE 1–1: PROCESS WHEEL: 93 WELDING AND ALLIED PROCESSES (CONTINUED)

Ultrasonic welding	USW	Flow welding	FLOW
Soldering (S)		Induction welding	IW
Dip soldering	DS	Laser beam welding	LBW
Furnace soldering	FS	Percussion welding	PEW
Induction soldering	IS	Thermit welding	TW
Infrared soldering	IRS	**Oxyfuel gas welding (OFW)**	
Iron soldering	INS	Air acetylene welding	AAW
Resistance soldering	RS	Oxyacetylene welding	OAW
Torch soldering	TS	Oxyhydrogen welding	OHW
Wave soldering	WS	Pressure gas welding	PGW
Resistance welding (RW)		**Thermal spraying (THSP)**	
Flash welding	FW	Arc spraying	ASP
Projection welding	PW	Flame spraying	FLSP
Resistance seam welding	RSEW	Plasma spraying	PSP
Resistance seam welding—high frequency	RSEW-HF	**Oxygen cutting (OC)**	
Resistance seam welding—induction	RSEW-I	Flux cutting	FOC
Resistance spot welding	RSW	Metal powder cutting	POC
Upset welding	UW	Oxyacetylene gas cutting	OFC-A
Upset welding—high frequency	UW-HF	Oxyfuel gas cutting	OFC
Upset welding—induction	UW-I	Oxygen arc cutting	AOC
Brazing (B)		Oxygen lance cutting	LOC
Block brazing	BB	Oxyhydrogen cutting	OFC-H
Diffusion brazing	DFB	Oxynatural gas cutting	OFC-N
Dip brazing	DB	Oxypropane cutting	OFC-P
Exothermic brazing	EXB	**Arc cutting (AC)**	
Flow brazing	FLB	Air carbon arc cutting	CAC-A
Furnace brazing	FB	Carbon arc cutting	CAC
Induction brazing	IB	Gas metal arc cutting	GMAC
Infrared brazing	IRB	Gas tungsten arc cutting	GTAC
Resistance brazing	RB	Plasma arc cutting	PAC
Torch brazing	TB	Shielded metal arc cutting	SMAC
Twin carbon arc brazing	TCAB	**Other cutting**	
Other welding		Electron beam cutting	EBC
Electron beam welding	EBW	Laser beam cutting	LBC
Electron beam welding—high vacuum	EBW-HV	Laser beam cutting—air	LBC-A
Electron beam welding—medium vacuum	EBW-MV	Laser beam cutting—evaporative	LBC-EV
Electron beam welding—nonvacuum	EBW-NV	Laser beam cutting—inert gas	LBC-IG
Electroslag welding	ESW	Laser beam cutting—oxygen	LBC-O

make one metal different from another. Two physical properties easily recognized are color and weight. The melting temperature of metal is another physical property that is not so easily determined.

The price or cost of metal is another determining factor, but the price of metal is not a physical property. The immediate cost can be quickly acquired with a phone call to a local metal supplier. Cost is determined by the marketplace, where companies compete with each other.

Hardness is another physical property of metal. Hardness is the resistance a material has to penetration. Hardness can be judged with the help of a file, but more accurate tests are also available. The Equotip portable hardness tester measures the hardness of a metal by recording impact and rebound velocities of a carbide or diamond tip impelled by the force of a spring off the metal (Figure 1–20). A monitor provides the necessary hardness information with comparison graphs for equivalent values of hardness according to the **Brinnell, Vickers,** or **Rockwell C** standard.

The hardness of metal is important because the hardness number corresponds to a **tensile strength** number. Tensile strength is the resistance of a material to being pulled apart.

Strength is the ability of metal to resist **strain** (deformation). The more resistant a metal is to penetration, the greater its hardness, which can

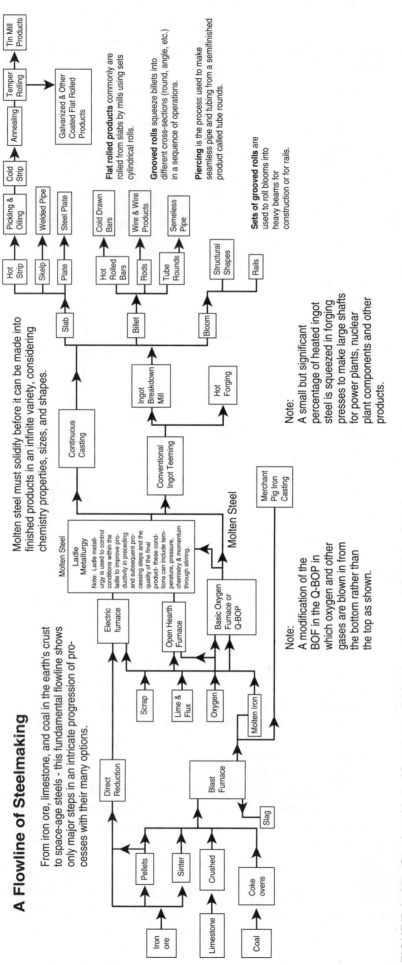

FIGURE 1–17 Steel making. *(Courtesy of American Iron and Steel Institute.)*

FIGURE 1–18 Alloying elements being added to a batch of steel. *(Courtesy of Jeffus, 1993.)*

FIGURE 1–19 Ingredients of a master chef.

indicate a higher tensile strength. Hardness is the one physical property that all metals have. Physical properties like hardness are important in the selection of metal for manufacturing any product.

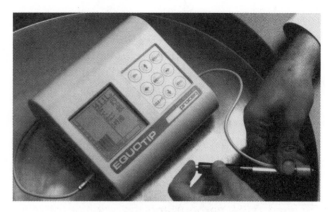

FIGURE 1–20 Equotip portable hardness tester. *(Courtesy of Foerster Instruments, Inc.)*

It could happen that a welder will be in the position of making the decisions: selecting the metal to be welded, deciding on the joint design, and choosing the welding process. In this case, knowledge of how a metal's physical properties are affected by welding becomes very important. Making the wrong decision could cause a welded joint to fail or could limit the useful life of the product.

CARBON STEELS

The amount of **carbon** in steel, determined during manufacturing at the steel-making plant, helps to determine its hardness. Hardness affects the strength of a steel. When carbon is increased, the hardness and strength increase also, but both the **ductility** (ability to bend or form) and the weldability decrease. **Low-carbon steel** contains 0.1% to 0.3% carbon (this is less than 1%). **Medium-carbon steel** contains 0.3% to 0.45% carbon. **High-carbon steel** contains 0.45% to 0.65% carbon, and **very high carbon steel** contains 0.65% to 1.5% carbon. Cast iron contains up to 4.5% carbon, though percentages may vary. The ductility and weldability of very high carbon steel and cast iron are not as great as those of low-carbon steel.

Low-carbon steel is also called **mild steel.** It is commonly welded, finding application in construction and fabrication. The greatest tonnage of steel manufactured is low-carbon steel. Because low-carbon steel cannot be hardened by heat treatment alone, it does not have application for tempered tools like punches and chisels. Unlike medium-carbon steel, low-carbon steel does not require additional **preheating** or **postheating** procedures to guard against weld failure due to the heat of welding. Thus, low-carbon steels are more easily welded than steels containing greater amounts of carbon.

Welding adds heat to every joint. As the amount of carbon within a batch of steel increases the hardness of that steel, there is an increased possibility of cracking as a result of the heat from welding. Generally speaking, the additional carbon (beyond 0.3% carbon) and the addition of other ingredients to a batch of steel can change the application of welding. By not using proper **welding procedure,** a medium-carbon steel can become **brittle** and crack from the heat of welding, whereas a low-carbon steel would not be affected using the very same welding procedure.

If a polished slice of steel is examined under a microscope, its **grain structure** can be observed. Like fingerprints, the grain structure separates one type of steel as different from other types of steel. Unlike fingerprints, however, the grain structure can be altered. The temperature produced by welding and the rate of cooling can produce changes that will help to determine the ultimate size of the grains.

A slice from a section of medium-carbon steel (0.3% to 0.45% carbon) taken out of the **heat-affected zone** (HAZ)—the area next to the weld—from a set of steel plates that were preheated 200° to 400° F before welding would show a grain structure consisting of **ferrite** and **pearlite** (Figure 1–21). A slice from another section of medium-carbon steel that was *not* preheated before welding would show a **martensite** grain structure (Figure 1–22). Martensite is the microstructure that forms when steel cools too rapidly. The much harder martensite is brittle enough to result in a failure of the joint in the heat-affected zone (Figure 1–23).

Martensite forms in steel when the cooling rate of a welded joint is so fast that a pearlite microstructure is unable to form. It would be like throwing the joint into water immediately after welding. In this case, however, the metal itself acts like water to quickly cool the welded joint. Preheating helps to slow the rate of cooling to prevent the formation of martensite when martensite is undesirable. In low-carbon steels (to 0.2% carbon) the cooling rate is usually not fast enough to let martensite

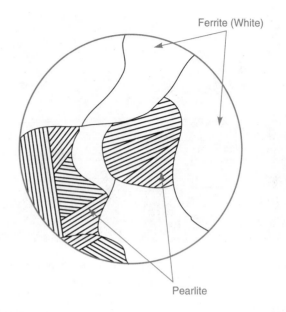

FIGURE 1–21 Ferrite and pearlite in heat-affected zone, preheated before welding.

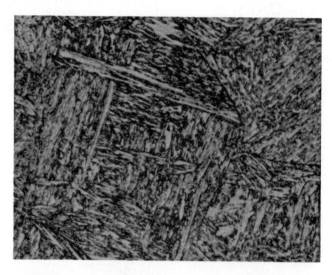

FIGURE 1–22 Martensite in heat-affected zone, not preheated before welding. *(Courtesy of Oak Ridge National Laboratories.)*

form. Thus, preheating is usually not necessary before welding most low-carbon steels.

Low-carbon steels have application in the structural members used for constructing skyscrapers and bridges. Medium-carbon steels are used in pressure vessels, axles, and machine parts. High-carbon steels find use in farm implements and some cutting tools. Unfortunately, high-carbon steels have poor weldability to greatly increase the possibility of cracking in the weld area.

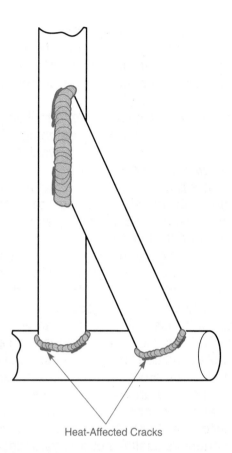

Heat-Affected Cracks

FIGURE 1–23 Failure in heat-affected zone, not preheated before welding.

1020 and 1025 are two examples of low-carbon steel. 10-- is the designation given by the **American Iron and Steel Institute** (AISI) and **Society of Automotive Engineers** (SAE) to carbon steels. The designations --20 and --25 indicate the percentage of carbon content—0.20% and 0.25% in these examples. 1040 and 1070 are two more types of carbon steel. 1040 is a medium-carbon steel with 0.40% carbon, and 1070 is a high-carbon steel with 0.70% carbon.

ALLOY STEELS

The American Iron and Steel Institute defines an **alloy steel** as (1) a steel in which one of the amounts: 1.65% manganese content, 0.60% silicon content, or 0.60% copper content are exceeded, or (2) a steel to which a definite minimum quantity of aluminum, **boron, chromium** (up to 3.99%), or a minimum quantity of any other alloying element is added to achieve a desired result.

Alloy steels are another group of steels that can be welded, but they require special attention, such as preheating or postheating procedures or the careful choice of welding **filler metals.**

The content information of metals is available in the product handbooks furnished by metal suppliers. For example, 4130 is an aircraft-quality (structural tubing) chromium-**molybdenum** steel with 0.30% carbon that requires a specified filler metal. Preheating and postheating treatment is necessary to improve **toughness** and to avoid cracking. 4130 is a **low-alloy steel** because it contains in the range of 0.80% to 1.10% chromium (Table 1–2).

A **high-alloy steel** is a steel in which chromium, manganese, or nickel content equals 12% or better. Notice the difference in these elements compared with low-alloy steels. Stainless steels are examples of high-alloy steels because their chromium content exceeds 12%. For example,

TABLE 1–2: AISI AND SAE NUMERICAL DESIGNATIONS FOR STANDARD LOW-ALLOY STEELS

Designation	Low-Alloy Steel
1330	Manganese, 1.60–1.90%
4130	Molybdenum, 0.15–0.25%; chromium, 0.80–1.10%
4320	Nickel, 1.65–2.00%; chromium, 0.40–0.60%; molybdenum, 0.20–0.30%
4620	Nickel, 1.65–2.00%; molybdenum, 0.20–0.30%
4720	Nickel, 0.90–1.20%; chromium, 0.35–0.55%; molybdenum, 0.15–0.25%
5140	Carbon, 0.38–0.43%; chromium, 0.70–0.90%
6150	Chromium, 0.80–1.10%; vanadium, 0.15% minimum
8615	Nickel, 0.40–0.70%; chromium, 0.40–0.60%; molybdenum, 0.15–0.25%
8720	Nickel, 0.40–0.70%; chromium, 0.40–0.60%; molybdenum, 0.20–0.30%
8822	Nickel, 0.40–0.70%; chromium, 0.40–0.60%; molybdenum, 0.30–0.40%
9259	Manganese, 0.75–1.00%; silicon, 0.70–1.10%; chromium, 0.45–0.65%

308 stainless steel has 19% to 21% chromium, 10% to 12% nickel, and a small amount (2%) of manganese (Table 1–3).

CAST IRON

The amount of carbon in steel is small in proportion to the amount of carbon in cast irons. There are four basic types of cast iron: **white, gray, malleable,** and **nodular.** White cast iron is not readily welded. Gray cast iron is weldable but should be kept cool by short welds cooled to the touch, **peening** to relieve stress. Gray cast iron can also be welded after preheating, then cooled slowly. Large gray iron castings, placed in a bed of coals or a temporary oven to preheat, can be welded in place.

Malleable and nodular (ductile) cast irons require heat treatment after welding to bring back their original physical properties. **Braze welding** can be used on both malleable and gray cast irons to control heat input without preheating or postheating. If the heat is properly controlled during the braze welding process, stress will not be enough to crack the casting.

White cast iron is used in boxcar wheels, rollers, and dies. Gray cast iron has application in woodstoves, motor blocks, cylinder heads, and brake drums. Malleable cast iron is used in shock impact tools, pipe, and valve parts, and nodular cast iron has application in pressure valves and pipes. Although brittleness is one physical property of some cast irons, toughness is another physical property as these applications indicate. Keep in mind that how a metal is used can be a very good indication of what the metal is. This helps the welder determine the welding procedure to follow in a repair.

HEAT

All metals melt at different temperatures. Temperature is a very important condition of welding.

The heat produced during welding becomes even more critical with higher amounts of carbon in the steel. The heat of welding brings changes in both the properties of the steel being welded and the properties of the weld itself.

Although the temperature in a molten pool of metal can approach 3,000° F, the temperature of the base metal (the metal being welded) a short distance away could be only 400° F. The cooling of metal as it is being welded can affect the physical properties of the metal. Remember that martensite forms when carbon steels are cooled rapidly from the range of 400° to 600° F.

Rapid cooling can result in carbon steels with greater hardness but less ductility. Slow cooling can result in carbon steels with less hardness but more ductility. Greater hardness can cause a brittle condition with possible cracking—an undesirable result of welding. However, hardness can also be a desirable condition in the manufacture of cutting tools and other products that require hardened surfaces. Less hardness can also lead to the premature failure of parts. Although the formation of martensite as a by-product of welding is undesirable, martensite itself is not always undesirable.

The cooling of a weld is affected by many factors, including the welding procedure and the physical properties of the metal being welded. The welding procedure consists of, among other things, the welding process, the selection of filler metal, the welding current, the number of passes, the preheating or postheating requirements, if any, and the joint design.

The physical properties of metal include the kind of metal being welded: Is it steel, aluminum, or something else? The type of steel or aluminum is determined by the ingredients that determine the tensile strength. For example, the amount of carbon in steel or the amount of magnesium in aluminum determines tensile strength. The thickness of the base metal is another physical property that can influence the choice of welding process. An aluminum welded joint will cool much more quickly than a steel welded joint of the same thickness. More heat has to be generated during welding to maintain enough heat to melt aluminum; the procedure may require preheating the joint before welding.

Finally, welding causes expansion by heating and contraction by cooling. Whereas cooling can

TABLE 1–3:	NUMERICAL DESIGNATIONS FOR STAINLESS STEELS	
Series	Alloy Elements	Characteristic
200	Chromium-nickel-manganese	Nonmagnetic
300	Chromium-nickel	Nonmagnetic
400	Chromium-molybdenum	Magnetic

create stress on a welded joint, controlled cooling can help to reduce stress and prevent strain. Strain, a deformation of the metal due to stress, can lead to weld failure. Experience and understanding of what is happening during welding will aid the skilled welder. Understanding the effects of heat enables welders to use it to their advantage.

Heat Treatments

A variety of **heat treatments** are used in many different applications. The focus in this unit is on **hardening** and **tempering** tools made of carbon steel.

Hardening can be accomplished by first raising the temperature of the steel above the **critical range** and then cooling rapidly. The critical range is the peak temperature at which a rapid rate of cooling results in the formation of martensite (an oxyacetylene torch or an oven can be used). Hardening tools like a chipping hammer or a cold chisel can be accomplished by heating the tool's taper end to a cherry red (1,300° to 1,425° F, depending on carbon content) and **quenching** it immediately in water. The tool should be moved about in the water until the steaming stops.

The martensite formed during hardening may be too hard for most applications. The tool would then require a second hardening heat treatment, just below the critical range this time, again quenching the tool immediately in water. This second heat treatment results in a softer microstructure combining with the harder martensite. This treatment may be necessary on the mushroomed end of a chisel, for example, where dangerous pieces of steel might break off.

Tempering is a process by which hardened steel is toughened. Once the tapered end of a tool has been hardened, the scale formed on the metal is ground or filed off the tapered end. A somewhat polished surface makes it easier to see the necessary color changes if using a torch; within an oven the temperature can be controlled. The tool is heated above the polished taper, and the color changes are noted, watching for a change from blue to purple on the polished surface. As purple moves to the edge of the taper, the tool is quenched immediately. The purple color that indicates the toughened edge forms around 525° F (Figure 1–24).

STRUCTURAL SHAPES AND SPECIFICATIONS OF METAL

Metal from the mill comes in a variety of shapes and sizes (Figure 1–25). Choosing the structural shape of metal is important in design. The quality of the **weldment** may depend on the shape of the metal being welded. A weldment is an assembly of welded parts.

Metal suppliers usually publish catalogs that list the structural shapes they carry as well as information on **material specifications.** Material specifications aid the customer in making a purchase. These specifications include dimensions such as length and width and measurements such as weight per foot. The weight per foot of structural metal can help to determine the final weight of a welded structure without going to the trouble of using a scale. Just add up the total length in feet of all the metal used in the structure by shape, and multiply each by its weight per foot.

PHYSICAL PROPERTIES OF METAL

Metal is characterized by its physical properties. The general strength of a metal is its resistance to mechanical forces without breaking. The physical properties of a metal determine its use. Each property is measured against a mechanical force. The following is a list of some properties that all metals have.

Tensile strength: Resistance to being pulled apart.
Impact strength: Resistance to sudden force without fracture.
Fatigue strength: Resistance to changing forces or **loads.**
Hardness: Resistance to penetration or denting.
Compression: Resistance to being crushed.
Toughness: Resistance to failure from a constant force.
Ductility: Resistance to deformation after stretching.

As already discussed, welding can change these physical properties of a metal, with a good

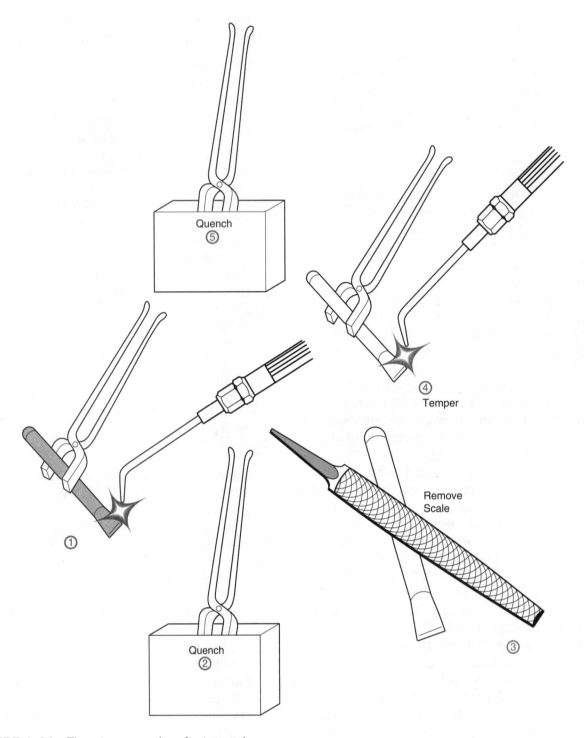

FIGURE 1–24   Five-step procedure for tempering.

or a bad result. For example, proper heating and cooling can increase the hardness of some metals, while an improper welding procedure can cause brittle kinds of cast iron to crack. Knowing the physical properties of the base metal helps in choosing the welding process and in selecting the filler metal (metal added to a joint).

Skilled welders understand welding and its effects on the weldment are valuable. The techno-

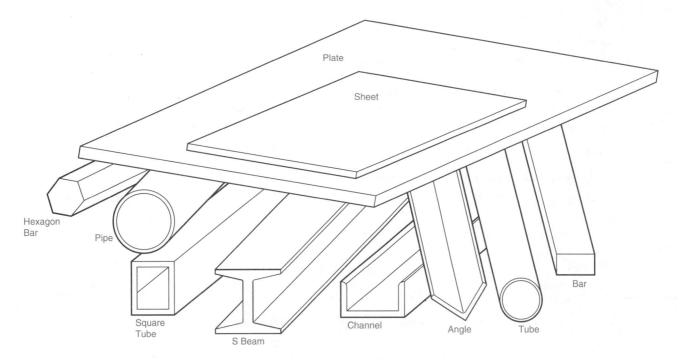

FIGURE 1–25 A variety of structural shapes of metal.

logical world of the twenty-first century will need welders with the skill and the knowledge of welding. Do not be afraid to spend some time reviewing this unit again, especially after completing the other units of this book.

REVIEW

1. What is welding?
2. Describe the outlook for jobs in welding.
3. What is skilled welding?
4. Name the five basic welded joints.
5. What is the American Welding Society?
6. Name one welding process other than oxyacetylene welding.
7. What metal is made from iron ore? From bauxite ore?
8. What other jobs, besides welding, could a welder be required to perform?
9. How does carbon change or affect steel?
10. Name three material specifications.

Create Three Questions

1.
2.
3.

Related Math and English Questions

1. Which number in each of the following pairs is larger?
   a. 1% or 0.1
   b. 2% or 0.45
   c. 3% or 0.035
   d. 4% or 0.45
2. Write a paragraph using the following terms: *strength, hardness, strain, carbon, steel, ductility.*
3. Explain the connection between martensite and cooling.

For Further Thought

1. What can be done to make welding more environmentally friendly?
2. Which of the five basic joints is used most often in automobile construction? Explain.
3. Are the standards established by the American Welding Society a good idea? Why or why not?
4. Aluminum is lighter than steel, which gives it some advantages. If you were building a bicycle frame, could aluminum tubing with the same dimensions (thickness and diam-

eter) as steel be used as a substitute for steel? Explain.

5. Imagine you are doing a repair job. Without using a file or torch and without even touching the broken part, how can you determine whether the metal is steel or cast iron?

SUGGESTED ACTIVITIES

1. Write a paper covering some aspect of welding. Use the library for research.

2. Compare the physical properties of different metals—weldability, hardness, strength, and so on.

3. Heat metal for color, and based on a chart, determine its temperature.

4. File different kinds of metal to check for hardness.

5. Bend a paper clip, and notice what happens.

6. Use 1045 bar steel to make cold chisels.

7. Try to identify different kinds of metal.

OXYACETYLENE WELDING (OAW)

"When I show it to him he nods and slowly goes over and sets the regulators for his gas torch. Then he looks at the tip and selects another one. Absolutely no hurry. He picks up a steel filler rod and I wonder if he's actually going to try to 'weld' that thin metal."

—Robert M. Pirsig, *Zen and the Art of Motorcycle Maintenance*

GOAL

- Develop an ability to do oxyacetylene welding safely, to make the mastery of other welding processes easier.

QUESTIONS

- What is oxyacetylene welding?
- What equipment is required?
- How does the equipment work?
- What does the weld pool look like?
- What is a quality weld?

SAFETY FIRST

Safety will be given particular attention throughout this book. Understanding safety begins with **commonsense** practices. Remember that if you do not watch out for yourself, you are leaving your safety to others. Accidents happen when you are not in control. You put yourself, the work environment, the equipment, and other people in jeopardy.

OXYACETYLENE PROCESSES

Oxyacetylene welding is popular in maintenance **welding.** However, because other welding processes are faster, the oxyacetylene **torch** is generally used for cutting **steel.** This unit covers oxyacetylene welding because of its importance in helping the beginning student learn welding (Figure 2–1).

The end product of oxyacetylene welding is the same as that of any other **arc welding** process. An important difference is that just two knobs on the oxyacetylene torch body control the heat input; arc welding machines have an assortment of switches and dials to do exactly the same thing. Because oxyacetylene welding is slower than other types of welding, it is easier to observe what is happening within the **weld pool** of molten metal. By turning the knobs to adjust the volume of heat, the **welder** can take immediate action to make any necessary corrections. The welder can quickly make the connection between the volume of heat and the size of the weld pool. As you will discover, there is nothing mysterious about watching a flame slowly create a weld pool of molten metal.

Oxygen

The **oxygen** required for oxyacetylene welding can be delivered in different ways. One way is to have the oxygen piped into the welding area from a large cylinder of liquid oxygen or a **cylinder manifold** (several cylinders connected together) in the gas storage area. One cylinder of liquid oxygen, 6 feet high and 1 1/2 feet in diameter, can deliver 3,000 cubic feet of oxygen or the equivalent of 12 K-size cylinders (244 cubic feet) of oxygen. Liquid oxygen is becoming popular because its convenience saves time and effort (Figure 2–2).

The second way to deliver oxygen is a portable high-pressure cylinder. This is the most readily

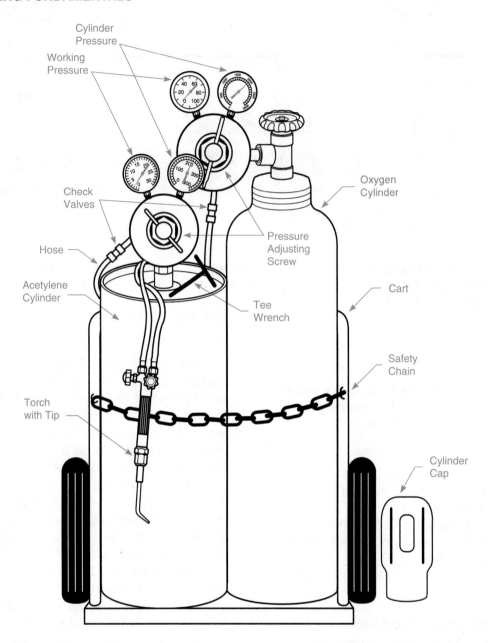

FIGURE 2–1 Oxyacetylene welding equipment.

recognizable way of delivering oxygen to the workplace. Oxygen (never say "air") is stored in different-size seamless cylinders that have been heat-treated and tested before going into operation. The oxygen in these cylinders is under normal pressure of 2,200 **psi** (pounds per square inch), measured at 70° F. These cylinders should never be moved without their protective caps in place (Figure 2–3). These caps avoid the possibility of damage to the **valve** on the cylinder during transport. The sudden release of unregulated oxygen pressure can propel a cylinder like a missile that is capable of going through concrete walls.

Acetylene

Acetylene is a gas compound of **hydrogen** and carbon. When water is added to calcium carbide, the gas is released. For industrial purposes, acetylene is stored in containers that are not as highly pressurized as those required for oxygen. A full cylinder of acetylene has a normal pressure of 250 psi, measured at 70° F.

These cylinders are filled with porous **calcium silicate** to absorb **acetone.** Acetone is a liquid chemical used to stabilize acetylene gas. Free acetylene must never be allowed to accumulate at

FIGURE 2–2 Liquid oxygen cylinder.

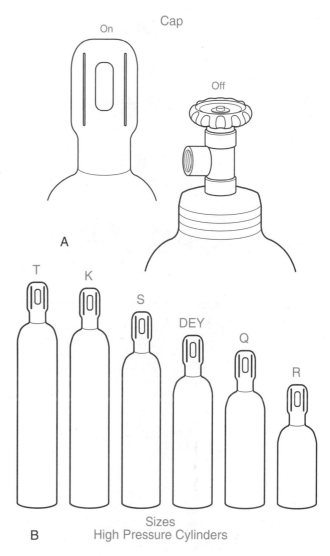

FIGURE 2–3 (A) Cylinders with the caps on and off. (B) High-pressure cylinders of various sizes.

pressures beyond 15 psi. At pressures greater than 15 psi, acetylene becomes unstable, creating the possibility of an explosion. *All* gas cylinders should be stored in the upright position, but especially acetylene. A leak at the cylinder valve on an acetylene cylinder laying on its side might allow acetone to flow out, causing acetylene to become unstable.

Cylinder Valves

There are different styles of valves. The oxygen cylinder valve must be opened completely to seat. The acetylene cylinder valve may be opened by hand, though most require a special wrench. The acetylene valve need not, and should not, be opened completely; one full turn is enough (Figure 2–4).

A

B

FIGURE 2–4 Cylinder valves: (A) Acetylene. (B) Oxygen. *(Photos courtesy of Jeffus, 1993.)*

OXYACETYLENE WELDING EQUIPMENT

Oxyacetylene welding requires its equipment be in good working order. Proper maintenance should keep the equipment performing up to its capability. Understanding how each piece of equipment works should aid in the welding.

Regulator

A **regulator** on a cylinder controls the release of gas under pressure. Regulators used with oxygen and acetylene should have two gauges. One

gauge gives cylinder pressure. The second provides a reading of line pressure in the hose. Some manufacturers mark the danger zone beyond 15 psi in red on the working-pressure gauges for acetylene (Figure 2–5).

There are two basic types of regulators. The **single-stage regulator** is very popular. It reduces the pressure coming from the cylinder in one step. The **two-stage regulator** reduces the pressure coming from the cylinder in two steps. The two-stage regulator does not require any readjustment of line pressure as cylinder pressure drops during long periods of operation. **Manual** welding or cutting operations do not require the benefits of a two-stage regulator.

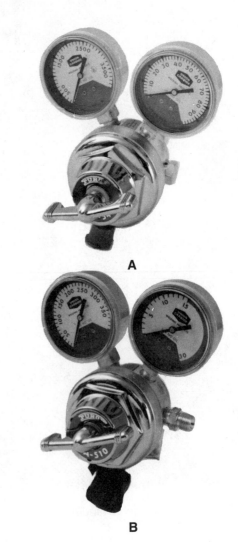

A

B

FIGURE 2–5 Gauges: (A) Oxygen. (B) Acetylene. *(Photos courtesy of Griffin, Roden, Jeffus, and Briggs, 1984.)*

Hoses

Top-quality grade RM hoses (with an oil-resistant cover) designed especially for oxyacetylene welding equipment should not be used at pressures over 200 psi. Although these hoses are tough, they are *not* indestructible. Use them with care to avoid damage from sharp objects or molten metal. The oxygen hose is **color-coded** green, and the acetylene hose is color-coded red.

Fittings

Brass fittings are used on oxyacetylene welding equipment. Hoses, regulators, and cylinder valves should be connected by first using finger pressure before tightening them with an adjustable wrench. This method will avoid thread damage should the threads be misaligned. Acetylene fittings have **left-handed threads** so that they cannot be confused with oxygen fittings. Acetylene fittings can sometimes be recognized by a single groove machined into the nut. Kits containing brass fittings are available from welding supply companies for repairing hoses that have been damaged or need to be lengthened.

Reverse-Flow Check Valves

Reverse-flow (one-way) **check valves** are highly recommended additions to the standard oxyacetylene welding equipment. They prevent one gas from flowing into the other gas line, mixing the gases. Check valves are attached to the outlet ports on the regulators or at the torch connections. Does your equipment have reverse-flow check valves?

Check valves are not designed to stop a **flashback.** A flashback occurs when the flame burns back inside the torch, usually producing a squealing noise. A flashback could indicate that something is wrong with the torch. After experiencing a flashback, turn off the torch immediately, and allow it to cool before relighting. If it happens again, the equipment requires service. A flashback can destroy welding equipment.

A flashback is different from a **backfire,** which results in a loud pop from the torch. A backfire may occur if you touch the tip to the work, try to use a dirty tip, set the pressure incorrectly, or overheat the tip, to name four causes. The torch can be relighted after a backfire.

Manufacturers of Oxyacetylene Welding Equipment

There are many manufacturers of oxyacetylene welding equipment. The purchase decision should depend on the availability of service for repairing, replacing, or adding equipment. Along with the equipment, manufacturers should provide information (**operator's manual** and charts) to aid welders in making the proper selection of **welding tip, cutting tip, heating tip,** or **gouging tip** for any given operation. The regulated pressure for welding and cutting and the correct tip size are not the same for every operation. Note that each manufacturer has its own numbering system for determining tip size (Figure 2–6). Following the manufacturer's suggestions will get the maximum efficiency from the equipment and will prevent the needless waste of gas.

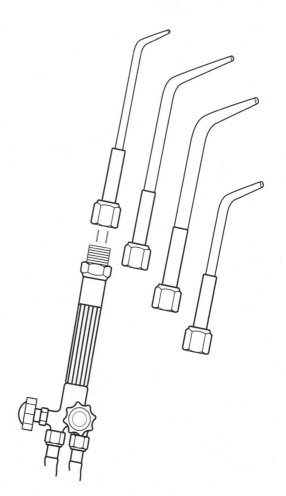

FIGURE 2–6   A variety of tip styles and sizes for one torch.

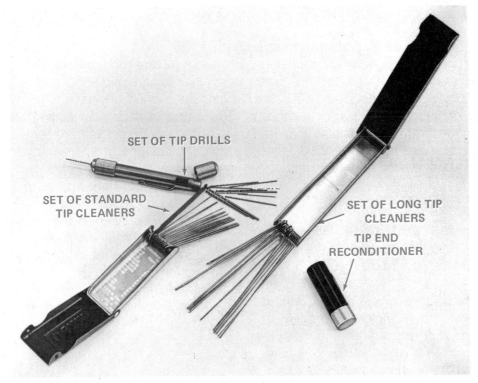

FIGURE 2–7 Two sets of tip cleaners. *(Courtesy of Jeffus, 1993.)*

Tip Cleaners

Tip cleaners are a necessary tool for oxyacetylene welding or cutting. A tip will never accomplish the job if it is partially plugged. Both the volume of heat and the focus of the flame are affected. Having a set of tip cleaners is only part of the solution. You must also know how to use them. Using tip cleaners incorrectly can damage a good tip (Figure 2–7).

Begin by locating a tip cleaner that has a much smaller diameter then the drill hole in the tip. Place the tip cleaner in the hole, and pull it straight out. Never twist the tip cleaner inside the tip; this will change the shape of the hole. Remember that the tip cleaner is of a harder metal than the tip. Find a tip cleaner of the next larger size, and follow the procedure just described. Continue in this fashion until you reach the tip cleaner with a diameter that is too large. If the tip is still plugged, repeat the entire procedure. No tip will last forever, but some welders have received a lifetime of service from one set of tips. Take care of the equipment, and make it a habit. Now is a good time to learn the good habits that will last you a lifetime.

To lengthen the life span of the tip cleaners themselves, always hold them properly. Keep the tip cleaner between your fingers, and slowly push it into the drill hole, holding the tip cleaner at the tip. Never push while holding the tip cleaners by the casing. Let no more than 1/4 inch of tip cleaner extend beyond your fingers. The idea is to keep the tip cleaner from bending (making the tip cleaner useless) or breaking off in the tip (Figure 2–8).

Goggles

Goggles must be worn whenever welding or cutting. The dark lens is necessary to filter out

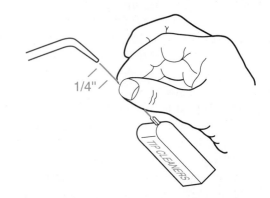

FIGURE 2–8 Correct way to use tip cleaners.

ultraviolet rays and infrared rays, which could damage the eyes. Those who choose not to wear goggles are asking for eye trouble later in life. This unsafe practice is not worth possible eye damage. The dark lens should be marked with the number 5 or 6, which is sufficient for most welding and cutting operations. The manufacturers of lenses for welding and cutting produce charts that should be available from any welding supply company. A chart gives the number of filter lenses based on light intensity.

The dark filter lens should be sandwiched between plastic cover lenses to protect both the inside and the outside of the filter lens. The first thing you may need to do before welding is to check the lenses for cracks (replace the goggles if cracked). If the lenses are dirty, clean them with a damp tissue or cloth. This is a good opportunity to take the goggles apart and become familiar with its parts (Figure 2–9).

Sparklighter

A sparklighter is the safe way of lighting a torch. It frees one hand so that you can hold the torch firmly

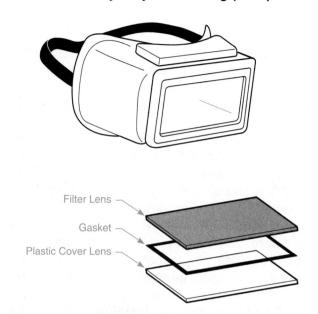

Filter Lens
Gasket
Plastic Cover Lens

FIGURE 2–9   Goggles and lenses.

while lighting without removing your gloves. There are several styles of sparklighters on the market. The two sparklighters pictured in Figure 2–10 use flints that can be replaced when worn down.

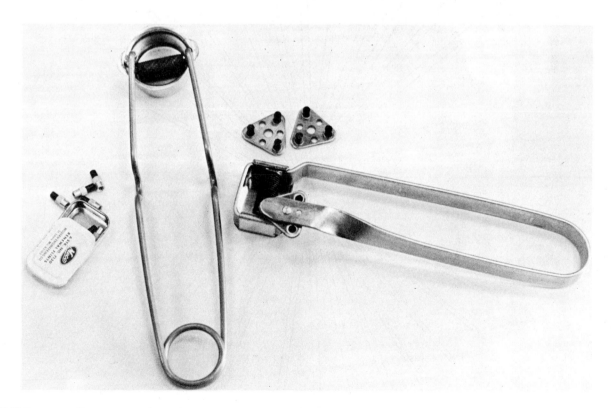

FIGURE 2–10   Two types of sparklighters with flints. *(Courtesy of Jeffus, 1980.)*

Steel Rods for Gas Welding

Melting a welding rod into the weld pool is not always done. Sometimes the pieces forming the joint are melted together without using a **welding rod.** The welding rod provides extra metal for the joint, creating a stronger weld.

Welding rods are designed to meet the specifications provided by the **American Welding Society.** Some rods are made for oxyacetylene welding only. The size of rods is measured by the diameter (Figure 2–11). The smallest size is 1/16 inch, and the largest is 3/8 inch. Two of the more popular sizes are 3/32 inch and 1/8 inch. **Copper**-coated steel rods help to prevent rust. **Stainless steel,** copper base, and **aluminum** are other popular rods that are available for oxyacetylene welding.

If **gas tungsten arc welding** is being done along with oxyacetylene welding, it is a good idea to use one rod for both welding processes. Under the American Welding Society specification A5.2, a welding rod that falls into the class RG 60 can be used for both oxyacetylene welding and gas tungsten arc welding.

THE OXYACETYLENE TORCH

The oxyacetylene torch consists of a torch body and a welding tip. The two knobs of the torch body control the flow of oxygen and acetylene into the mixing chamber of the tip. It is important that the torch be used for the purpose of welding as recommended by the manufacturer.

Torch Body

A tip used for welding can be screwed off and replaced with a cutting attachment. The one item

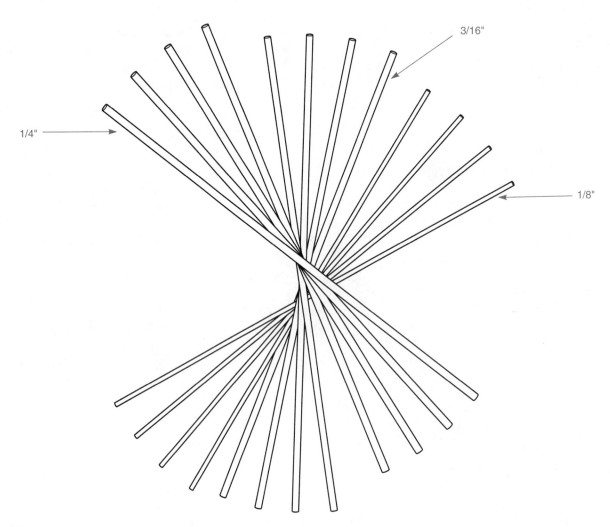

FIGURE 2–11 Welding rods of various diameters.

that does not change is the torch body, which consists of two valves (Figure 2–12). One valve controls the acetylene, and the other controls the oxygen. If the knobs are so tight that they cannot be opened and closed smoothly, they should be repaired. Likewise, if the knobs are loose, they should also be repaired. Any problem that makes it difficult to keep the flame constant should be corrected.

Welding Tips

One tip does not work for all welding jobs. Match the tip to the thickness of metal. No two manufacturers produce identical products, but all tip manufacturers provide charts with information about their tips. These charts can be put on the wall. When changing a tip, never let go of the torch body until another tip is in place. The threads on the torch can be damaged if exposed to the impact of a fall.

Become familiar with the welding tips and how they are designed. If the tip consists of an O-ring (washer), examine it for wear (Figure 2–13). Be sure the tip fits tightly on the torch. Leaks can cause a backfire. Give attention to your equipment, and you will develop appreciation for it. The completion of a job well done depends on equipment being kept in good working order.

Holding the Torch

The recommended method for holding the torch gives you greater flexibility to perform welding in all situations. Examine Figure 2–14 carefully. Do not hold the torch as though it were a pencil (Figure 2–15). The fingers will tire faster and the position of the torch can become awk-

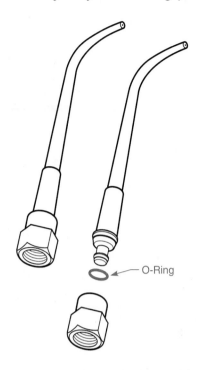

FIGURE 2–13 Two welding tips, one with an O-ring.

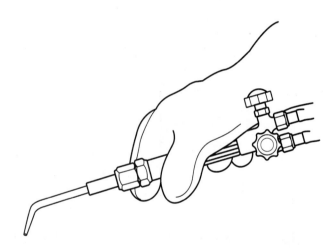

FIGURE 2–14 Correct position for holding torch.

ward with the weight of the equipment above the wrist.

SETTING UP WELDING OR CUTTING EQUIPMENT

Follow this step-by-step procedure for setting up welding or cutting equipment.

1. Make sure that the cylinders are **securely** fastened so they cannot fall. Chain them

FIGURE 2–12 Torch body with acetylene and oxygen valves. *(Courtesy of Albany Calcium Light Co., Inc.)*

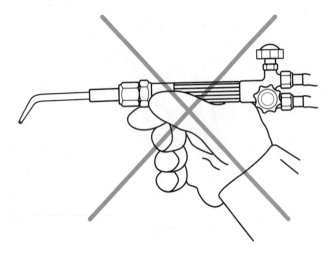

FIGURE 2–15 Incorrect position for holding torch.

to a portable cart or some other object. Never allow cylinders to stand freely (Figure 2–16).

2. Remove the protective caps from the cylinders, check for damage, and clean if necessary. Crack the cylinder valves to purge (remove) foreign particles. To do this, rapidly open and close each cylinder valve, directed away from anyone standing in the immediate area (Figure 2–17).

3. Attach clean and dry regulators (without lubricants) by hand, then tighten them with an adjustable wrench. Be sure that the pressure-adjusting screws are turned out to the left on each regulator. This helps to avoid sudden pressure, which might damage the regulator, when opening the cylinder valves (Figure 2–18).

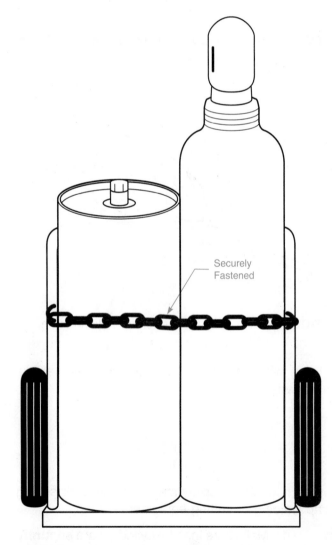

FIGURE 2–16 Cylinders securely fastened to a cart.

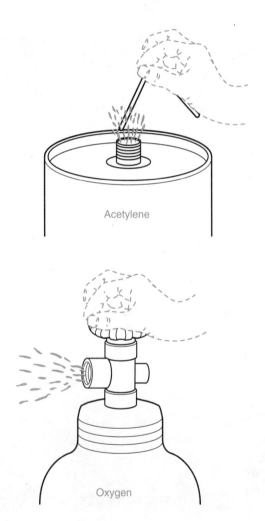

FIGURE 2–17 Cracking a cylinder valve. *(Courtesy of Victor Equipment Co.)*

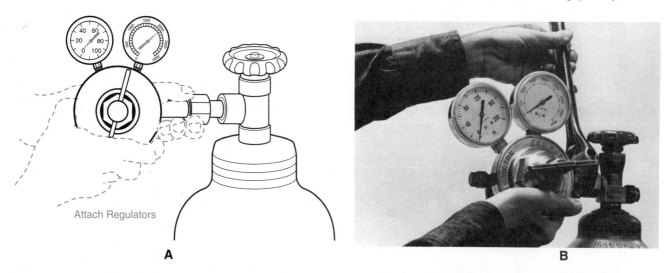

A   **B**

FIGURE 2–18   (A) Attaching the regulators. (B) Using a wrench to tighten the nut. *(Photo courtesy of Jeffus, 1993.)*

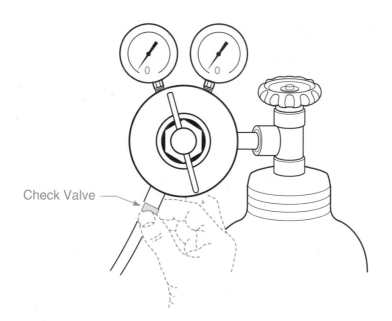

FIGURE 2–19   Connecting the hoses.

4. Connect the hoses if you are putting together new equipment. Add check valves, if available. Screw the valves on by hand before tightening them with an adjustable wrench (Figure 2–19).
5. Connect the torch body by hand, tighten it with an adjustable wrench, and screw on the welding or cutting attachment (Figure 2–20).
6. Open the oxygen cylinder valve. Stand to one side away from the regulator, and open the valve slowly at first. Open the valve completely to seat the valve. It is

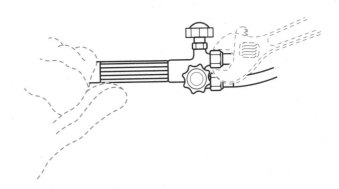

FIGURE 2–20   Connecting the torch body.

good practice to stand away from the regulator because a regulator malfunction could cause a serious injury (Figure 2–21).

7. Open the acetylene cylinder valve one-quarter turn or until the pressure registers on the working-pressure gauge. If a T-wrench is used, it must always remain on

the cylinder so that you can quickly close the valve in an emergency (Figure 2–22).

8. Consult the manufacturer's chart for the brand of equipment in the shop to find the proper pressure settings for both the oxygen (Figure 2–23) and acetylene (Figure 2–24) regulators. Pressure settings depend on the thickness of the metal to be welded or cut. Once the metal thick-

FIGURE 2–21 Opening the oxygen cylinder valve. *(Courtesy of Jeffus, 1993.)*

TIP PRESSURE SIZE THICKNESS

Metal Thickness	Tip Size	Oxygen Pressure Min./Max.	Acetylene Pressure Min./Max.
Up to 1/32"	000	3/5	3/5
1/64" to 6/64"	00	3/5	3/5
1/32" to 5/64"	0	3/5	3/5
3/64" to 1/8"	1	3/5	3/5
3/64" to 1/8"	2	2/5	2/5

Manufacturer's Settings

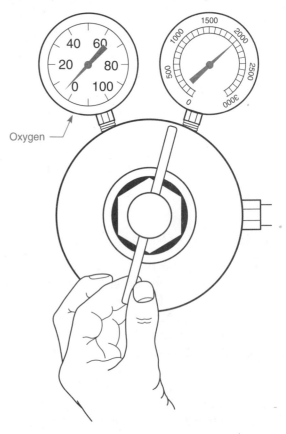

FIGURE 2–23 Adjusting oxygen pressure to the manufacturer's recommended setting.

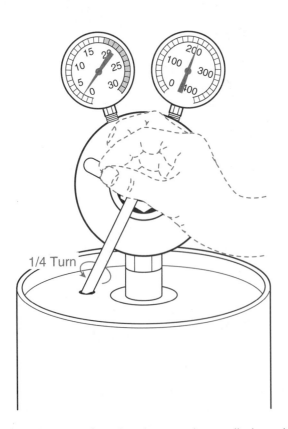

FIGURE 2–22 Opening the acetylene cylinder valve one-quarter turn.

TIP PRESSURE SIZE THICKNESS

Metal Thickness	Tip Size	Oxygen Pressure Min./Max.	Acetylene Pressure Min./Max.
Up to 1/32"	000	3/5	3/5
1/64" to 6/64"	00	3/5	3/5
1/32" to 5/64"	0	3/5	3/5
3/64" to 1/8"	1	3/5	3/5
3/64" to 1/8"	2	2/5	2/5

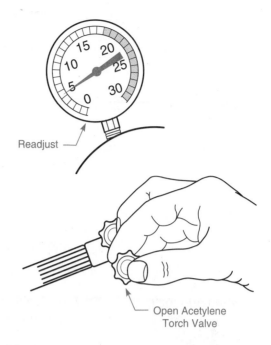

FIGURE 2–25 Opening and adjusting the acetylene torch valve.

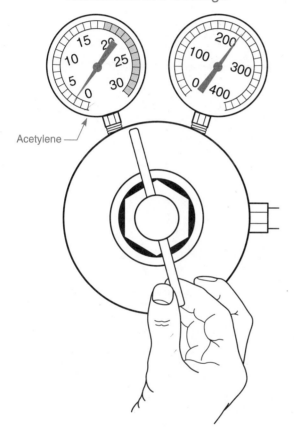

Manufacturer's Settings

FIGURE 2–24 Adjusting acetylene pressure to the manufacturer's recommended setting.

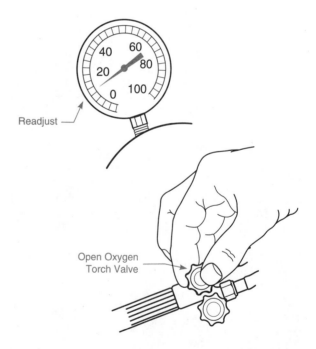

FIGURE 2–26 Opening and adjusting the oxygen torch valve.

ness has been determined, you can find the proper tip and pressure settings on the chart.

9. Open the acetylene torch valve, and turn the pressure-adjusting screw to the recommended working pressure (Figure 2–25). Close the torch valve.

10. Open the oxygen torch valve, and turn the pressure-adjusting screw to the rec-

ommended working pressure. Close the torch valve (Figure 2–26).

If you suspect a leak, apply a nonpetroleum-based liquid to all fittings, and watch for bubbles

(Figure 2–27). Make repairs or adjustments, if needed. Never use petroleum-based products around any welding equipment.

SHUTTING DOWN WELDING OR CUTTING EQUIPMENT

Follow this procedure to shut down welding or cutting equipment.

1. With the torch flame out, close the acetylene and oxygen cylinder valves (Figure 2–28).
2. Open the acetylene torch valve until both gauges (cylinder and line) read zero pressure (Figure 2–29).
3. Open the oxygen torch valve until both gauges (cylinder and line) read zero pressure (Figure 2–30).
4. Turn out the pressure-adjusting screws on both regulators (Figure 2–31).

Close Cylinder Valves

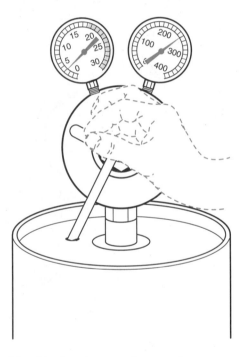

FIGURE 2–28 Closing the acetylene and oxygen cylinder valves.

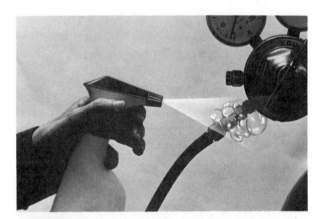

FIGURE 2–27 Checking for leaks at connections and valves. *(Photos courtesy of Jeffus, 1993.)*

5. Wrap up the hoses properly, and store the equipment (Figure 2–32).

SAFETY REMINDERS

1. Be sure the shop is clear of all **flammable** and volatile substances. Paints, solvents,

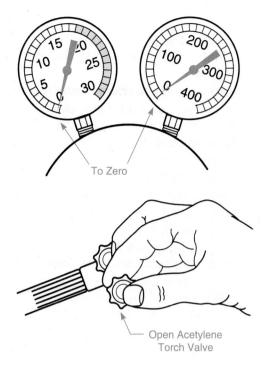

To Zero

Open Acetylene
Torch Valve

FIGURE 2–29   Opening the acetylene torch valve.

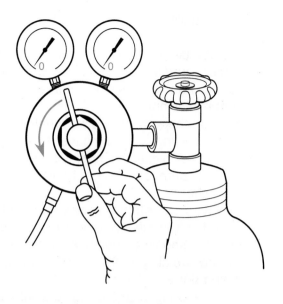

FIGURE 2–31   Turning out the pressure-adjustment screw.

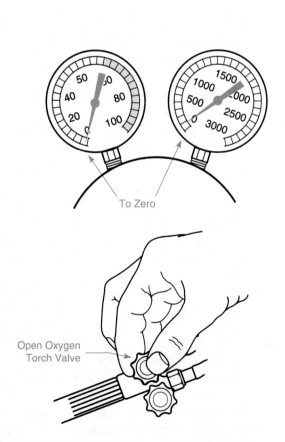

To Zero

Open Oxygen
Torch Valve

FIGURE 2–30   Opening the oxygen torch valve.

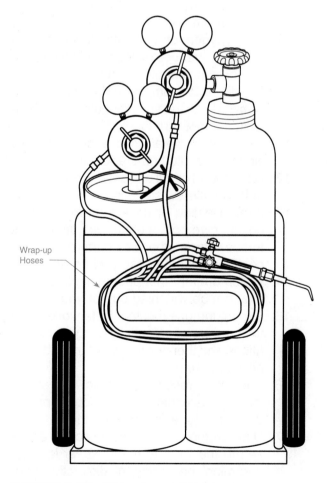

Wrap-up
Hoses

FIGURE 2–32   Wrapping up the hoses.

and gasoline have no place in a welding shop.

2. Do not use oil, grease, or any petroleum-based products around oxyfuel fittings. Oxygen under pressure is subject to violent explosions.

3. Be sure the work area is well ventilated to allow welding or cutting gases to escape.

4. Wear cotton clothing, not synthetic fibers that will burn, and leather boots, not athletic shoes.

5. Always use **safety glasses** and leather **welding gloves.** No one without safety glasses should be permitted in any shop where welding, cutting, or grinding is being done.

6. Do not bring cigarette lighters into the welding shop.

7. Always wear goggles with a filter lens when welding or cutting. Infrared rays can cause headaches and may lead to cataracts in later life.

8. Always remain in control of your equipment. Never lay down a lit torch.

9. Use oxygen only for its intended purposes. Never use oxygen to blow off people or equipment; do not use it as a substitute for compressed air.

10. Know where the fire extinguishers in your welding shop are located.

11. Only handle metal while wearing welding gloves to avoid burns and cuts.

12. Do not weld or cut on containers that have held volatile substances.

13. Avoid welding or cutting on metals that could give off potentially dangerous fumes. Some otherwise harmless plastic coatings become deadly when heated.

14. Watch out for other people because they may not be watching out for you.

15. Expect the unexpected. Safe workers try to be aware of potentially dangerous situations and then act to correct them before accidents happen.

16. Know the procedures to follow in case of injury or accident in the shop. If procedures do not exist, develop them.

17. Use common sense, and pay close attention to the work at hand. Never let daydreaming, horseplay, or distraction control your behavior.

REVIEW

1. When do accidents happen?
2. What is the pressure of a full cylinder of oxygen?
3. What is the maximum working pressure of acetylene? Free acetylene?
4. Why do acetylene fittings have left-handed threads?
5. What is a flashback?
6. What information do manufacturers provide about their equipment?
7. How are regulated working-pressure settings determined?
8. Why are goggles recommended?
9. How far should you open the acetylene valve when setting up the equipment?
10. What kinds of products must never come in contact with oxygen?

Create Three Questions

1.
2.
3.

Related Math and English Questions

1. Which welding rod is larger: 3/32″ or 1/8″?
2. Which welding rod is smaller: 3/8″ or 1/16″?
3. Write a paragraph on the proper procedure for cleaning a welding tip.
4. Write a paragraph describing the most difficult part of the shop exercises.

For Further Thought

1. Why not buy calcium carbide instead of leasing acetylene cylinders?
2. Why use brass fittings on gas cylinders when steel fittings would be less expensive?
3. Can scrap steel **plate** be used for filler metal instead of welding rods? Explain.
4. Suppose that oxyacetylene welding does not produce enough heat to weld together some 1/2-inch-thick pipe. If no other welding process is available, what can be done to help?
5. The head of a bolt breaks off inside a cast iron block. How can oxyacetylene welding be of help?

SUGGESTED ACTIVITIES

1. Discuss safety.
2. Set up and shut down the oxyacetylene equipment.
3. Locate the fire extinguishers in your shop.
4. With supervision, stage a mock injury to practice the prescribed shop procedure; involve the school nurse.

UNIT 2: EXERCISE 1

Oxyacetylene Welding (OAW): Lighting the Torch

1. Follow the manufacturer's recommendations for gas pressure settings and welding tip selection based on metal thickness.
2. Using a sparklighter (never a match or cigarette lighter), point the torch away from nearby objects or people, open the acetylene torch valve slightly, and light the torch (Figure 2–33).
3. Add acetylene until the black carbon smoke clears, but do not add so much that the flame separates from the end of the tip.
4. Slowly open the oxygen torch valve, adding oxygen until the acetylene flame shrinks into the different stages of the **carburizing flame** (Figure 2–34).
5. When the carburizing flame disappears into a bright inner **cone,** you have reached the **neutral flame.** The difference between the neutral flame and the **oxidizing flame** is slight. For some applications, you will make the necessary adjustment to maintain a small acetylene feather and avoid the oxidizing flame altogether. If the bright inner cone forks into two or more flames, the tip needs to be cleaned (Figure 2–35).
6. Adding slightly more oxygen to the neutral flame will produce an oxidizing flame. An excessive oxidizing flame can become quite loud as more oxygen is added. Continuing to add oxygen will eventually extinguish the flame (Figure 2–36).
7. Turn the acetylene off first to shut down the equipment. Although this can cause the torch to pop, it will avoid smoke and **carbon** buildup at the end of the tip.
8. Turn off the oxygen.

Three Basic Flames

By now you should have adjusted your torch through the three basic flames used in oxyacetylene welding. The carburizing flame is used in welding **high-carbon steel, aluminum,** and stainless steel. The neutral flame is used in welding **cast iron** and **low-carbon steel.** The oxidizing flame has application in the **braze welding** of most steels and cast iron.

Flame adjustment can be critical. Excessive amounts of oxygen or acetylene in the flame can lead to weld failure. The **physical properties** of the **base metal** can be affected by the amount of oxygen or carbon deposited in the weld. The oxidizing flame will cause the weld pool to foam when welding **mild steel,** and the carburizing flame will cause the weld pool to boil when welding mild steel. Avoid these two situations.

FIGURE 2–33   Using a sparklighter to light the torch. *(Courtesy of Jeffus, 1993.)*

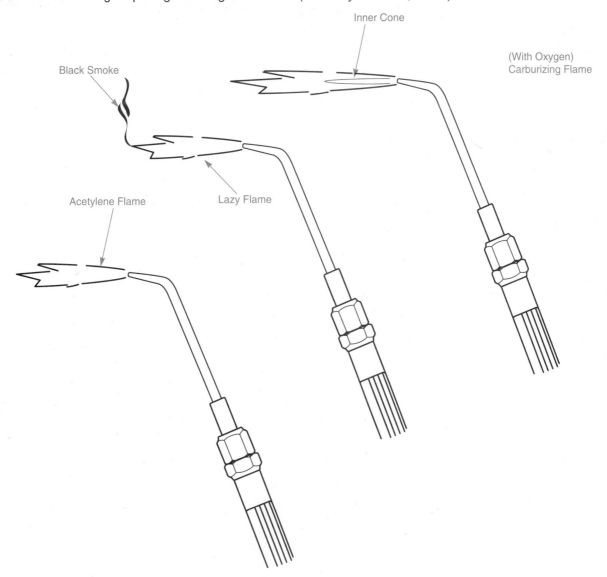

FIGURE 2–34   Different acetylene flames.

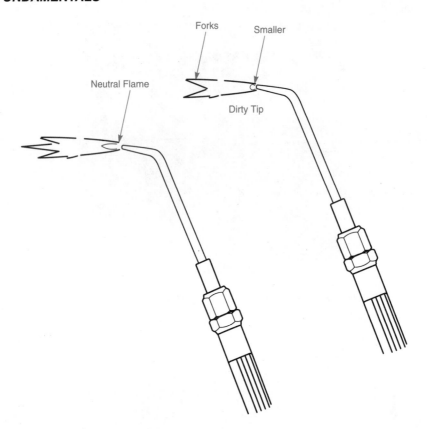

FIGURE 2–35   Neutral flame and flame produced by a dirty tip.

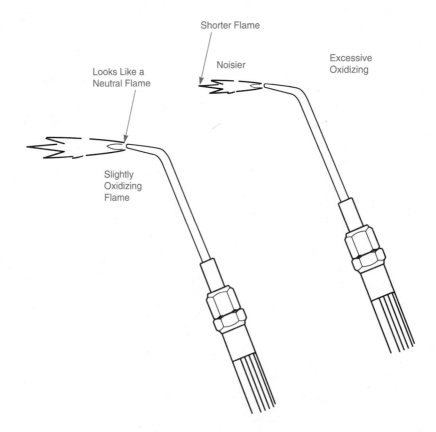

FIGURE 2–36   Oxidizing flames.

UNIT 2: EXERCISE 2

Oxyacetylene Welding (OAW): Weld Pool without Welding Rod

Root penetration is an important concept in welding. In most **joint designs,** 100% root penetration is required. Root penetration is 100% when there is complete **fusion.** In a **butt joint,** the **weld** melts through the base metal to the other side, and the **root bead** can be seen when the base metal is turned over.

You might need to prepare the base metal for welding to achieve 100% root penetration. **Plate** steel (more than 3/16 inch thick) may require **beveling,** which is one method used to achieve complete root penetration. With both **sheet** steel and plate steel, any **mill scale** or **rust** should be removed by grinding before completing the welding exercises.

This welding exercise is intended to develop your coordination and to familiarize you with the weld pool. The weld pool is basic to all of the welding exercises covered in this book.

Gauge numbers are introduced in this unit. A gauge number is a measurement of metal thickness. A single gauge number covers a variety of standards (different decimal numbers) in use to measure a variety of metal thicknesses. In the Manufacturer's Standard, to measure sheet steel a 5 gauge is 0.220 inches thick, and in the Brown and Sharpe, to measure aluminum and brass a 5 gauge is 0.1819 inches thick. The following is a short list of gauge numbers with approximate fractional equivalents.

```
 3 gauge = 1/4 inch
11 gauge = 1/8 inch
16 gauge = 1/16 inch
22 gauge = 1/32 inch
28 gauge = 1/64 inch
```

Work angle and **travel angle** are used to describe welding techniques throughout this book (Figure 2–37). Work angle refers to the angles of the tip and the welding rod in relation to the base metal. Work angle is always 90°. Travel angle refers to the angles of the tip and the welding rod in relation to the direction of travel. The angles given for all of the instructions in this unit refer to travel angle (Figure 2–38).

Necessary Material and Equipment

Safety glasses	Welding gloves	Tip cleaners	Goggles
Tongs or pliers	Sparklighter	2 firebricks, or C-clamp	Welding tip
16-gauge steel, 2″ × 4″	Protective clothing	fixture	

Instructions

1. Adjust the flame to neutral.
2. Keep the tip at a 45° angle with the inner cone just above the weld pool.
3. Heat a spot on the steel beyond red, causing the metal to melt. Begin travel after establishing the weld pool (molten metal).
4. Never let the weld pool dry up; always move forward with a slight movement from side to side.
5. Turn the base metal over, and inspect the bottom. A **convex** (raised) **weld bead** pattern on the underside of the base metal indicates a steady travel speed.
6. Holes in the base metal indicate too much heat. The volume of gas could be too high, or the travel speed could be too slow.
7. No root penetration indicates not enough heat, or travel speed is too fast.
8. If the tip is partially plugged, use the tip cleaners to clean the **orifice** (hole), removing the metal particles that are plugging the tip.
9. Check with the instructor for evaluation, if necessary.
10. Practice this exercise until you meet the standards established for the course.

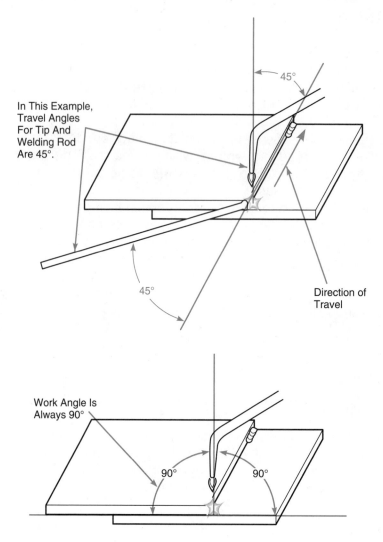

In This Example, Travel Angles For Tip And Welding Rod Are 45°.

45°

45°

Direction of Travel

Work Angle Is Always 90°

90°

90°

FIGURE 2–37   Work angle and travel angle.

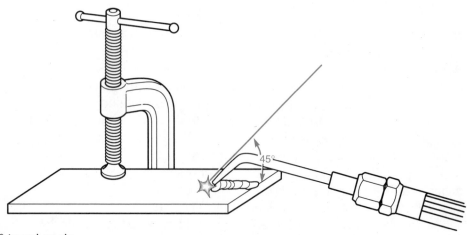

45°

FIGURE 2–38   45° travel angle.

UNIT 2: EXERCISE 3

Oxyacetylene Welding (OAW): Weld Pool with Welding Rod

Always direct the inner cone of the flame at the base metal. The welding rod should melt into the weld pool. **Feed** the rod at a constant rate; do not dab the rod in and out of the weld pool. Before the rod becomes short enough to heat up your welding gloves, weld another 3-foot length of rod onto the short rod so that nothing is wasted.

Root penetration is important in this exercise. If 16-gauge metal is not used, it will be very difficult to achieve complete root penetration. This exercise is intended to develop coordination with both hands. For all exercises, follow the manufacturer's recommendations for tip size and pressure settings.

Necessary Material and Equipment

Safety glasses	Welding gloves	Protective clothing	Goggles
Tongs or pliers	Sparklighter	Tip cleaners	Welding tip
16-gauge steel, 2″ × 4″	2 firebricks, or C-clamp fixture	1/16″ welding rod	

Instructions

1. Adjust the flame to neutral.
2. Hold the welding tip and welding rod at 45° angles (Figure 2–39). If the tip is not held at the proper angle, it will plug easily from the sparks given off by the welding.
3. Position the inner cone of the flame just over the weld pool.
4. Place the welding rod on the leading edge of the weld pool, dragging it on the metal.
5. If the torch pops, the tip may be plugged (dirty).
6. Popping may also be caused by **overheating** the tip as the flame is deflected back at the torch or by using an improper flame adjustment.
7. Completely fill the **crater** that forms at the end of the weld bead.
8. Check with the instructor for evaluation, if necessary.
9. Practice this exercise until you meet the standards established in the course.

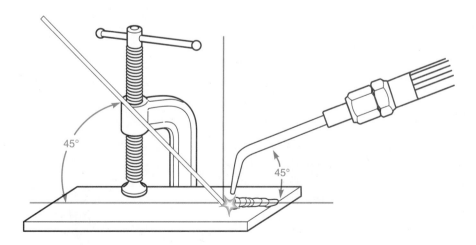

FIGURE 2–39   Tip and welding rod held at 45° angles.

UNIT 2: EXERCISE 4

Oxyacetylene Welding (OAW): Corner Joint without Welding Rod, Fillet Weld in Flat Welding Position

The **corner joint** is popular in welding applications for which no welding rod is required. The two pieces of base metal used in the corner joint are melted or fused together, without a gap between them. Complete root penetration is important for welding a strong corner joint. Remember to melt enough of the base metal into the weld to prevent the corner joint from cracking during testing (Figure 2–40).

Necessary Material and Equipment

Safety glasses	Welding gloves	Protective clothing	Goggles
Tongs or pliers	Sparklighter	Tip cleaners	Welding tip
16-gauge steel, 2″ × 4″	Firebrick, or C-clamp fixture		

Instructions

1. Adjust the flame to neutral.
2. Hold the welding tip at a 45° angle.
3. **Tack weld** two pieces of steel at the beginning and at the end of the corner joint.
4. Travel at a steady pace, keeping the width of the weld pool constant. The entire length of the bead should have a consistent width.
5. Turn over the finished corner joint, and check for complete root penetration.

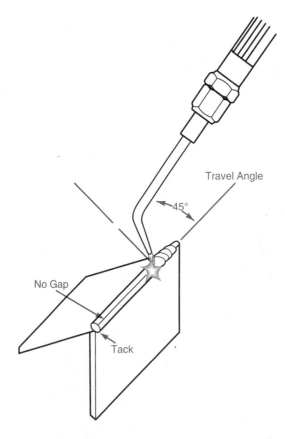

FIGURE 2–40   Corner joint with no welding rod in flat position.

6. Allow to cool. Test by hammering the joint flat.
7. Look for **cracks** between the base metal and the weld. Redo the weld if you find less than 100% root penetration, excessive buildup, or cracks in the weld.
8. Check with the instructor for evaluation, if necessary.
9. Practice this exercise until you meet the standards established in the course.

UNIT 2: EXERCISE 5

Oxyacetylene Welding (OAW): Corner Joint with Welding Rod, Fillet Weld in Flat Welding Position

Many times it is not enough to melt two pieces of metal together. Filler metal is usually added to give greater **strength** to the weld or to prevent **melt-thru.** The use of filler metal requires a gap or spacing between the two pieces of base metal. This gap is referred to as the **joint root** opening. It aids in welding, making a corner joint with complete root penetration. Be careful, though: Too large a joint root opening will result in melt-thru; and not enough of a joint root opening will cause **poor root penetration.**

At this point, you should be able to easily handle a 3-foot welding rod. Wearing heavy welding gloves, let the rod slide through the glove fingers to grip higher up. Do not fight the welding rod if it sticks in the weld. Use the flame to free the end of the rod. Be sure to fill the crater in completely to avoid cracks (Figure 2–41).

Necessary Material and Equipment

Safety glasses	Welding gloves	Tip cleaners	Goggles
Tongs or pliers	Sparklighter	Firebrick, or C-clamp	Welding tip
16-gauge steel, 2″ × 4″	Protective clothing	fixture	1/16″ welding rod

Instructions

1. Adjust the flame to neutral.
2. Hold the welding tip and welding rod at 45° angles.
3. Using an extra piece of metal as a spacer, set a 1/16-inch joint root opening.
4. Tack weld two pieces of steel at the beginning and at the end of the corner joint.

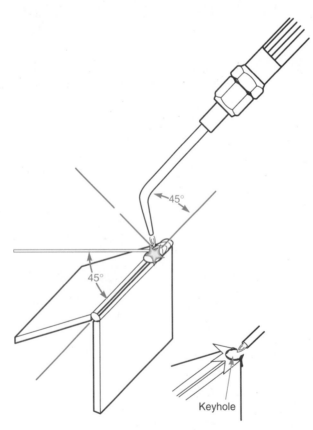

FIGURE 2–41   Corner joint with welding rod in flat position.

5. Melt the welding rod into the weld pool, mixing filler metal with base metal.
6. Turn over the finished corner joint and check for complete root penetration.
7. Allow to cool. Test by hammering the joint flat.
8. Look for cracks. Redo the weld if you find less than 100% root penetration, excessive buildup, or cracks in the weld.
9. Check with the instructor for evaluation, if necessary.
10. Practice this exercise until you meet the standards established in the course.

UNIT 2: EXERCISE 6

Oxyacetylene Welding (OAW): Butt Joint, Square-Groove Weld in Flat Welding Position

Filler metal is necessary to **build up** or reinforce a weld. A joint root opening between the pieces of base metal to be joined allows the formation of a **keyhole.** A keyhole is the partial melting away of each edge of the joint at the front of the weld pool. Continually adding filler metal into the keyhole as it forms gives a visual indication of complete root penetration. Remember that traveling too fast and adding too much filler metal both reduce root penetration (Figure 2–42).

Necessary Material and Equipment

Safety glasses	Welding gloves	Protective clothing	Goggles
Tongs or pliers	Sparklighter	Tip cleaners	Welding tip
16-gauge steel, 2″ × 4″	Firebrick, or C-clamp fixture	1/16″ welding rod	

Instructions

1. Adjust the flame to neutral.
2. Hold the welding tip and welding rod at 45° angles.
3. Tack weld two pieces of steel at the beginning and at the end of the butt joint.
4. Set enough of a joint root opening to prevent the gap from closing up.
5. Place the welding rod at the front edge of the weld pool, moving the tip from side to side.

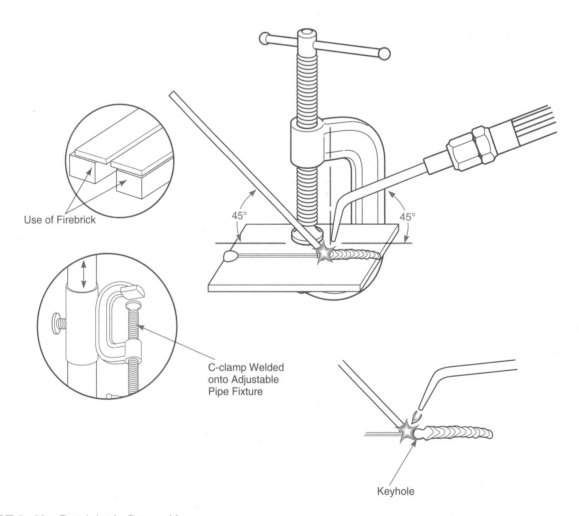

Use of Firebrick

C-clamp Welded onto Adjustable Pipe Fixture

45°   45°

Keyhole

FIGURE 2–42 Butt joint in flat position.

6. Add filler metal just fast enough to prevent melt-thru on one side of the butt joint or the other. Fill the crater completely at the end of the weld bead.
7. Turn over completed butt joint and check for complete root penetration.
8. Allow to cool. Bend-test the weld. Place the weldment in a vice, positioning the weld just above the jaws. Hammer the bottom of the butt joint over, and squeeze together in the vice.
9. Look for cracks on the underside of the butt joint. Redo the weld if you find cracks or less than 100% root penetration.
10. Check with the instructor for evaluation, if necessary.
11. Practice this exercise until you meet the standards established in the course.

UNIT 2: EXERCISE 7

Oxyacetylene Welding (OAW): Lap Joint Fillet Weld in Horizontal Welding Position

A filler rod is used to reinforce this **fillet weld** in the horizontal position. Although a **lap joint** requires complete root penetration, this does not mean that you should weld through the bottom piece. The heat of the torch should be directed toward the bottom piece because the edge of the top piece will heat more quickly (Figure 2–43). This will prevent the top piece from melting away, causing **undercut**. Undercut is a welding **defect** that results when the base metal becomes thinner alongside the weld.

If the weld pool becomes too large, causing the top piece to burn away, **flash off.** This means taking the torch away from the weld pool for a moment, allowing the weld pool to cool down. Flashing off is one method of controlling the heat.

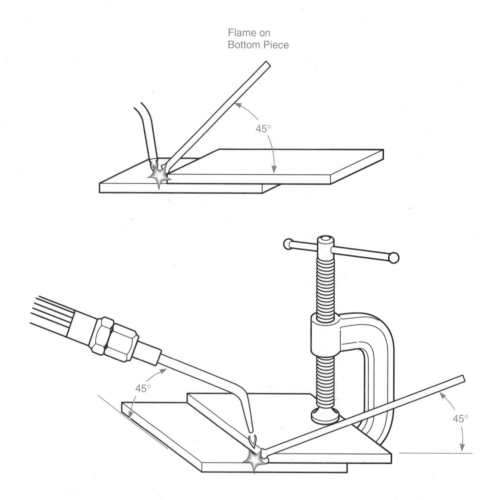

FIGURE 2–43 Lap joint in horizontal position.

Necessary Material and Equipment

Safety glasses	Welding gloves	Protective clothing	Goggles
Tongs or pliers	Sparklighter	Tip cleaners	Welding tip
16-gauge steel, 2″ × 4″	Firebrick, or C-clamp fixture	3/32″ welding rod	

Instructions

1. Adjust the flame to neutral.
2. Keep the welding tip and the welding rod at 45° angles.
3. Tack weld the pieces of steel tightly together.
4. Melt plenty of filler metal into the center of the weld pool. Fill the crater completely at the end of the weld bead.
5. Examine the completed lap joint for undercut (underfill along the top edge of the weld).
6. Allow to cool. Test the weld in a vice, positioning the weld just above the jaws. Hammer the top piece over.
7. Look for cracks and bending along any undercut. Redo the weld if you find cracks or undercut.
8. Check with the instructor for evaluation, if necessary.
9. Practice this exercise until you meet the standards established in the course.

UNIT 2: EXERCISE 8

Oxyacetylene Welding (OAW): Tee Joint Fillet Weld in Horizontal Welding Position

A larger volume of gas is required for the **tee joint.** More volume is necessary to heat the bottom piece. Concentrate the heat on the bottom piece. Too much heat on the top piece will cause melt-thru. Weld only one side of the tee joint to allow for testing (Figure 2–44).

Necessary Material and Equipment

Safety glasses	Welding gloves	Protective clothing	Goggles
Tongs or pliers	Sparklighter	Tip cleaners	Welding tip
16-gauge steel, 2″ × 4″	Firebrick, or C-clamp fixture	3/32″ welding rod	

Instructions

1. Adjust the flame to neutral.
2. Tack weld the pieces of steel together.
3. Remelt the tack welds, blending them into the weld bead.
4. Keep the welding rod in the middle of the weld pool, heating both top and bottom pieces. Fill the crater completely at the end of the weld bead.
5. Allow to cool. Test the weld by clamping the bottom piece in a vice, and hammer the top piece toward the weld.
6. Squeeze the two pieces together.
7. Look for cracks and incomplete fusion. Redo the weld if it is cracked.
8. Check with the instructor for evaluation, if necessary.
9. Practice this exercise until you meet the standards established for the course.

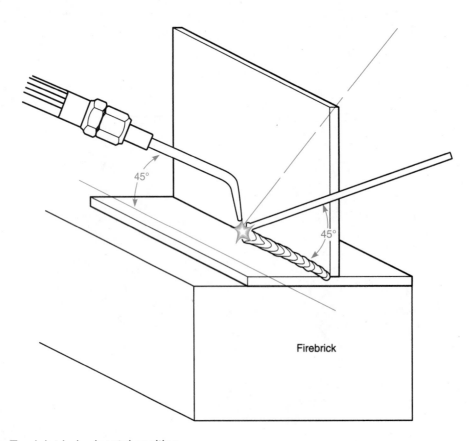

FIGURE 2–44 Tee joint in horizontal position.

UNIT 2: EXERCISE 9

Oxyacetylene Welding (OAW): Corner Joint Fillet Weld in Vertical Welding Position

The position of the torch and the welding rod is important for welding **uphill** (against gravity). The effect of gravity must be overcome. Note that the welding rod should be angled away from the heat of the torch. Keep the welding rod in the weld pool, and direct the heat toward the edges of the joint. Always keep the inner cone of the flame just off the weld pool. Move the tip slightly back and forth, traveling up the joint. Move at a steady speed to keep the weld pool at a constant size (Figure 2–45).

Necessary Material and Equipment

Safety glasses	Welding gloves	Protective clothing	Goggles
Tongs or pliers	Sparklighter	Tip cleaners	Welding tip
16-gauge steel, 2″ × 4″	Firebrick, or C-clamp fixture	1/16″ welding rod	

Instructions

1. Adjust the flame to neutral.
2. Check Figure 2–45 for welding tip and welding rod angles.
3. Feed the rod into the weld pool from above, and gravity will do the rest.
4. Position the joint for vertical welding (up to a 15° slope is allowable).
5. Fill the crater completely at the end of the weld bead.
6. Examine the weld for complete root penetration.
7. Allow to cool. Test by hammering flat.
8. Look for cracks. Redo the weld if you find less than 100% root penetration, excessive buildup, or cracks in the weld.
9. Check with the instructor for evaluation, if necessary.
10. Practice this exercise until you meet the standards established for the course.

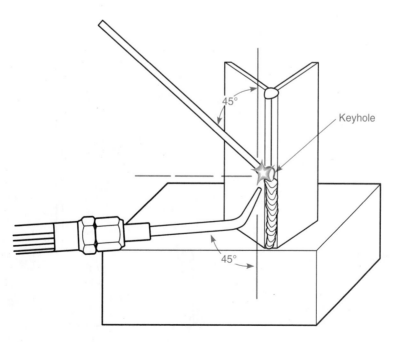

FIGURE 2–45 Corner joint in vertical position.

UNIT 2: EXERCISE 10

Oxyacetylene Welding (OAW): Tee Joint Fillet Weld in Vertical Welding Position

Watch the position of the torch and the welding rod when traveling uphill. The effect of gravity must be considered whenever welding in the vertical position. Note that the flame of the torch should be directed upward. Concentrate the flame at the middle of the piece, where more heat is required. When that piece begins to melt as the weld pool forms, add filler metal, and adjust the position of the tip. Then concentrate the heat on both pieces. Move at a steady pace to keep the weld pool at a constant size (Figure 2–46).

Necessary Material and Equipment

Safety glasses	Welding gloves	Protective clothing	Goggles
Tongs or pliers	Sparklighter	Tip cleaners	Welding tip
16-gauge steel, 2″ × 4″	Firebrick, or C-clamp fixture	1/16″ welding rod	

Instructions

1. Adjust the flame to neutral.
2. Review Figure 2–46 for welding tip and welding rod angles.
3. Position the tee joint for welding so it cannot fall.
4. Examine the completed weld for good fusion between the pieces. Look also for consistency in the size of the weld bead and the absence of any undercut. You should have filled in the crater completely.
5. Allow to cool. Test the weld by hammering the top (vertical) piece toward the weld held in a vice.
6. Look for cracks and melt-thru on the back side of the tee joint.
7. Redo if the weld is defective.
8. Check with the instructor for evaluation, if necessary.
9. Practice this exercise until you meet the standards established for the course.

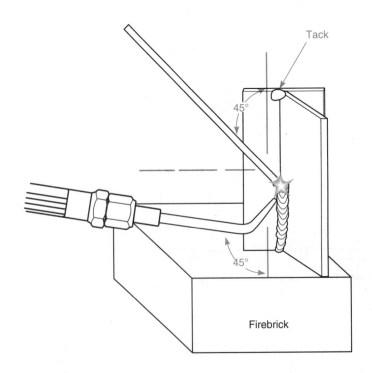

FIGURE 2–46   Tee joint in vertical position.

UNIT 2: EXERCISE 11

Oxyacetylene Welding (OAW): Butt Joint (Exhaust Tubing), Square-Groove Weld in 2G Position

After cutting two pieces of tubing (the thickness is not important), set one piece of tubing on top of the other without a root opening. Tack weld the pieces together at two or more points, depending on the diameter of the tubing. While the tubing is positioned vertically, complete the welding around the tubing from the horizontal position. Examine the inside for complete root penetration (Figure 2–47).

Necessary Material and Equipment

Safety glasses	Welding gloves	Protective clothing	Goggles
Tongs or pliers	Sparklighter	Tip cleaners	Welding tip
Hacksaw	2 pieces of exhaust tubing, 2″ long	Firebrick, or C-clamp fixture	1/16″ welding rod

Instructions

1. Adjust the flame to neutral.
2. Begin the weld pool at the edge of a tack weld.
3. Feed the welding rod into the weld pool from above, using gravity. Move around the tubing or turn the tubing to complete the welding.
4. Keep an even speed of travel along the tubing, feeding the rod into the weld pool at a constant rate.
5. Examine the weld for appearance. Is the weld undercut on the top edge? Is the size of the weld bead consistent?
6. Look for complete root penetration along the inside of the tubing.
7. Redo the weld if the root penetration is incomplete.
8. Check with the instructor for evaluation, if necessary.
9. Practice this exercise until you meet the standards established by the course.

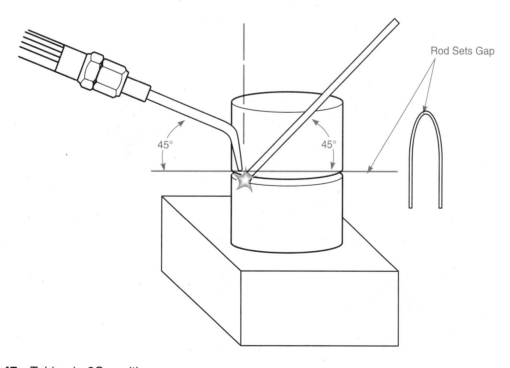

FIGURE 2–47 Tubing in 2G position.

UNIT 2: EXERCISE 12

Oxyacetylene Welding (OAW): Butt Joint (Exhaust Tubing), Square-Groove Weld in 5G Position

This exercise requires a means of holding the tubing in position once the tubing has been tack welded together. Once again, cut two pieces of 2-inch long tubing (the thickness is not important). You will complete the welding uphill. Practice on tubing develops the kinds of skills required for pipe welding (Figure 2–48).

Necessary Material and Equipment

Safety glasses	Welding gloves	Protective clothing	Goggles
Tongs or pliers	Sparklighter	Tip cleaners	Welding tip
Hacksaw	2 pieces of exhaust tubing, 2" long	C-clamp fixture	1/16" welding rod

Instructions

1. Adjust the flame to neutral.
2. Tack weld the pieces of tubing together at two or more places, depending on the diameter of the tubing.
3. Place the tubing in a fixture, as indicated by Figure 2–48, and begin welding from the bottom up.
4. Feed the rod from above into the weld pool, using gravity.
5. Remelt the tack welds into the weld pool, stop at the top, and begin the same procedure from the other side without moving the tubing.
6. Examine the completed weld for appearance, consistent size, and complete root penetration on the inside of the tubing.
7. Redo if necessary.
8. Check with the instructor for evaluation, if necessary.
9. Practice this exercise until you meet the standards established in the course.

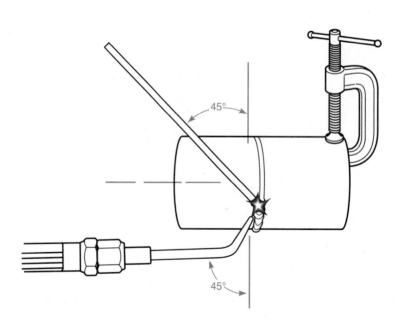

FIGURE 2–48   Tubing in 5G position.

UNIT 2: EXERCISE 13

Oxyacetylene Welding (OAW): Practice on Other Projects

This would be a fine opportunity to do some repair welding. After all, most people who make a living welding, do so joining together functional products that make living better. Let's put some welding to the test of holding up under the stress of real life.

Metals other than steel can also be welded together by the heat of a flame. If no welding project is possible, this exercise is an opportunity to try welding other metals (if available). Try to stretch the learning experience by undertaking any activity that can increase your skill level.

Necessary Material and Equipment

Safety glasses	Welding gloves	Protective clothing	Goggles
Sparklighter	Tip cleaners	Welding tip	Welding project

Oxyacetylene Welding: Other Metals

Gray cast iron.   Oxyacetylene welding (using cast iron welding rods and cast iron welding **flux**) is not the best choice for gray cast iron. Most castings are too large to go through the necessary **preheating** and **postheating** designed to prevent them from cracking. An oven of sufficient size or a charcoal fire may be out of the question. If a casting cannot be heated in a few minutes with a heating tip, another welding process should be considered. For many projects, braze welding (Unit 7) will do the job.

Aluminum.   Aluminum repair by oxyacetylene welding requires aluminum welding rods (1100, 4043, or 5356), aluminum welding flux, and lots of practice. Welders who are skilled at using the torch can weld aluminum as thin as .040 thickness. If there is much welding to be done, gas tungsten arc welding (Unit 6) and gas metal arc welding (Unit 5) are better choices. In general, gas tungsten arc welding is preferred for sheet aluminum, and gas metal arc welding is preferred for aluminum plate.

Stainless steel.   Very little stainless steel welding is done with the oxyacetylene welding process. Suitable stainless steel filler rods and a stainless steel flux are required. Because stainless steel is a poor **conductor** of heat, subject to distortion and **oxidation,** special care must be taken in setup and in welding.

OXYACETYLENE (OFC-A) AND OTHER CUTTING PROCESSES

"Cleanest welding place I've ever seen. Great high trees and deep grass line an open space in back, giving a kind of village-smithy appearance."

—Robert M. Pirsig, *Zen and the Art of Motorcycle Maintenance*

GOAL

- Develop an ability to cut metal safely using some popular cutting processes

QUESTIONS

- What are some of the processes used for cutting metal?
- How is the choice of a cutting process made?
- What equipment is required?
- How does the equipment work?
- What is quality cutting?

SAFETY FIRST

Before using any cutting process, be sure the area is clear of all **flammable** and **volatile** materials. Sparks from cutting can travel more than 30 feet. Be careful not to cut through a hose, **electrode lead,** or **workpiece lead,** which should be removed from the cutting area. Pay attention: Your nose may be the first to sense something is on fire. Know where the fire extinguishers are located, and know the shop procedures for responding to an emergency. Review safety guidelines and the **operator's manual** before using any piece of equipment discussed in this unit.

CUTTING METAL

The fabrication of almost all products requires the use of one or more methods for preparing metal for **welding.** A hacksaw, band saw, abrasive cut-off saw, and shear are four methods of cutting metal that a welder may have to use. Time is an important consideration when choosing a method for cutting through metal.

A hacksaw (Figure 3–1) can make a straight cut through a piece of 2-inch pipe, but if several straight cuts are required, a band saw or an abra-

FIGURE 3–1 Hacksaw. *(Courtesy of Bill Johnson)*

sive cut-off saw (Figure 3–2) would be faster and not as tiring. However, a band saw or an abrasive cut-off saw would not be used if the job required cutting a sheet of **steel** in half. A shear designed for cutting sheet metal (Figure 3–3) will do the job, but it is not made to cut through 1-inch-thick plate. If a shear for cutting plate (Figure 3–4), or appropriate equipment for cutting some other structural shape, is not available, one of the cutting processes covered in this unit can be substituted. These other methods are defined by the **American Welding Society** as cutting processes.

The job of selecting a cutting process to use when preparing metal for welding might fall on the shoulders of the **welder.** The knowledge of and the ability to use these cutting processes will be helpful for welders in making this decision.

OXYACETYLENE CUTTING (OFC-A)

The **oxyacetylene cutting** process is the popular method for cutting **ferrous** metal when the electricity required for other cutting processes is not available. Other gases, such as **propane** and natural gas, which are cheaper than acetylene, are also combined with oxygen to produce the chemical reaction needed to cut ferrous metal. Gases like propane and natural gas require different accessories. Also, because the temperature of their flames is not as high as acetylene's, these two gases require more time to reach preheat temperature for cutting. The oxyacetylene cutting process is faster and more versatile than other flame cutting processes.

The **oxyacetylene flame** used to cut ferrous metal really does not cut at all. The flame acts chemically to speed up the **oxidizing** process that occurs naturally as metal **rusts.** The **torch** concentrates the oxidizing action of the flame at the targeted area until it reaches the **kindling temperature.** At a kindling temperature of 1,400° to 1,600° F, there is a chemical reaction that allows rapid oxidation of metal. A jet of **oxygen** can then be used to blow the oxidized metal away from the **kerf** in a controlled manner.

Refer to Unit 2 for safety procedures and requirements for setting up the oxyacetylene cutting

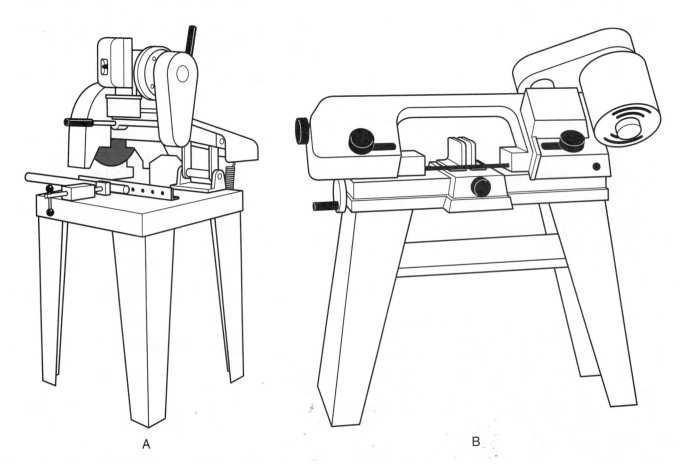

A   B

FIGURE 3–2   (A) Abrasive cut-off saw. (B) Horizontal band saw.

A

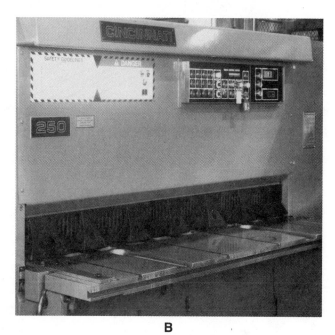

B

FIGURE 3–3   (A) Electric shear. *(Courtesy of Milwaukee Electric Tool Corp., Brookfield, Wisconsin.)* (B) Sheet metal shear. *(Courtesy of Cincinnati Incorporated, Cincinnati, Ohio).*

FIGURE 3–4   Heavy-duty shear for cutting plate. *(Courtesy of Cincinnati Incorporated, Cincinnati, Ohio).*

equipment. Although several models of cutting torches are available, there are basically only two types. One type of cutting torch is designed only for cutting (Figures 3–5 and 3–6). It is longer and heavier but much more comfortable to use than the **combination torch** (Figure 3–7). With a 75° torch, the torch body is raised above the plate. This provides for smooth cuts without interference. With the combination torch, the cutting attachment can be easily removed and replaced by a **welding tip** or a **heating tip** (Figure 3–8). Need should determine the type of cutting torch provided in any shop. Some shops have both types.

Consult the manufacturer's chart for specifications in choosing the proper size **cutting tip** and the correct pressure settings. The cutting tip size should meet the thickness of the steel to be cut.

FIGURE 3–5   Torch for cutting only. *(Courtesy of Victor Equipment Company.)*

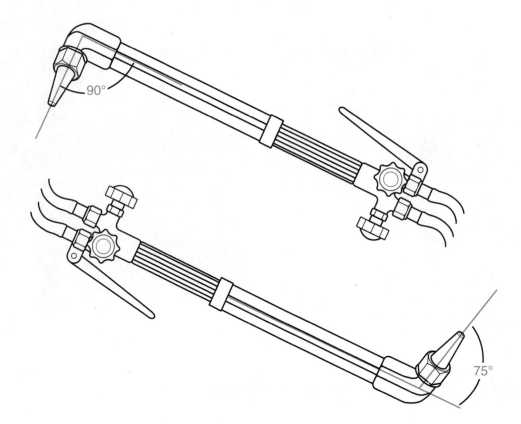

FIGURE 3–6 75° and 90° cutting torches.

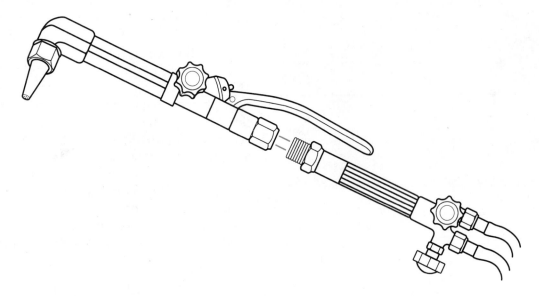

FIGURE 3–7 Combination torch.

Note that the size of the preheat holes in the tip vary with the thickness of the steel; the holes for cutting through thicker steel have larger diameters.

A set of tip cleaners is as important to the cutting operation as is having a selection of cutting tips to choose from (Figure 3–9). Tip cleaners should be used correctly with an up-and-down motion. Do not use a side-to-side or circular motion, which would enlarge the **orifices.** Tip cleaners keep the cutting tip clean and improve your ability to get the job done. Remember that oxyacetylene cutting is for use on ferrous metal (metal containing iron).

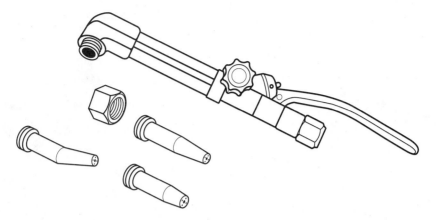

FIGURE 3-8   Torch head with various tips.

FIGURE 3-9   Using a tip cleaner. *(Courtesy of Jeffus, 1993.)*

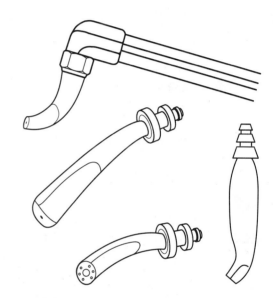

FIGURE 3-10   Gouging tips.

Gouging tips are used to remove a **weld** or to remove **base metal.** They aid repair and ensure complete **root penetration** in welding. Gouging tips are available in different sizes (Figure 3–10).

An adjustable wrench is as important as the cutting tips themselves (Figure 3–11). The wrench is necessary to remove and tighten tips securely. You should already be familiar with the use of the wrench for removing or tightening different pieces of equipment when setting up or breaking down the oxyacetylene cutting process. To remove a cutting tip easily, hold the torch out and place the wrench on the torch with the wrench handle parallel to the floor. Grab the handle, pushing downward with the heel of the hand (Figure 3–12).

Heating tips are used to increase the volume

FIGURE 3-11   Adjustable wrench. *(Courtesy of Bill Johnson)*

of heat to an area of metal (Figure 3–13). It may be necessary to **preheat** metal before welding or cutting. Heating tips are also used to deform metal in order to bend, straighten, or flatten it. The use of heat correctly applied is helpful in loosening nuts, bolts, or seized parts, although these procedures do not necessarily require a heating tip.

Many   shops   are   equipped   with   electric-

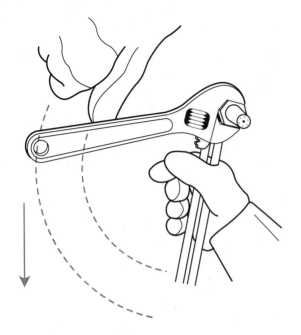

FIGURE 3–12 Removing tip from cutting torch.

FIGURE 3–13 Heating tip. *(Courtesy of Jeffus, 1993.)*

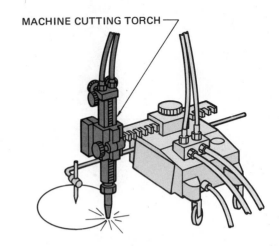

MACHINE CUTTING TORCH

FIGURE 3–14 Portable flame-cutting machine. (Courtesy of Chemetron Corporation.)

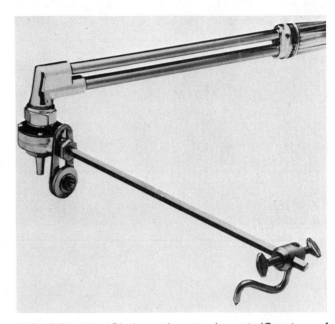

FIGURE 3–15 Circle-cutting attachment. *(Courtesy of Jeffus, 1993.)*

powered cutting machines designed for use with cutting torches (Figure 3–14). Some of these machines can cut straight lines and complicated patterns. There are also attachments to help the welder **manually** cut a **bevel** or a circle (Figure 3–15). Although these pieces of equipment are necessary in some situations, a person with the ability to use a cutting torch **freehand** is much more versatile than any machine.

For assistance with cutting a straight line, a piece of metal can be clamped in position for support to make an even kerf (cut) (Figure 3–16). Once you have acquired the skill to do freehand cutting, you will find that cutting machines and other aids are welcome additions. Skilled welders depend on their knowledge and ingenuity as well as on the equipment.

Using the **neutral flame,** hold the torch with the inner **cones** of the cutting tip just over the steel (Figure 3–17). Watch the color change, and wait until the red in the steel turns yellow. Push in the lever to send a jet of oxygen into the metal, blowing oxidized steel out of the kerf.

Cast irons are ferrous metals that can be flame-cut. The kerf is not usually as smooth as when cutting through steel. To cut through cast iron requires a slower travel speed and a **carburizing flame.** The torch is worked slowly from side to side. Select a tip of the size that would be used for steel of the same thickness.

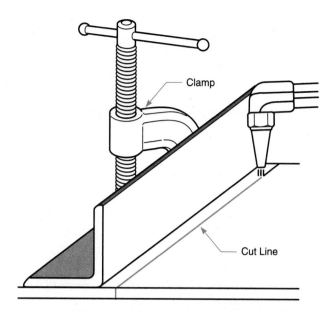

FIGURE 3–16 Metal guide for cutting straight lines.

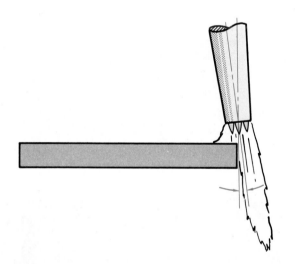

FIGURE 3–17 Position of torch for cutting steel. *(Courtesy of Larry Jeffus.)*

AIR CARBON ARC CUTTING (CAC-A)

Air carbon arc cutting is a versatile process. Because air carbon arc cutting does not require oxidation, it can be used to cut any kind of metal. Air carbon arc cutting requires the following equipment: a **power source, compressed air,** a specially designed torch (with an electrode holder), and carbon-graphite **electrodes.** An electrode is a **consumable** or a **nonconsumable** through which

electricity flows to create an **arc** that melts metal. The torch is connected to compressed air with a second fitting for connecting it to the electrode holder of the welding machine. The torch has an on/off button to control the air flow to its own electrode holder (Figure 3–18).

The power source for use in air carbon arc cutting should consist of a welding machine with an **amperage** setting that allows for a continuous welding (100%) **duty cycle.** The duty cycle is the recommended percentage of time a welding machine should be under the load of welding. A 60% duty cycle means that the welding machine is rated for welding six minutes out of every 10 at the maximum amperage setting. Information about the duty cycle should be attached to the welding machine.

Most welding machines under 400 **amperes** are not large enough to take the constant load of air carbon arc cutting. To determine the maximum amperage setting for continuous welding on any welding machine, multiply the duty cycle percentage by the maximum amperage of the given welding machine. For example, a welding machine with a maximum amperage of 200 multiplied by its 20% duty cycle puts its rated continuous welding ceiling at 40 amperes (200 × 0.20 = 40). As shown in Table 3–1, 40 amperes is much too low for the three electrode sizes indicated. Thus, a welding machine used for air carbon arc cutting must be over 400 amperes.

Although air carbon arc cutting has some application for cutting metal, this process is not as fast as other cutting processes. The electrode is positioned in the holder, not extending more than 5 inches. This allows air flow to push molten metal

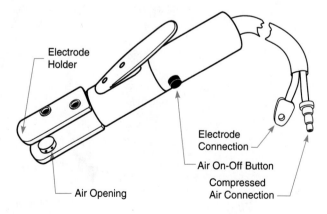

FIGURE 3–18 Air carbon arc cutting torch.

TABLE 3–1: INFO FOR AIR CARBON ARC CUTTING

	Electrode Size		
Current	5/32	3/16	1/4
Direct current electrode positive **(DCEP)**	90–150 amperes	150–200 amperes	200–400 amperes
Alternating current **(AC)**	* * * * * * *	150–200 amperes	200–300 amperes

out of the kerf (Figure 3–19). The kerf is usually not as smooth, making a finishing process, such as grinding, necessary in some cases. Air carbon arc cutting is very popular for use in gouging out welds or defects in castings that must be welded.

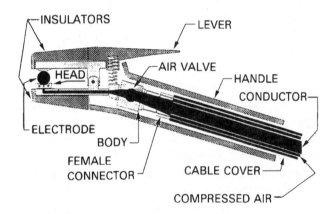

FIGURE 3–19 Placing electrode in the torch. *(Courtesy of American Welding Society.)*

PLASMA ARC CUTTING

Plasma arc cutting (PAC) is used for cutting **nonferrous** metals, like **aluminum,** and ferrous **high-alloy steels,** like **stainless steel,** which cannot be cut by the oxyacetylene cutting process. Less distortion occurs when cutting sheets of steel with plasma arc cutting than with oxyacetylene cutting (Figure 3–20).

When first developed, the equipment needed for the plasma arc cutting process was very expensive and required a large power source and special gas mixtures. Advanced technology has reduced the size of the power source, lowered the purchase price, and improved the torch design (Figure 3–21). As a result, the popularity of plasma arc cutting has been growing. Today, plasma arc cutting can be done with compressed air. New equipment, weighing less than 50 pounds in total, can cut cleanly through 1-inch-thick material (Figure 3–21).

In plasma arc cutting, the heat of an arc is

FIGURE 3–20 Plasma arc cutting. *(Courtesy of Jeffus, 1993.)*

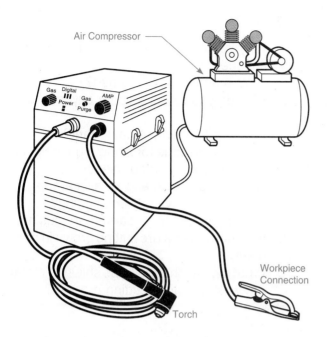

FIGURE 3–21   Plasma arc with air compressor.

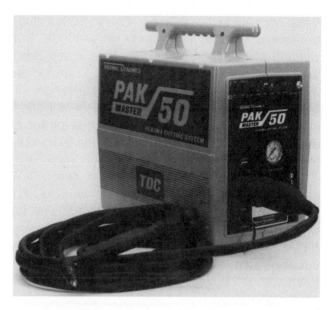

FIGURE 3–22   Versatile power source for plasma arc cutting. *(Courtesy of Thermal Dynamics.)*

concentrated to melt metal, which is removed by the force of air. The heat generated within by plasma arc cutting can reach temperatures approaching 60,000° F. Compare this temperature with that of an oxyacetylene cutting **oxidizing flame,** which approaches a temperature of 6,300° F. It is important to remember, though, that temperature alone does not make the critical difference. If temperature were so important, a tip of just one size could do the job in all cutting processes. The volume of heat generated at the cut is the major factor in all cutting operations, and the volume of heat for plasma arc cutting depends on the amount of amperage generated at the cut.

The minimum equipment required for plasma arc cutting consists of a power source (Figure 3–22), a torch, and compressed air. Plasma arc cutting has an advantage over any oxygen/fuel cutting process because of its versatility. Plasma arc cutting can cut a variety of metals. The torch itself is made up of a gas-shielding **nozzle,** a tip, and an electrode. These are attached to the torch body (Figure 3–23). The parts of the torch can be changed for different applications.

ARC CUTTING (AC)

Arc cutting is the process of cutting through metal with the heat generated between an elec-

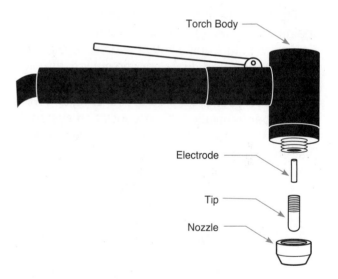

FIGURE 3–23   Plasma torch.

trode and the base metal. Arc cutting is a convenient method for cutting metal when a more appropriate cutting method is not available. **Shielded metal arc welding, gas metal arc welding,** and **gas tungsten arc welding** are three welding processes that can be used for arc cutting. Special carbon electrodes have been designed for cutting with the shielded metal arc welding process.

A **flux**-covered electrode made for shielded metal arc welding can also be used for shielded metal arc cutting (SMAC). E6010 and E6011 electrodes work better than some other types, but welding electrodes should never be considered the primary method for cutting steel. Cranking up the amperage setting well beyond the range for welding will turn flux-covered welding electrodes into cutting electrodes. The welder uses the force of the arc and the **manipulation** of the electrode to push molten metal out of the kerf.

Current flow and gas flow must also be increased when using gas metal arc welding and gas tungsten arc welding equipment for gas metal arc cutting (GMAC) and gas tungsten arc cutting (GTAC). Gas tungsten arc cutting can cut through stainless steel and aluminum when a heavy-duty torch with a current setting of greater than 200 amperes is used.

SAFETY REMINDERS

Remember the hazards that must be guarded against when using any of the cutting practices discussed in this unit. Protect yourself against **ultraviolet radiation,** fumes during cutting, the high temperature of molten metal, noise, and high voltage. Protect your eyes, protect your ears, and work with proper ventilation. If the ventilation system is inadequate, wear a mask with the proper filtration system. Some procedures might even require breathing oxygen through a respirator (Figure 3–24). Wear welding gloves and proper clothing, eye protection, and earplugs. Stay dry. Never depend on your employer for your safety; the primary responsibility is always your own.

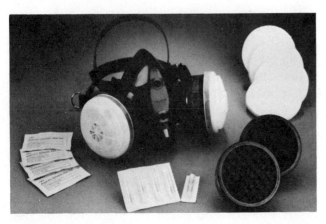

FIGURE 3–24 Respirator. *(Courtesy of Paint-Safe Products.)*

REVIEW

1. Why is the selection of a cutting method important?
2. How is sheet metal defined?
3. How do you select a tip for oxyacetylene cutting?
4. Why are tip cleaners important?
5. Give two reasons for using a gouging tip.
6. When is a heating tip used?
7. Name one advantage of air carbon arc cutting over oxyacetylene cutting?
8. Give one use for air carbon arc cutting.
9. Why has plasma arc cutting grown in popularity?
10. What simple change turns shielded metal arc welding into a cutting process?

Create Three Questions

1.
2.
3.

Related Math and English Questions

1. A 30% duty cycle means that a welding machine can operate at maximum amperage for how many minutes out of every 10?
2. What is the maximum continuous welding amperage of a 200-amperage welding machine with a 30% duty cycle?
3. Write a paragraph explaining duty cycle.
4. Can a 200-amperage welding machine be used for air carbon arc welding? Explain your answer.

For Further Thought

1. Besides plasma equipment and shears, name another method that could be used to cut a 4-foot by 8-foot sheet of 16-gauge stainless steel in half?
2. A cast iron fitting was repaired using **braze welding,** but the fitting did not hold. Arc welding will be used the second time. What must be done to prepare the **joint** for arc welding?
3. An experienced welder picked up the shop's oxyacetylene cutting torch but, making a test cut, found **slag** on the bottom of the cut that would require unnecessary

grinding on the actual project. What could be done to remedy the problem?

4. When there is a choice between a band saw and a cutting torch, why is the band saw preferred?

SUGGESTED ACTIVITIES

1. Discuss any safety issues of concern.

2. Display any equipment used with the oxy-acetylene process, such as a hose repair kit.

3. Examine an assortment of tips used for heating, cutting, and gouging.

4. Note the duty cycles of different welding machines, and compute the continuous duty cycle amperage for each one.

UNIT 3: CUTTING EXERCISE 1

Oxyacetylene Cutting (OFC-A): Neutral Flame

Begin by setting up the equipment as explained in Unit 2. Check the number on the cutting tip, and compare that number with the manufacturer's recommendations for the thickness of steel to be used in Exercise 2 (Figure 3–25). Set the pressure of the regulators as required by the chart for the thickness of steel.

Necessary Material and Equipment

Safety glasses	Welding gloves	Protective clothing	Goggles
Sparklighter	Adjustable wrench	Tip cleaners	

Instructions for Combination Torch

1. If using a combination torch (a torch designed for both welding and cutting), open completely the primary oxygen **valve** on the torch body.
2. Open the acetylene valve and light torch, adding acetylene until the flame separates from the cutting tip. Turn down the acetylene until the flame just returns to the tip.
3. Open the secondary oxygen valve on the cutting attachment, adding oxygen until a neutral flame is reached (Figure 3–26).

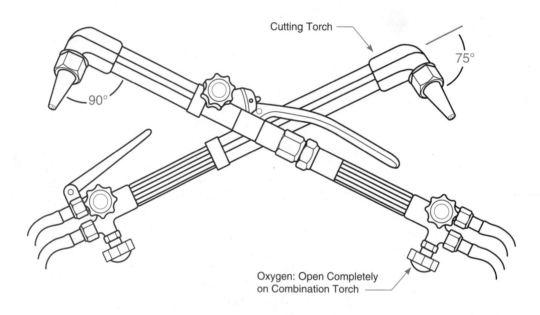

Cutting Torch

75°

90°

Oxygen: Open Completely
on Combination Torch

CUTTING TIP CHART

Metal Thickness	Tip Size	Oxygen Pressure	Acetylene Pressure
1/8"	1	20 to 25	3 to 5
1/4"	2	20 to 25	3 to 5
3/8"	3	25 to 30	3 to 5
1/2"	4	30 to 35	3 to 5

BRAND X

FIGURE 3–25 Preparing to light the torch.

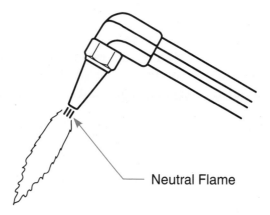

FIGURE 3–26   Neutral flame.

4. Push in the lever for the jet of oxygen. Readjust at the same time the secondary oxygen valve to neutral, if necessary.
5. Examine the jet of oxygen coming from the center hole of the tip. Two thin white lines should extend out from the tip (Figure 3–27). If they do not, the center hole is partially plugged. Clean with tip cleaners as required.
6. To extinguish the flame, turn off the acetylene valve first.
7. Turn off the secondary oxygen valve at the cutting attachment.

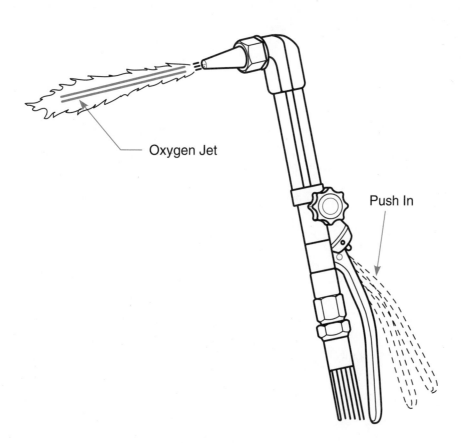

FIGURE 3–27   Oxygen jet.

Instructions for Cutting Torch

1. Open the acetylene valve, and light torch, adding acetylene until the flame separates from the tip. Turn down the acetylene until the flame just returns to the cutting tip.
2. Open the oxygen valve, adding oxygen until a neutral flame is reached (see Figure 3–27).
3. Push in the lever for the jet of oxygen. At the same time, readjust the oxygen to neutral, if necessary.
4. Examine the jet of oxygen coming from the center hole of the tip. Two thin white lines should extend out from the tip (see Figure 3–28). If they do not, the center hole is particially plugged. Clean it with tip cleaners as required.
5. To extinguish the flame, turn off the acetylene valve first.
6. Turn off the oxygen valve.

UNIT 3: EXERCISE 2

Oxyacetylene Cutting (OFC-A): Cutting Straight Lines

Relax and find a comfortable position from which to begin the kerf (Figure 3–28). Try to steady yourself against some stationary surface to keep your motion smooth throughout the cutting exercise. Be aware of where the molten metal is falling (a cutting table should be used). Never allow molten metal to fall on the floor (it will damage concrete), to touch the hoses, or to land on you.

Remember if the correct tip size is used at the proper pressure settings and the tip is not plugged, your travel speed across the metal is the most likely cause of your problem. Aside from the positioning of the torch, you are either traveling too fast or too slow.

Necessary Material and Equipment

Safety glasses	Welding gloves	Protective clothing	Goggles
Sparklighter	Adjustable wrench	Soapstone	Straightedge
Punch	Ball peen	Tip cleaners	Chipping hammer
Scrap steel, 1/8″–1/2″			

Instructions

1. With the soapstone and a straightedge, mark off several lines from the edge of the steel, 1/2 inch apart.
2. Use the punch and hammer to mark along the lines every inch so that some evidence of the lines will remain when the soapstone disappears. Have everything ready before lighting the torch.
3. Light the torch. Begin by opening the acetylene valve, and use a sparklighter. Then open the acetylene valve until the flame separates from the end of the tip, bringing the flame back to a point where the flame appears to be touching the end of the tip.
4. Open the oxygen value, adjusting to a neutral flame.
5. Push in the lever to release a jet of oxygen, readjusting to neutral if necessary. Note the two

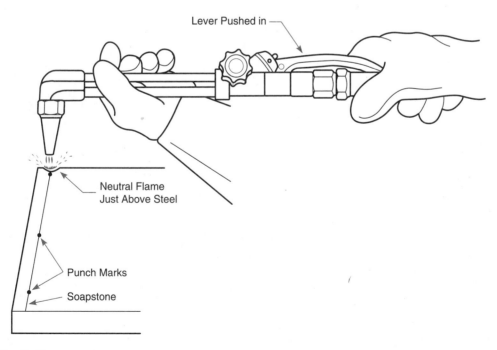

Lever Pushed in

Neutral Flame
Just Above Steel

Punch Marks

Soapstone

FIGURE 3–28   Cutting.

white lines for the jet of oxygen. These two lines should extend for some distance beyond the tip. If they are not sharp (clear), use tip cleaners to clean out the center orifice.

6. Find a comfortable position that allows you to easily roll over the arm that is providing the stability for cutting.

7. Preheat the beginning of the cut until molten (appears yellow), using the neutral flame. Push in the lever when the steel reaches kindling temperature. While pushing in the lever for the jet of oxygen, move along the line of kerf at an even speed. Do not move so fast as to overrun the molten metal, which would require stopping to preheat once more.

8. Keep the preheat flames (inner cones) just above the plate, with torch head at a 90° angle, directing the oxygen jet downward.

9. Practice until you are comfortable enough with the torch to provide satisfactory results.

10. Turn off the torch, turn the steel over, and check the slag (oxidized metal). If you carried out the cutting process successfully, you should be able to easily knock off any visible slag along the kerf with the chipping hammer. Never use the torch as a hammer, and never lay a lighted torch down!

11. Repeat the cutting process until smooth. Make cuts from both the right and the left. Try another thickness of steel that requires a different tip size. Thicker steel slows the travel speed across the line of cut. Thinner steel quickens the travel speed; use a sharp angle nearly parallel for cutting the thinner sheet steel.

12. Check with the instructor for evaluation.

13. Practice this exercise until you meet the standards established for the course.

UNIT 3: EXERCISE 3

Oxyacetylene Cutting (OFC-A): Bevels

The bevel is very important when welding plate where 100% root penetration is required. The bevel is an angle put on the edge of metal plate. In many applications, two pieces of plate welded together will not achieve complete root penetration without a bevel. This can be readily seen by turning the **weldment** over. A beveled edge helps to ensure a **quality weld** with 100% root penetration for butt joint.

Be sure to match the tip size and regulated pressure settings recommended by the torch manufacturer based on the width of the face on the bevel, not the thickness of the steel (Figure 3–29).

Necessary Material and Equipment

Safety glasses	Welding gloves	Protective clothing	Goggles
Sparklighter	Adjustable wrench	Soapstone	Straightedge
Punch	Ball peen	Tip cleaners	Chipping hammer
Scrap steel, 1/2″ or greater thickness			

Instructions

1. With the soapstone, mark a line (no more than 6 inches) with the straightedge. Mark the line from the edge, a width less than the thickness of the plate. This will result in a **groove angle** less than 90° and more in line with **welding procedures.** Punch the line every inch.
2. Light the torch. Begin by opening the acetylene valve; then use the sparklighter. Open the acetylene valve until the flame separates from the end of the tip. Reduce the gas volume until the flame appears to be touching the end of the tip.
3. Open the oxygen valve, adjusting to a neutral flame.
4. Push in the lever to release a jet of oxygen, readjusting to neutral if necessary. Release the lever.

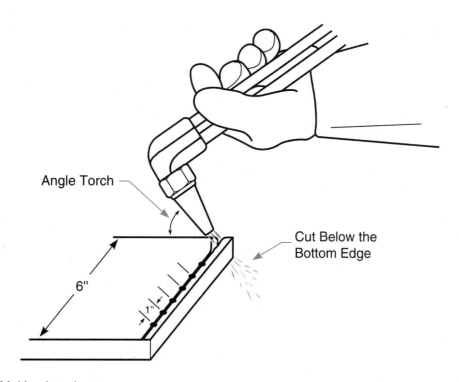

Angle Torch

Cut Below the Bottom Edge

6″

1″

FIGURE 3–29 Making bevel cuts.

5. Find a comfortable position and angle the torch head in line with the bottom edge of the plate. Be in position to watch the cut of the kerf as the torch moves along the plate away from your body. Keep the inner cones of the neutral flame off the plate.

6. Preheat the base metal to kindling temperature (appears yellow) before pushing in the lever. After starting the cut, move the torch along the plate to completion.

7. Use the chipping hammer to knock away any slag that has formed.

8. Make several bevels, looking for quality work.

9. Check with the instructor for evaluation, if necessary.

10. Practice this exercise until you meet the standards established in the course.

UNIT 3: EXERCISE 4

Oxyacetylene Cutting (OFC-A): Holes

It is sometimes necessary to pierce holes and circles when precise openings are not required. This exercise is not difficult once you are familiar with the cutting torch. Begin by setting up the equipment as explained in Unit 2. Match the tip size and regulator pressure settings from the manufacturer's chart against the thickness of the steel. Light the torch, and check the jet of oxygen to see if any of the orifices are partially plugged. Relax, find a comfortable position, and begin by making a few piercing holes. When cutting circles, start the hole at the center, working the kerf to the outside where the circle has been marked by soapstone (Figure 3–30).

Necessary Material and Equipment

Safety glasses	Welding gloves	Soapstone	Circle pattern (piece of
Sparklighter	Adjustable wrench	Tip cleaners	pipe)
Punch	Ball peen	Goggles	Chipping hammer
Scrap steel, 1/8″–1/2″	Protective clothing		

Instructions

1. Light the torch. Begin by opening the acetylene valve, then use the sparklighter. Open the acetylene valve until the flame separates from the end of the tip. Reduce the volume until the flame appears to be touching the end of the tip.
2. Open the oxygen valve, adjusting to a neutral flame.
3. Push in the lever to release a jet of oxygen, readjusting to neutral if necessary.
4. Relax and get comfortable. Preheat the spot to be pierced.
5. Angle the tip slightly out of 90°, and push the oxygen lever at kindling temperature (appears yellow). A slight angle at first will keep molten metal from shooting back at the tip before piercing completely through the metal.

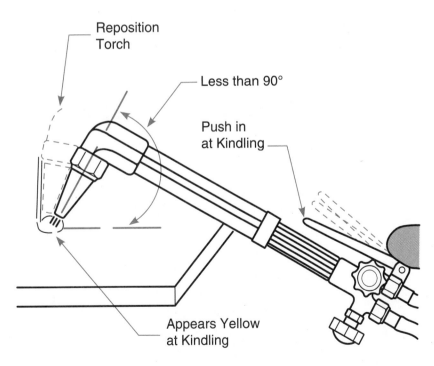

FIGURE 3–30 Piercing holes.

6. Upon cutting through the steel, reposition the torch instantly to 90°, and cut a small hole. Make several.
7. Extinguish the torch properly. Turn off the acetylene valve first.
8. Use a pattern (a round object) to make several circles, and mark them with a punch. Put the circles close together to save steel.
9. Light the torch, and adjust the flame to neutral.
10. Pierce a hole into the center of the circle before making a kerf to the outside.
11. Cut several circles. Concentrate on making them as round as possible. A shaft of a given diameter can be used for fit to measure your work.
12. Check with the instructor for evaluation, if necessary.
13. Practice this exercise until you meet the standards established in the course.

UNIT 3: EXERCISE 5

Oxyacetylene Cutting (OFC-A): Gouging

Begin by setting up the equipment as explained in Unit 2. Replace the cutting tip with a gouging tip. Adjust the pressure regulator as recommended by the manufacturer for the tip selected, based on the tip number. With practice, you should be able to gouge out weld metal without damage to the weldment. In many situations, welds can be gouged out without ever raising the rest of the weldment to the kindling temperature, in effect, peeling welds away.

This exercise can be used to cut old welds out of pieces of steel that have been welded together. On the job, gouging out defective welds and cracks in a weldment are the primary uses for the gouging tip. Recycling discarded weldments gives additional life to pieces of steel that have been relegated to the scrap pile (Figure 3–31).

Necessary Material and Equipment

Safety glasses	Welding gloves	Protective clothing	Goggles
Sparklighter	Adjustable wrench	Gouging tip	Tip cleaners
Chipping hammer	Steel with old welds to be removed		

Instructions

1. Set up this exercise so that flying sparks go in a safe direction.
2. Light the torch. Begin by opening the acetylene valve, then use the sparklighter. Open the acetylene valve until the flame separates from the end of the tip. Reduce the volume until the flame appears to be touching the end of the tip.
3. Open the oxygen valve, adjusting to a neutral flame.
4. Push in the lever to release a jet of oxygen, readjusting to neutral, if necessary.
5. Direct the flame in a safe direction. Bring to kindling temperature, and push in the oxygen lever.
6. Move the **molten pool** along the weld or crack, gouging out metal.
7. Check with the instructor for evaluation, if necessary.
8. Practice this exercise until you meet the standards established in the course.

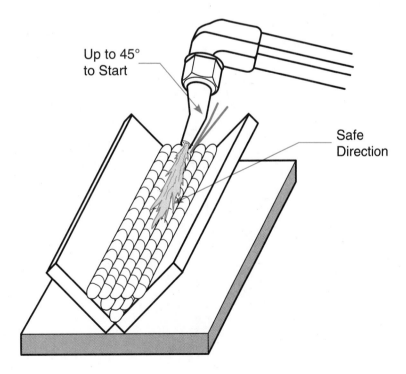

Up to 45°
to Start

Safe
Direction

FIGURE 3–31   Gouging.

UNIT 3: EXERCISE 6

Air Carbon Arc Cutting (CAC-A)

Begin by setting up the equipment. Follow the manufacturer's recommended procedure, as laid out in the operator's manual for the torch. Match the carbon-graphite electrode size to the amperage setting on the welding machine. Choose electrodes for use with a direct current (DC) power source or an alternating current (AC) power source.

Be sure that all connections are **secure** with air pressure (80 to 100 psi). Earplugs are recommended for this exercise. Direct the sparks in a safe direction. This exercise is suggested for gouging out old welds from scrap metal or for gouging out completed welds in Unit 4 (Figure 3–32).

Necessary Material and Equipment

Safety glasses	Earplugs	Welding gloves	Welding machine
Welding helmet	Carbon-graphite	Air arc cutting torch	
Metal (ferrous or	electrodes	Protective clothing	
nonferrous)			

Instructions

1. Connect the **workpiece connection** to the work or a metal structure, like the worktable that is in contact with the work.
2. Place an electrode into the holder. No more than 5 inches of the electrode should extend beyond the holder. Line up the electrode with the air openings of the holder so that the airstream will blow the molten metal away.
3. Angle the electrode up from the surface of the work approximately 30°.
4. Push in the button on the torch.
5. Lightly touch the electrode to the surface of the work. Move along the surface of the work at the travel speed necessary to keep the arc going, gouging out metal. A slight motion from side to side will remove more metal. A steeper angle will cut deeper into the surface of the work.
6. Use the sound of the arc to help you judge the travel speed.

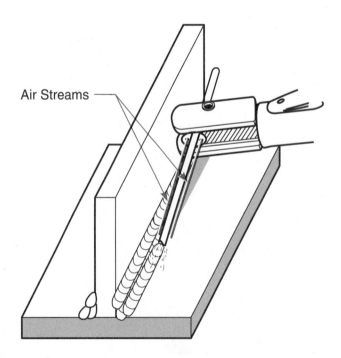

Air Streams

FIGURE 3–32 Air carbon arc cutting.

7. Do not let the electrode become too short. This can damage the holder. Stop, push the button in, and readjust the electrode. Begin again.
8. Several passes may be required to remove all of the necessary metal. Continue until the work is completed. The idea is to be as smooth and consistent with each pass as possible.
9. Check with the instructor for evaluation, if necessary.
10. Practice this exercise until you meet the standards established in the course.

UNIT 3: EXERCISE 7

Plasma Arc Cutting (PAC): Straight Lines

Begin by setting up the equipment. Follow the manufacturer's recommended procedure, as laid out in the operator's manual for the torch. For some power sources, **nitrogen** gas can be substituted for compressed air using the same equipment.

With a straightedge and soapstone, draw several lines. The current **output** and orifice diameter of the nozzle in combination with travel speed will result in a dross (slag). A dross is oxidized metal formed on the underside of the kerf. Select a nozzle with an orifice diameter as small as possible within the current setting, but avoid too small a diameter, which can damage the nozzle. With safety in mind, never use plasma arc cutting while in contact with a wet surface or while in or under water (Figure 3–33).

Necessary Material and Equipment

Safety glasses	Earplugs	Metal (ferrous or	Compressed air
Welding helmet	Plasma cutting console	nonferrous)	Protective clothing
Soapstone	torch	Welding gloves	Straightedge

Instructions

1. Connect the workpiece connection to the work or to a metal structure, such as a worktable, that is in contact with the work.
2. Check the torch and the consumable electrode. Make sure that the torch is set up for the job at hand, matching metal thickness with orifice diameter and current output.
3. With the straightedge and soapstone, make several lines. Mark them with a punch every inch or so. Put the lines close together to save metal.
4. Find a comfortable position, and switch on the torch. No preheat is required, so begin travel across the surface of the metal.
5. Make several cuts.
6. Check with the instructor for evaluation.
7. If cutting on steel, try various thicknesses. Also try nonferrous metals to gain experience. Use this opportunity to make bevel cuts and piercing holes.
8. Once again, check with the instructor for evaluation, if necessary.
9. Practice this exercise until you meet the standards established in the course.

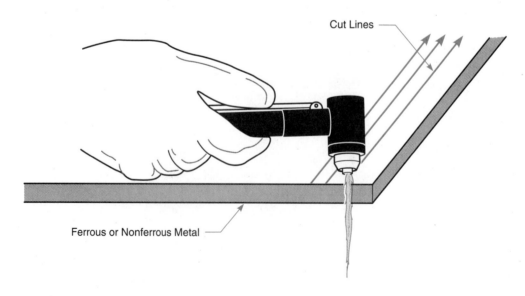

FIGURE 3–33   Plasma arc cutting.

UNIT 3: EXERCISE 8

Shielded Metal Arc Cutting (SMAC)

Field welding is welding in any situation away from the shop. The use of welding electrodes for cutting purposes may be necessary when no other practical cutting process is available. Rough cuts through nuts, bolts, and structural steel shapes can be made by turning up the amperage on the shielded metal arc welding machine. Although special cutting electrodes are manufactured to serve this purpose, welding electrodes can be used. This exercise calls for 6011 welding electrodes, but any electrode will do. Some work better than others. This exercise is one example of tapping into the ingenuity that every welder should develop (Figure 3–34).

Necessary Material and Equipment

Safety glasses	Welding gloves	Protective clothing	Welding helmet
Shielded metal arc welding machine	6011 electrodes	Scrap steel	

Instructions

1. Depending on the size of the electrode and the type of electrode, double the amperage setting recommended for welding.
2. Work the electrode quickly in and out of the molten pool, forcing the molten metal out of the kerf by both the force of the arc and the pressure of the electrode itself in the molten pool.
3. Adjust the amperage setting, if necessary. For example, overheating an electrode can result in an electrode sticking to the steel as the covering on the electrode drops off. Replace with another electrode and begin again, lowering the amperage.

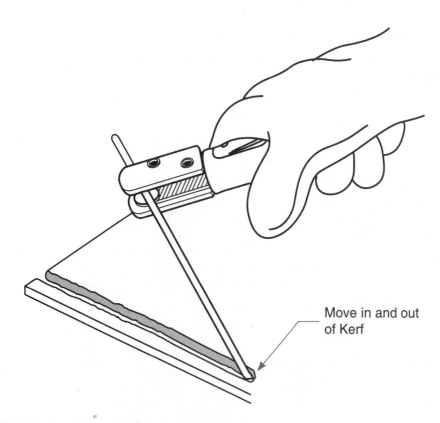

Move in and out of Kerf

FIGURE 3–34   Shielded metal arc cutting.

4. Repeat the procedure with another type of electrode or another diameter of electrode. Note the difference of using one electrode rather than another.
5. Check with the instructor for evaluation, if necessary.
6. Practice this exercise until you meet the standards established in the course.

SHIELDED METAL ARC WELDING

"But a person who does machining or foundry work or forge work or welding sees 'steel' as having no shape at all. Steel can be any shape you want if you are skilled enough, or any shape but the one you want if you are not."
—Robert M. Pirsig, *Zen and the Art of Motorcycle Maintenance*

GOALS

- Develop an ability to do shielded metal arc welding safely
- Pass destructive tests that meet the specifications set forth in the American Welding Society's *Structural Welding Code*

QUESTIONS

- What is shielded metal arc welding?
- What equipment is required?
- How does the equipment work?
- What are some of the problems in making quality welds?

SAFETY FIRST

Safety is important when using electric **welding** equipment. Safe electric welding practices prevent injury and save lives. Twenty-one suggestions for electric welding safety are given toward the end of this unit. This list is just the beginning in developing a safe attitude around welding equipment. Always keep safety in mind as you familiarize yourself with this welding process (Figure 4–1).

ELECTRICITY

Shielded metal arc welding (SMAW) begins with electric power. **Current, voltage,** and **amperage** are three electrical terms used in welding. Current is the flow. Amperage is a measurement

FIGURE 4–1 Welder at work. *(Courtesy of Miller Electric Mfg. Co., Appleton, Wisconsin.)*

of current. Voltage is a measurement of the force (or pressure) causing the flow. These three terms are at the foundation for an understanding of electricity.

Alternating current (AC) and **direct current** (DC) are the two kinds of current. Simply stated, AC is the back-and-forth motion of current, as described by the term *alternating*. AC is the standard

frequency for the United States. The current changes direction 120 times per second within an electrical circuit, so the standard frequency is 60 **hertz** (cycles) per second.

AC is the **input** used to power familiar household appliances, such as toasters and hair dryers. It is also the input used to power some welding machines. More important, AC is used for the **output** in the arc during shielded metal arc welding.

Unlike AC, DC is current flow in one direction. DC operates electrical systems on automobiles, RVs, and power boats. DC is also used for the output in the arc during shielded metal arc welding through the use of a **rectifier.** A rectifier converts AC into DC.

Choice of Current

Three choices of current may be used as output in electric welding. AC is one. Because AC current changes direction many times within the electrical circuit, AC is different from the other two currents available for welding. The less-expensive **welding machines** on the market are designed for AC.

The two other choices of current involve DC. They are **direct current electrode positive** (DCEP) and **direct current electrode negative** (DCEN). With DC welding, when the **electrode** holder is attached to the positive terminal and the **workpiece connection** (ground clamp) is attached to the negative terminal, the current is said to be **electrode positive.** The terminals may

be marked on the welding machine by a plus (+) for positive and a minus (−) for negative.

By shifting the connections **manually** or by throwing a switch on the welding machine, DC current can be reversed. When the **electrode lead** (cable) attached to the electrode holder is changed to the negative terminal and the workpiece connection is changed to the positive terminal, the current is said to be **electrode negative.** The more-expensive welding machines offer all three choices of welding current: AC, DCEP, and DCEN.

Note that DCRP and DCEP are identical and that DCSP and DCEN are identical. The notation "DCEP" informs the **welder** that the procedure calls for attaching the electrode holder to the positive terminal (electrode positive). The notation "DCEN" informs the welder that the procedure calls for attaching the electrode holder to the negative terminal (electrode negative).

Electrode positive is another name for the non-standard term *reverse polarity.* Electrode negative is another name for the nonstandard term *straight polarity.* Because DCEP and DCEN describe the welding terminal connections directly, DCEP and DCEN are the abbreviations that will be used throughout this book (Figure 4–2).

Other differences between AC and DC are important. With AC welding, the arc goes out and restarts itself every time the current switches direction. This charactistic of the arc is not present with DC welding. This is one reason that **striking** (starting) an ARC under DC is much easier than

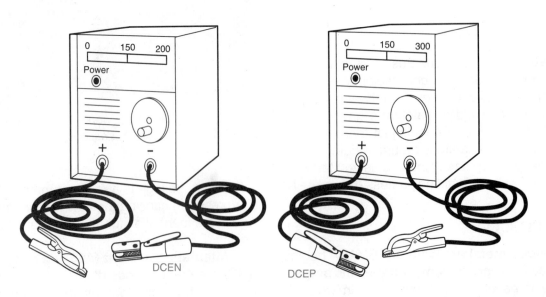

FIGURE 4–2 Power sources labeled DCEP and DCEN.

under AC. Also, during DC welding, the arc is usually more stable unless **arc blow** becomes a problem. Arc blow is an erratic arc that results in sputtering and **spatter.** It is caused by a concentration of magnetic forces. Simply put, arc blow means that the arc refuses to go where the welder wants it to go.

Finally, with all other welding conditions being equal, DCEP results in deeper **root penetration** than AC or DCEN. Although some electrodes are designed for both AC and DC welding, many electrodes are designed for use only with DC welding or only with AC welding. Understanding the differences between the three choices of current available for welding will help make you a better welder.

POWER SOURCE

There are five basic types of **power sources.** The **transformer,** one type, is a small and inexpen-

sive AC machine (Figure 4–3). The high voltage and low amperage from power lines is transformed into the low voltage and high amperage necessary for output during welding.

A second type of power source is basically a transformer with a rectifier. The rectifier is a device for changing AC to DC (Figure 4–4). Although some rectifier welding machines are designed for DC only, others have both AC and DC capabilities. A welding machine designed for both AC and DC welding is versatile and can be easily converted to **gas tungsten arc welding.** (This welding process is covered in Unit 6.)

The **motor generator,** the third type of power source, produces DC only (Figure 4–5). It generally requires **three-phase power** and is activated by AC power supplied by power companies. This is a sturdy machine with low maintenance, though some are quite noisy because of a rotating armature.

The **engine-driven generator,** the fourth type of power source, can be taken anywhere (Figure

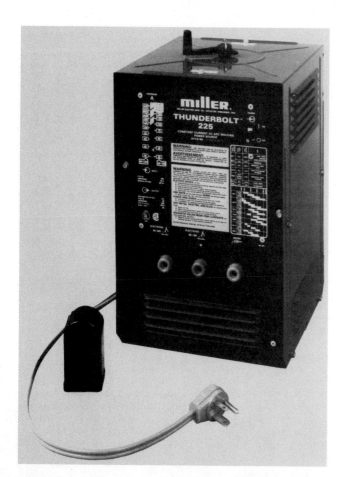

FIGURE 4–3 Transformer welding machine. *(Courtesy of Miller Electric Manufacturing Company, Appleton, Wisconsin.)*

FIGURE 4–4 Transformer/rectifier multiple process. *(Courtesy of Aftek, Inc.)*

FIGURE 4–5 Motor generator power source. *(Courtesy of Lincoln Electric Co., Cleveland, Ohio.)*

4–6). It can be lifted by helicopter to remote locations. The engine-driven generator is activated by an internal combustion engine. Some machines can also double as electric power supplies. Both the motor generator and the engine-driven generator produce their own welding current.

The **inverter** is the final type of power source

(Figure 4–7). The latest generation of inverters have the potential to combine shielded metal arc welding, **gas metal arc welding,** gas tungsten arc welding, **flux cored arc welding,** and air carbon arc cutting in one machine. A special feature on some units allows the inverter to be easily converted from line voltage of 208 to 230 to 460 single-phase power or three-phase power. These options can be combined in a package that weighs less than 80 pounds.

Future Technology

In recent years, there has been a revolution in welding equipment, with the development of the inverter. Today's new technology revolves around electronics, which have reduced the size of welding machines, and high-strength plastics, which protect the electronics. The new plastics are tough and corrosion-resistant, act as insulators by not conducting electricity, and are lighter than metal

FIGURE 4–6 Engine-driven generator. *(Courtesy of Lincoln Electric Company, Cleveland, Ohio.)*

FIGURE 4–7 Inverter power source. *(Courtesy of Thermal Dynamics, St. Louis, Missouri.)*

casings. Thanks to the use of the microprocessor (a miniature computer), interface boards have become the guts of the welding machine. Conventional welding machines, which contain the added pounds of winding required to step down the line voltage coming into the machine for output during welding, will become a thing of the past.

By pushing buttons like those found on calculators, the welder can easily set the welding machine precisely with the information programmed into the machine. With these machines, the welder can store welding parameters and retrieve them for use later.

Where is the technology heading? Manufacturers will continue to introduce a variety of new welding machines. The next generation of power sources will offer greater capacity (even higher amperages) in lightweight packages. Welders will be able to convert these power sources easily from one application to another by using a computer to punch up the program of the welding process desired.

SHIELDED METAL ARC WELDING

Shielded metal arc welding (SMAW) is a welding process in which a **flux**-covered electrode is melted into a liquid pool of metal. An arc that generates enough energy to melt metal is produced by current jumping across a gap. The air in the gap offers high resistance to the arc, creating intense heat. Temperatures within the **weld pool** can range from 6,000° to 10,000° F. The metal of the electrode melts together with the **base metal** to form the **weldment.**

Personal requirements and electrodes aside, the equipment used for this welding process is quite minimal. Perhaps this is one reason that the electrode holder and the workpiece connection are often neglected. It is important to keep the equipment in good repair. A damaged, worn-out electrode holder or workpiece connection should be repaired or replaced (Figure 4–8). A high-quality electrode holder can be repaired easily by replacing worn insulators. Finally, there is no excuse for allowing frayed **lead** connections to affect the quality of the work.

Shielded metal arc welding is a convenient and a very mobile welding process. It can easily be set up for use in extreme conditions that would limit the effectiveness of other welding processes.

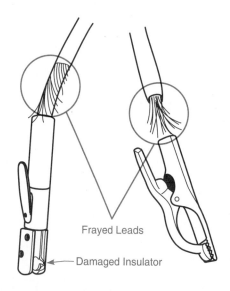

Frayed Leads

Damaged Insulator

FIGURE 4–8 Damaged electrode holder and workpiece connection.

A helicopter can lift an engine-driven shielded metal arc welding machine into a remote location. This equipment can operate in climates that would create problems for other welding processes. Learning this process requires training and involves a considerable amount of time, more than is necessary for **oxyacetylene welding** or any other manual welding process.

PROTECTIVE CLOTHING

Shielded metal arc welding is one type of electric welding with special protective attire. Sparks, **slag,** molten metal, and rays given off by the arc can result in serious burns. Screening the body from the harmful effects of the arc requires special attention. Safety glasses and earplugs are an absolute must in all shop conditions. Beyond these required safety measures, cotton clothing is recommended over synthetics, which can melt to the skin or spread a fire quickly over the body. Leather jacket, sleeves, and chaps protect clothing from damage by sparks and protect the body from burns. A helmet with special filter lenses protects the head and eyes. Safety glasses must be worn under the helmet. Earplugs will safeguard your hearing from the loud noises of the shop environment. Welding gloves protect your hands and wrists. Leather boots with **steel** toes help to protect your feet and toes from burns and from injury (Figure 4–9).

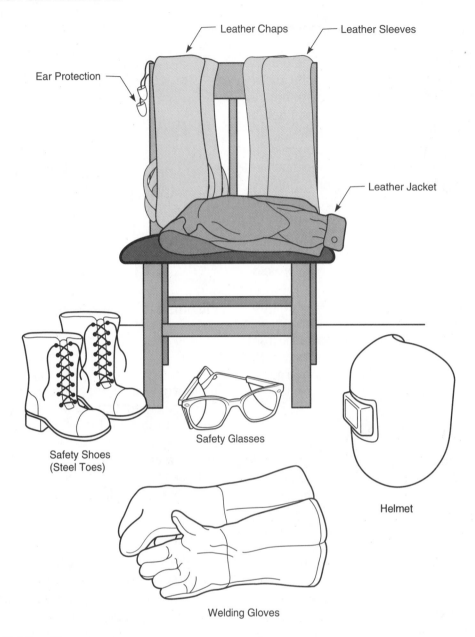

FIGURE 4–9   Safety attire.

ELECTRODES

Shielded metal arc welding requires electrodes. An electrode is the **filler metal** added to the weld. Bare electrodes (no flux) were first introduced, but they were not found to be very effective for most applications of shielded metal arc welding. They create an unstable arc that is difficult to start and keep going, and contaminants like **oxygen** and **nitrogen** from the air make the welds brittle and weak. The air, which will affect unprotected welding, consists of about 21% oxygen and 78% nitrogen. The remaining 1% is made up of several other gases mixed with air pollutants. Bare electrodes are still used today in some welding applications.

The addition of flux covering to bare wire provides protection during the welding process. The flux burns and produces a gaseous shield that protects the weld pool from the air. The flux also helps to stabilize the arc, making welding much easier. The lighter flux rises to the surface of the weld pool created by the arc, forming a coating called slag. Slag also protects the **weld** from the air until it cools down. It is then chipped off the weld with a slag hammer and swept away with a wire brush. Weld failure can result from not removing all the slag before making another **pass** over the **weld bead.**

There are many different kinds of electrodes (Figure 4–10). Each kind has special characteristics that lead to a specific result. For example, the presence of iron powder in the flux increases the deposition rate (the amount of metal added to the weld). Some electrodes are easier to use than others. They can be dragged across the weld surface, with the flux maintaining the **arc length.** The arc length is the distance from the tip of the electrode to the surface of the weld pool.

Just any electrode will not do. Matching the electrode with the base metal to be welded is a very important part of the welding operation. Other things to be considered in choosing the electrode are the welding position, the **joint design,** the thickness of the metal, and the type of current. Some electrodes cannot be used for all three current selections. Skilled welders understand that the selection of an electrode goes beyond picking up the closest electrode lying around.

Identification of Electrodes

There are many ways to identify electrodes. Nicknames, colors, and painted spots are three.

The classification system developed by the **American Welding Society** is the best way. This numbering system allows welders to use similar **carbon steel** electrodes produced by different manufacturers without a loss of quality. Knowing the American Welding Society's electrode number will save you time and possible trouble on the job in case a supplier does not have your favorite brand.

Besides identifying electrodes, the Amercan Welding Society's numbering system provides valuable information. Look at four examples:

E6011
E6013
E7024
E11018

The *E* means electrode. The first two numbers (or the first three if it is a five-digit number) refer to the minimum **tensile strength.** Remember, tensile strength is a measure of a material's resistance (in pounds per square inch) to being pulled apart (Figure 4–11). To find the tensile strength, in pounds per square inch, add three zeros to the first

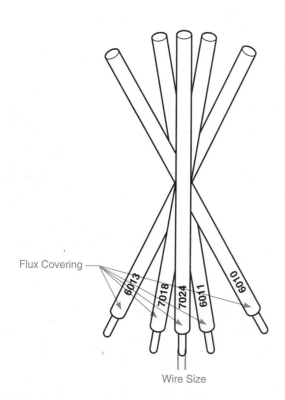

Flux Covering

6013 7018 7024 6011 6010

Wire Size

FIGURE 4–10   Electrodes.

FIGURE 4–11   Tensile testing machine. *(Courtesy of Jeffus, 1993.)*

two (or three) numbers. For example, the tensile strength of the electrodes mentioned earlier are as follows:

Electrode	Tensile Strength
E6011	60,000 psi
E7024	70,000 psi
E6013	60,000 psi
E11018	110,000 psi

The third number provides information about the welding position in which the electrode can be used. Number 1 means that the electrode can be used in all four welding positions. Number 2 means that the electrode can be used in the **flat** and **horizontal** welding positions only. Number 4 means that the electrode can be used in the flat, horizontal, overhead, and **downhill** welding positions. For example, E6011, E6013, and E10018 are all-position welding electrodes. E7024 is an electrode for welding in the flat and horizontal positions. E7048 is an electrode for flat, horizontal, overhead, and downhill welding positions.

The last number provides much information. It tells the kind of electric current to be used with a given electrode. It also provides information

about the ingredients in the flux, the degree of penetration (deep, medium, shallow), and the amount of slag and the shape of the weld bead. Here are some examples:

- E6011 is popular for work on dirty or **rusty** steel that is difficult to clean before welding. This electrode can be used with AC or DCEP. It has deep penetration with a **concave** weld bead and thin slag coating.
- E6013 is popular on clean steel. This electrode can be used with all three currents. It has shallow penetration with a **convex** bead and moderate slag coating.
- E7024 contains iron powder in the flux. This electrode can be used with all three currents. It has medium penetration with a convex bead and heavy slag coating.
- E11018 has iron powder in the flux. This electrode can be used with AC or DCEP. It has shallow penetration with a convex bead and moderate slag coating.

Table 4–1 lists the type of covering and the welding current for various types of electrodes, and Table 4–2 describes the penetration, shape of the bead, and slag.

TABLE 4–1: CHARACTERISTICS OF CARBON STEEL ELECTRODES

Fourth Digit	Type of Covering	Welding Current
0	Cellulose sodium	DCEP
1	Cellulose potassium	AC, DCEN, DCEP
2	Titania sodium	AC, DCEN
3	Titania potassium	AC, DCEN, DCEP
4	Iron powder titania	AC, DCEN, DCEP
5	Low-hydrogen sodium	DCEP
6	Low-hydrogen potassium	AC, DCEP
7	Iron powder iron oxide	AC, DCEN, DCEP
8	Iron powder low hydrogen	AC, DCEP

TABLE 4–2: CHARACTERISTICS OF CARBON STEEL ELECTRODES

Fourth Digit	Penetration	Shape of Bead	Slag
0	Deep	Concave	Thin
1	Deep	Concave	Thin
2	Medium	Convex	Moderate
3	Shallow	Convex	Moderate
4	Medium	Flat	Heavy
5	Medium	Convex	Moderate
6	Medium	Convex	Moderate
7	Medium	Flat	Heavy
8	Medium	Convex	Heavy

Sizes of Electrodes

The size of an electrode is determined by the diameter of the bare wire. The flux covering is not used to determine size because the thickness of the flux varies among the different electrodes. The selection of the size is based on the welding position, among other things. For example, when welding in the **vertical** or **overhead** position, gravity becomes a factor that limits the size of the electrode. A smaller-diameter electrode requires a lower amperage. A lower amperage results in a smaller weld pool, which is less affected by gravity in the vertical and overheat positions than a larger weld pool.

The size of an electrode also determines the amount of fill. A larger-diameter electrode is preferred when a large amount of fill is needed. Electrodes with an 1/8-inch diameter are probably the most popular. The smallest electrode has a 1/16-inch diameter, and the largest electrode manufactured has a 5/16-inch diameter (Figure 4–12).

Varieties of Electrodes

Although **mild steel** welding electrodes are very popular, there are other varieties, too. **Low-alloy** electrodes are used on high-tensile-strength steels. **Stainless steel** electrodes are used to arc weld stainless steel and for special applications. Bare **cast iron, nickel,** and nickel alloy electrodes are used to arc weld cast iron. **Aluminum** electrodes are used to arc weld aluminum, and **copper** alloy electrodes are used to arc weld copper alloys. Hard-surfacing electrodes are used to **build up** and overlay metal that is subject to abrasion (wear).

Care of Electrodes

Electrodes should be kept dry. **Low-hydrogen electrodes** (E7015, E7016, E7018) must be kept in dry conditions. A holding oven heated to between 150° to 300° F will keep electrodes safely in a moisture-free environment (Figure 4–13). Redrying wet electrodes can restore weldability in some applications, but any critical reconditioning of electrodes should be done in accordance with the electrode manufacturer's recommendations. Table 4–3 provides guidelines for storing some types of electrodes in an oven. Unless specified by the manufacturer, other electrodes should not be stored at the elevated temperatures indicated in Table 4–3. Some electrodes require a degree of moisture in the flux.

Some manufacturers that do welding have pol-

TABLE 4–3: OVEN STORAGE

AWS Electrode Number	Storage Temperature
E6010	a
E6011	a
E6012	a
E6013	a
E6020	a
E7024	150° F
E7016	300° F
E7018	300° F
Stainless Steels	300° F

ªOven storage not recommended.

FIGURE 4–12 Electrodes of various sizes.

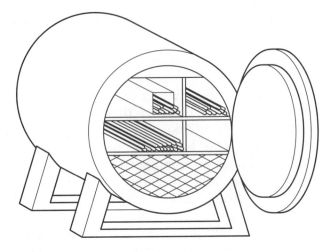

FIGURE 4–13 Commercial electrode oven.

icies concerning the handling of welding electrodes. For example, one manufacturer's policy states that low-hydrogen electrodes that are not taken from sealed containers or ovens must not be used. General guidelines are available for all varieties of electrodes. Welding suppliers who follow the American Welding Society numbering system should provide this information.

WELDING PROBLEMS

If the results of your welding do not meet the standards established in your class, review the following points. Have you matched the current setting on the welding machine to the electrode? Are you maintaining the proper **arc length,** the proper travel speed, and the proper **electrode angles**? There are two electrode angles: **work angle** and **travel angle** (Figure 4–14).

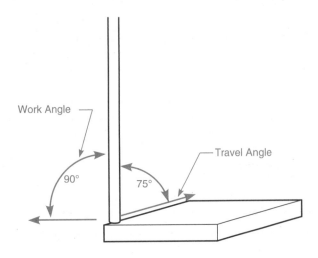

FIGURE 4–14 Work angle and travel angle.

Arc Blow

Arc blow is movement of the arc out of its intended path. Arc blow is caused by an unbalanced condition, and it creates a concentration of magnetic fields. With arc blow, the welder may be unable to control where the weld bead is deposited. The bead does not flatten out, and the slag becomes difficult to remove. Spatter (drops of metal surrounding the bead) may be excessive.

Generally, arc blow is a problem when welding with DC current. It becomes more intense as amperage increases. Arc blow can occur in deep groove welds, when welding into or out of corners, or when working on magnetized metal.

Here are some suggestions to deal with arc blow. You can control it by switching to AC, especially above 250 amperes. Arc blow can sometimes be controlled by changing the placement of the workpiece connection or by changing the welding direction (from left to right or vice versa). Finally, you might be able to control arc blow by reducing the arc distance while welding.

Common Welding Problems

Several problems can cause the failure of a weldment (Figure 4–15). Poor root penetration can result from a travel speed that is too fast, a current setting that is too low, or the improper preparation of the assembly for welding. **Undercut** can result from a current setting that is too high or from an improper electrode angle. **Porosity** is the pinholes (trapped gas) that results from impurities in the base metal or in previous passes, improper preparation or cleaning for the next pass, improper arc length, improper current setting, drafty conditions, or traveling too fast.

SAFETY REMINDERS

Review the safety points listed below:

1. Wear safety glasses with side shields. Safety glasses protect your eyes from slag (a nonmetal byproduct of the weld pool) and from grinding particles. They also deflect **ultraviolet rays,** which is given off by the arc, and helps prevent arc flash. Safety glasses are so important

Top View

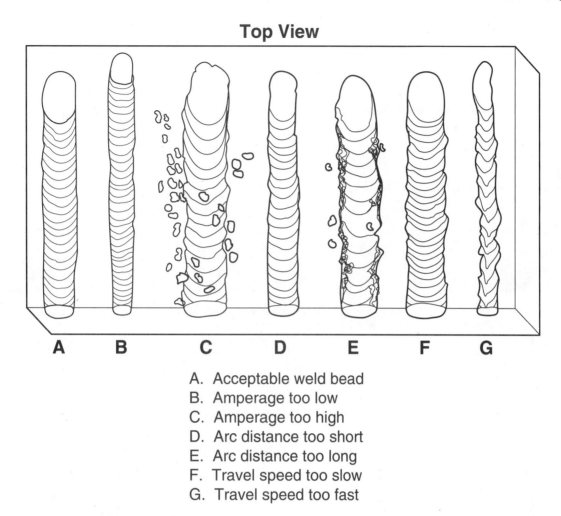

A. Acceptable weld bead
B. Amperage too low
C. Amperage too high
D. Arc distance too short
E. Arc distance too long
F. Travel speed too slow
G. Travel speed too fast

FIGURE 4–15   Common welding problems.

that no one should be in a welding shop without wearing them. However, safety glasses offer no protection against careless shop practices.

2. Use leather **welding gloves** to protect your hands and lower arms from burns and shock (Figure 4–16). Only gloves made for welding do an adequate job of handling heat without losing their shape. If you use tongs to handle hot metal, your welding gloves will last longer and not become stiff.

3. Wear cotton clothing, not synthetic fibers that will burn or melt to the skin, to shield exposed skin from arc rays, which are more intense than sunlight. A leather jacket, leather sleeves, and chaps, although expensive, are preferred because leather will shield your clothing from the sparks and spatter of welding (Figure 4–17).

4. Be sure the shop is clear of **volatile** and **flammable** materials (Figure 4–18). Gasoline and solvents have no place in a welding shop.

5. Do not bring cigarette lighters under pressure into the welding shop (Figure 4–19).

6. Always wear a welding helmet with the correct filter lens to protect against **arc flash** (Figure 4–20). Arc flash is a painful, but usually temporary, eye condition caused by the light of the welding arc. If you suspect an arc flash is more serious, get proper medical attention. Replace cracked lenses or damaged helmet. Check the shade number on the lens. Shielded metal arc welding requires a shade number of 10 to 14.

FIGURE 4–16 Welding gloves. *(Courtesy of Jeffus, 1993.)*

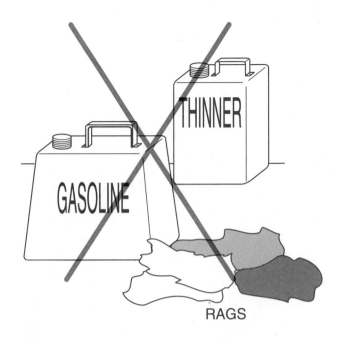

FIGURE 4–18 Volatile and flammable materials.

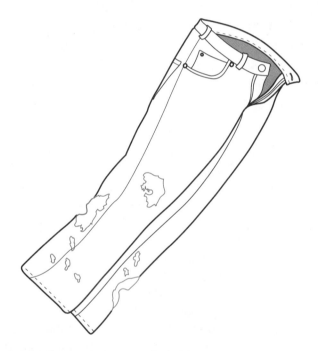

FIGURE 4–17 Damaged clothing.

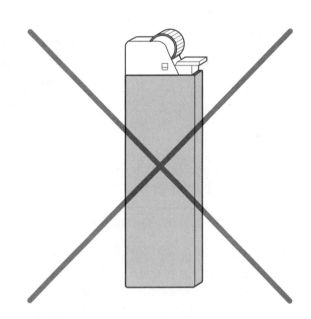

FIGURE 4–19 Cigarette lighter.

7. The welding shop should be equipped with fire extinguishers. Know their location and classification (Figure 4–21).
8. The welding shop should be well ventilated to remove smoke and fumes. Turn on the system.
9. Electricity can kill, so take the necessary precautions. Dry clothing, gloves, and a dry workplace help prevent shocks. If you stay dry while welding, you act as an insulator and not a conductor of electricity. Electricity always seeks the path of least resistance, so stay dry when welding to stay safe.
10. Wear steel-toe leather boots in work areas (Figure 4–22). They offer some protection against foot injury.

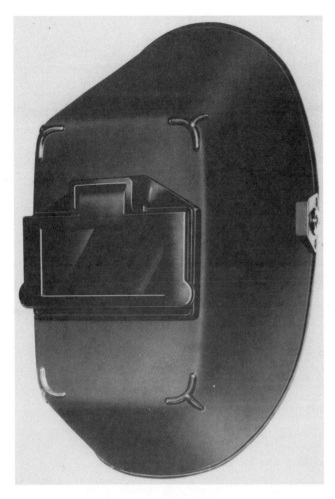

FIGURE 4–20 Welding helmet with flip-up lens. *(Courtesy of Kedman Co., Salt Lake City, Utah.)*

11. Dark clothing is preferred over white clothing because dark clothing absorbs light rays and white clothing deflects light rays. Arc flash can occur off aluminum and other sources of deflection.

12. Before turning on the welding machine, be sure to stretch out the electrode lead and the **workpiece lead** (Figure 4–23). Never wrap the leads around your body. Always place the workpiece connection on the work or as close to the work as possible for good electrical contact. The workpiece connection can be welded, clamped, or screwed into position.

13. Never weld on pressurized cylinders or on containers that have held flammable substances. A stray arc strike on a gas cylinder can cause an explosion.

14. As a common courtesy, always warn others in the area before striking an arc. Welding booths with curtains should be closed. Get in the habit of always beginning an arc strike at the point of the welding (Figure 4–24).

15. Be alert to strange smells. Visibility is limited under a welding helmet, and your sense of smell becomes important. Unusual odors could indicate a fire or a toxic substance, alerting you to possible danger.

16. Use common sense, and pay close attention to the work at hand. Never let daydreaming or distraction have its way.

Safety for Shielded Metal Arc Welding

17. Keep track of electrode stub ends, and dispose of them properly (Figure 4–25). They can puncture tires.

18. Remove the electrode before setting the electrode holder down, and turn off the welding machine if you must leave it unattended.

19. Replace the **insulator** on the electrode holder when it becomes worn out (Figure 4–26).

20. Replace a damaged electrode holder, and repair any frayed workpiece connections to ensure good electric contact.

21. A well-designed area to weld also contributes to your health and safety (Figure 4–27).

REVIEW

1. What do the letters SMAW stand for?
2. Name the three types of current used in welding.
3. What is the current called when an electrode holder is negative?
4. What is the purpose of the flux covering for electrodes?
5. Why must all of the slag be removed?
6. What determines the size of the electrode to be used?
7. In what welding positions can the E7018 electrode be used?

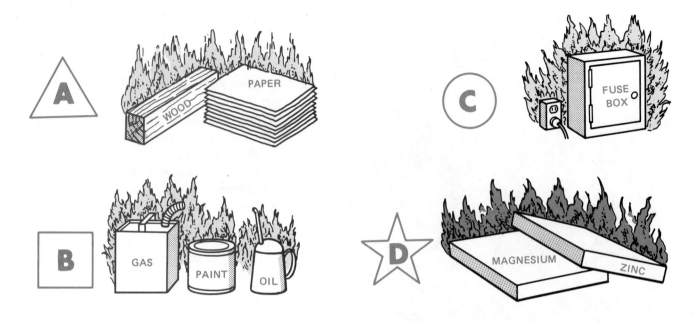

FIGURE 4–21   Classes of fire extinguishers. *(Courtesy of Jeffus and Johnson, 1988.)*

FIGURE 4–22   Steel-toe boots. *(Courtesy Jeffus and Johnson, 1988.)*

8. Name two things that can be done to control arc blow.
9. Why is wearing safety glasses so important?
10. Why is it important to stay dry when welding?

Create Three Questions

1.
2.
3.

Related Math and English Questions

1. What is the tensile strength of the following electrodes?
   a. E7014
   b. E8018
   c. E10018
2. Which size of electrodes produces the greater amount of fill?
   a. 3/32″ or 1/16″
   b. 1/4″ or 5/32″
3. What do the following abbreviations stand for?
   a. AC
   b. DCEN

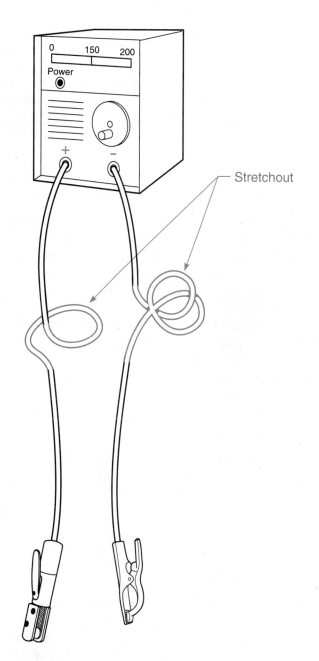

FIGURE 4–23   Twisted leads.

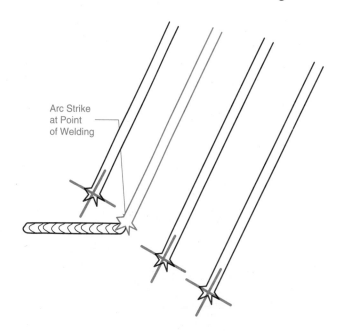

FIGURE 4–24   Correct and incorrect arc strikes.

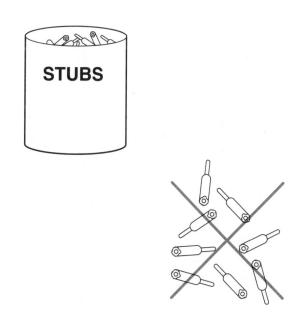

FIGURE 4–25   Proper disposal of electrode stubs.

c. AWS
d. DCEP
4. What was the hardest part of the shop exercises? Explain.

For Further Thought

1. What can sometimes be done to increase the amount of filler metal deposited?
2. Suppose you must weld a frame to support a motor. The 12 pieces of the frame, cut to length, are gathered in a pile. What should you do to ensure that the frame stays square and all the pieces fit without distortion?
3. Welder A has a box of 3/16″ electrodes, and Welder B has a box of 1/8″ electrodes. Which welder will deposit the most filler metal at the end of the day?
4. Which one of the four basic welding posi-

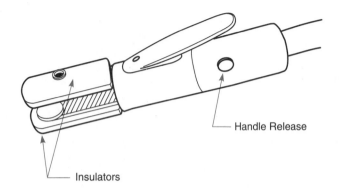

FIGURE 4–26   Electrode holder.

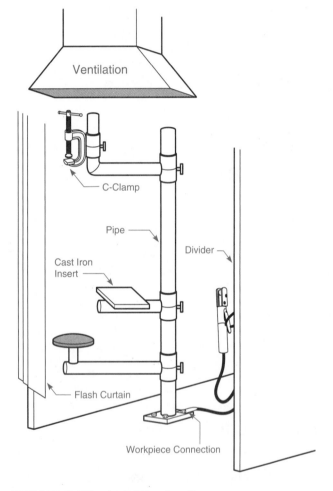

FIGURE 4–27   A welding booth.

tions results in the most smoke going into the face of the welder?

5. A welder must repair a **crack** on the bucket of a backhoe raised waist-high by hydraulic cylinders. Unfortunately, the workpiece connection does not reach to the bucket.

The welder decides to make the connection on the opposite side of the backhoe where it will reach. Having second thoughts, the welder then instructs an operator to turn the backhoe around. Now the connection can be put on the bucket where the welding is to be completed. Why was this maneuver necessary?

SUGGESTED ACTIVITIES

1. Discuss accidents that might occur if safety procedures are not followed.
2. Demonstrate the care and use of the different pieces of equipment used in welding.
3. Demonstrate the care and use of the welding machine.
4. Show and describe various types of welding electrodes.

UNIT 4 EXERCISES

Setting Up to Weld

The following list of reminders will become second nature to you in a very short time. If you take an interest in the details, you will gain the edge needed to make you the very best welder you can be.

1. Make sure that there is good electric contact at the workpiece connection. If necessary, grind the surface clean.
2. Check the welding helmet and lenses for cracks, and clean the lenses if dirty.
3. Relax and find a comfortable position to weld. Good welding techniques (electrode angles, amperage, and so on) depend on being comfortable (Figure 4–28).
4. Be conscious of the work angle and the travel angle. Refer back to Figure 4–14.
5. Keep both hands on the electrode holder. There will be plenty of time later to develop the one-handed technique.
6. If the weight of the electrode lead is causing stress on the hand holding the electrode holder during welding, lay the lead over an object in the area.
7. Keep one hand in front of your face when removing slag, and always chip away from your body. Better yet, keep your helmet down if it is designed with a flip filter lens.

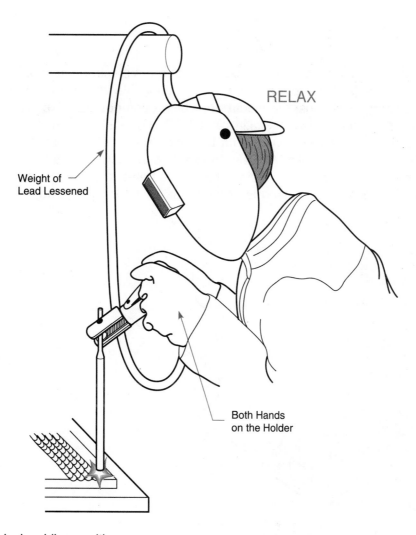

RELAX

Weight of
Lead Lessened

Both Hands
on the Holder

FIGURE 4–28   A typical welding position.

8. Metal is expensive, so *never* waste material. For example, do not stop welding one
joint and begin on another without completing the first.

Note that DCEP is recommended for these

exercises, but AC can always be used if it is available. In addition, 1/8-inch-diameter electrodes are
suggested for use in all of the following exercises
unless otherwise stated. Substitute, if necessary,
to develop skill with electrodes of other sizes.

UNIT 4: EXERCISE 1

Shielded Metal Arc Welding (SMAW): Weld Bead Formation

You begin every weld bead by placing an electrode into the electrode holder. By using a 135° angle, you can burn each electrode down to the stub end with very little waste. By using 45° angle, there will be waste unless you stop to readjust the electrode, which does not make sense.

The TAP method is recommended for striking an arc (Figure 4–29). This method takes a little practice, but you will appreciate it in the long run when it becomes necessary to put the electrode on a single spot. The **tap method** is used by skilled welders to keep the **arc strikes** in the welding area only. E6013 electrodes are chosen for this opening exercise. If the weld bead is made correctly, the slag coating can be easily removed. The amperage setting is provided as a starting point; adjustments will have to be made.

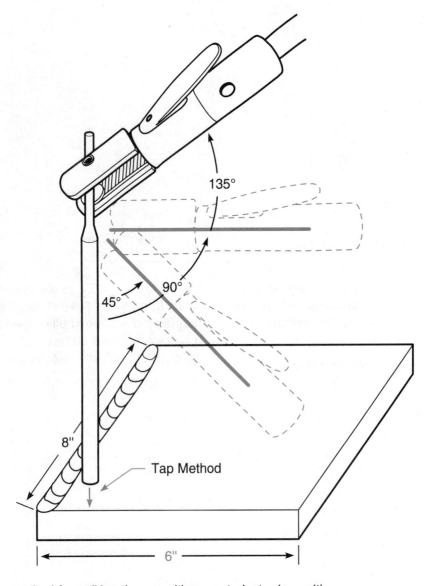

FIGURE 4–29   Tap method for striking the arc with correct electrode position.

Necessary Material and Equipment

Safety glasses	Helmet	Earplugs	1 piece 3/8″ steel, 6″ × 8″
Welding gloves	Slag hammer	Grinder	Amperage: 90–110
Protective clothing	Wire brush	E6013 electrodes, 1/8″	

Instructions

1. Clean the surface with the grinder.
2. Tap (strike) the electrode on the plate, raising it quickly. The electrode will stick if raised too slowly, and the arc will go out if raised too high. Maintain an arc length that equals the diameter of the electrode. If the electrode sticks, release it from the holder before raising your helmet.
3. If you are right-handed, move from left to right. If you are left-handed, move from right to left.
4. Practice laying weld beads in a straight line. Start the first weld bead along the back edge of the plate.
5. Run **stringers** to keep the weld beads in a straight line. A stringer is a weld bead that has been made without **weaving,** which tends to spread the heat (reducing the penetration). At this stage in your training, concentrate on being steady.
6. Continue to push the electrode into the weld pool to maintain the correct arc length while moving in the direction of travel.
7. Chip off the slag and brush the metal clean after each pass.
8. Check with the instructor for evaluation, if necessary, before moving on to the next exercise.

UNIT 4: EXERCISE 2

Shielded Metal Arc Welding (SMAW): Padding Plate in Flat Position, Surfacing Weld

Because vision is limited under the helmet, lay the first weld bead along the edge. Use the edge as a guide for keeping the first bead straight. Lay each weld bead so that it overlaps the previous bead by one-half its width (Figure 4–30). If the **plate** has been cut to the dimensions recommended for this exercise, only a stub of electrode should be left after traveling the entire 8-inch length. If some electrode remains, your travel speed was too fast. When you reach the edge, break the arc momentarily, then quickly restart the arc filling in the **crater.** This motion will prevent the edge of the plate from melting away.

This exercise is an example of the kind of welding used to build up worn parts. Some electrodes designed for buildup are of a quality to reduce wear.

Necessary Material and Equipment

Safety glasses	Helmet	Earplugs	1 piece 3/8″ steel, 6″ × 8″
Welding gloves	Slag hammer	Grinder	Amperage: 90–110
Protective clothing	Wire brush	E6013 electrode, 1/8″	

Instructions

1. Clean the surface with the grinder.
2. Fill the plate completely. If the electrode is used up before you reach the edge of the plate, practice tying in the end of one weld bead with the beginning of the next weld bead. The challenge is to make two beads appear as one.
3. The ripples provide a pattern like fingerprints in each weld bead. The ripples should be rounded. Ripples that come to a point indicate a travel speed that is too fast. High, narrow weld beads suggest that the arc length is too small, DCEN is being used, or the amperage is too low.
4. If the arc length is too long, the weld bead will flatten out, and spatter will result. Spatter is the small drops of metal that form along the weld bead. The ripples will also be uneven.
5. Look for straight weld beads with even ripples and consistent overlap.
6. Always tilt the electrode 5° to 10° in the direction of travel. This helps prevent slag from being trapped in the weld bead.
7. Check with the instructor for evaluation, if necessary.
8. Practice filling as many padding plates as needed to meet the standards established for the course.

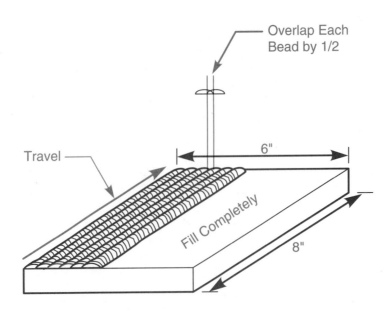

FIGURE 4–30 Laying beads.

UNIT 4: EXERCISE 3

Shielded Metal Arc Welding (SMAW): Padding Plate in Flat Position, Surfacing Weld with Different Electrodes

Because each type of electrode melts differently, it is important to become proficient with the various types of electrodes. The E6011 and E6010 electrodes, recommended for this exercise, burn more quickly and must be fed at a faster rate into the weld pool than the E7018 electrode. The E7018 requires a higher amperage setting, and its slag forms a scale that chips off easily (Figure 4–31).

Remember that the proper travel speed, amperage setting, arc length, and electrode angle are important for good welding.

Necessary Material and Equipment

Safety glasses	Slag hammer	Grinder	E6010 electrode, 1/8″
Welding gloves	Wire brush	E6011 electrode, 1/8″	1 piece 3/8″ steel, 6″ × 8″
Protective clothing	Earplugs	E7018 electrode, 1/8″	Amperage: 80–130
Helmet			

Instructions

1. Clean the surface with the grinder.
2. Use the slag hammer and brush on every weld bead before laying an overlapping bead.
3. Fill all craters at the edge of the plate.
4. Keep a tight (close) arc length to reduce spatter.
5. Run one **layer** of overlapping weld beads using E6011. Look for straight weld beads with even ripples and correct overlap.
6. Check with the instructor for evaluation, if necessary.
7. Practice as many padding plates with the E6011 as necessary to meet the standards established for the course.
8. Complete a second layer using the E7018, which will require a higher amperage setting.
9. Complete a third layer using the E6010. Must you make another change in the amperage setting?
10. Check with the instructor for evaluation, if necessary.
11. Practice this exercise until you meet the standards established in the course.

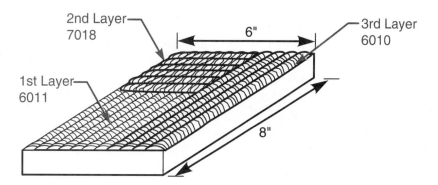

FIGURE 4–31   Bead patterns.

UNIT 4: EXERCISE 4

Shielded Metal Arc Welding (SMAW): Corner Joint in Flat Position, Fillet Weld

Angle iron (steel) is recommended for the inside **corner joint,** but two pieces of steel plate can be **tack welded** together. Like the padding plate exercises, stringer weld beads are laid so that they overlap one another. Note that no space should be left between the weld beads for trapping slag (Figure 4–32).

Necessary Material and Equipment

Safety glasses	Slag hammer	E6011 electrode, 1/8″	Padding plate
Welding gloves	Wire brush	E7018 electrode, 1/8″	Amperage: 80–150
Protective clothing	Earplugs	1 1″ × 1″ × 1/4″ angle ×	
Helmet	Grinder	8″ or 2 pieces of plate	

Instructions

1. Clean the surface with the grinder.
2. Tack weld the angle iron (or two pieces of plate 90° angle) to the scrap padding plate.
3. Lay the weld beads in each layer by moving from one side to the other, overlapping each weld bead by one-third to one-half its width.
4. Build up layer upon layer.
5. Remove all slag from each weld bead before laying another.
6. Make each layer as smooth and clean as the buildup in the padding plate exercises.
7. Complete three to five layers with the 1/8″ E6011 electrode.
8. Complete an additional two layers with the 1/8″ E7018.
9. Check with the instructor for evaluation, if necessary.
10. Practice this exercise until you meet the standards established in the course.

This completed weldment can be saved for use in practicing air carbon arc cutting. Alternatively, a bandsaw cutting completely through the weldment can be **etched** and examined for **defects.**

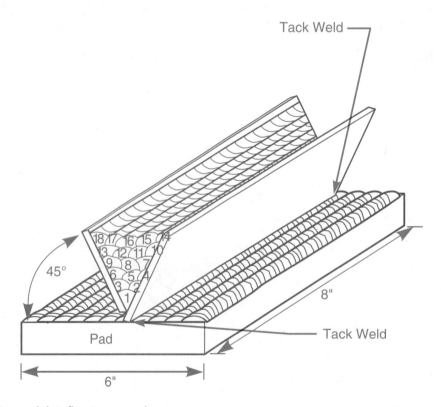

FIGURE 4–32 Corner joint, five to seven layers.

UNIT 4: EXERCISE 5

Shielded Metal Arc Welding (SMAW): Corner Joint in Flat Position, Fillet Weld

Learning to weld means repetition, practicing the same skills over again, trying for perfection. If becoming a competent welder is your goal, then practice is the only way of reaching that goal. The 5/32″ E7018 electrode used in this exercise is very popular in industry. Be sure to clean every weld bead between passes. Using the material from Exercise 4 will help conserve steel and provide greater mass. Cool to avoid overheating. The setup is shown in Figure 4–33.

Necessary Material and Equipment

Safety glasses	Helmet	E7018 electrode, 5/32″	Amperage: 130–160
Welding gloves	Slag hammer	1 piece 1/4″ angle, 8″ long	
Protective clothing	Wire brush	Padding plate	

Instructions

1. Tack weld the steel used in Exercise 4 to the scrap padding plate.
2. Lay weld beads in each layer by moving from one side to the other.
3. Build up layer upon layer. Note the higher amperage setting required for the E7018 electrode.
4. Make each layer as smooth and clean as the buildup in the padding plate exercises. Remember, a layer is a series of overlapping weld beads laid on a surface.
5. Complete three to five layers until you meet the standards established for the course.
6. Test by **visual inspection,** looking for a smooth surface with good overlap.
7. Check with the instructor for evaluation, if necessary.
8. Practice this exercise until you meet the standards established for the course.

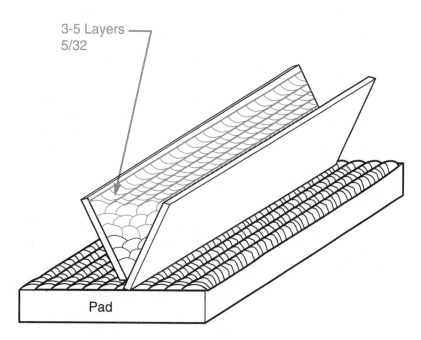

3-5 Layers
5/32

Pad

FIGURE 4–33   Three to five layers of 5/32″ electrodes.

UNIT 4: EXERCISE 6

Shielded Metal Arc Welding (SMAW): Butt Joint (Designed to Meet AWS *Structural Welding Code*) V-Groove Weld with Backing in Flat Position

The **V-groove weld** is the popular test given to qualify welders for jobs. Passing a V-groove weld test, which requires a **butt joint,** qualifies you for the four other joints in the flat position. Passing this test is evidence that you have achieved the skill necessary for moving on to the horizontal position (Figure 4–34).

Necessary Material and Equipment

Safety glasses	Wire brush	2 pieces 1/8″ to 1″ A36	**2 strong backs**
Welding gloves	Earplugs	steel, 3″ × 7″ (6″ if 1″	Grinder
Protective clothing	E7018 electrode,	thick)	Cutting machine, 22.5°
Helmet	1/8″	1 piece 1/4″ bar, 1″ × 8″	bevel
Slag hammer	Amperage: 130–150	(backing)	

Instructions

1. Follow Figure 4–34 in preparing the pieces for welding.
2. Grind the surfaces to be welded, including the **backing** material. Grind the top and bottom 1/4 inch beyond the weld area.
3. The fit up is important before tack welding the pieces together. The backing material should have complete contact with the V-groove.
4. The first weld bead requires a slight weave to fuse both pieces together with the backing. Remember to pause momentarily on the sides to avoid undercut.
5. Do not weld over the top edges of the V-groove until you reach the final layer of the weld. Use the edges as a guide for laying the first weld bead of each layer.

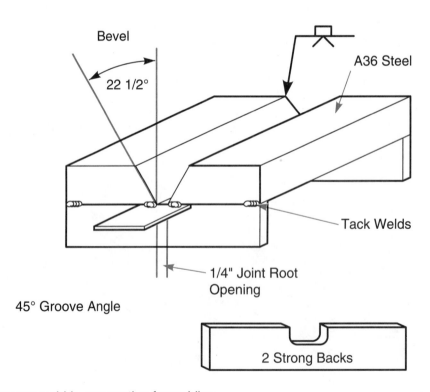

FIGURE 4–34 V-groove weld in preparation for welding.

6. Air cool. Do not cool in water.
7. Prepare for the **guided bend test,** as described in Unit 10.
8. Check with the instructor for evaluation, if necessary.
9. Pass the guided bend test before moving on.

UNIT 4: EXERCISE 7

Shielded Metal Arc Welding (SMAW): Butt Joint (Designed to Meet AWS *Structural Welding Code*) V-Groove Weld without Backing in Flat Position

The V-groove weld without backing will help prepare you for pipe welding. (The practice exercises necessary to develop the skill of pipe welding are covered in Unit 12.) In addition, many joint designs require a V-groove weld without backing (Figure 4–35).

The **keyhole** (a partial melting of the edges on both pieces) is the basis for success in this exercise. The keyhole is necessary for achieving **complete root penetration**.

Necessary Material and Equipment

Safety glasses	Slag hammer	Amperage: 80–110	2 strong backs
Welding gloves	Wire brush	2 pieces 1/8" to 1" A36	Grinder
Protective clothing	Earplugs	steel, 3" × 7" (6" if 1"	Cutting machine, 30° bevel
Helmet	E6010 electrode, 1/8"	thick)	

Instructions

1. Follow Figure 4–35 in preparing the pieces for welding.
2. Grind the surfaces to be welded. Grind the top and bottom 1/4 inch beyond the weld area.
3. Fit up, align, and tack weld.
4. The first bead (**joint root** pass) is critical. The success of this joint depends on the joint root pass. Use a **whipping** motion for the E6010 electrode. Whipping is a quick motion out of the weld pool and back in again, pushing the keyhole along. This is not a weaving motion.
5. Do not weld over the top edges of the V-groove until the final cover layer. Use the edges as a guide for laying the first weld bead of each layer. Always chip away the slag and brush clean between passes.
6. Prepare for the guided bend test, as described in Unit 10.
7. Check with the instructor for evaluation, if necessary.
8. Pass the guided bend test before moving on.

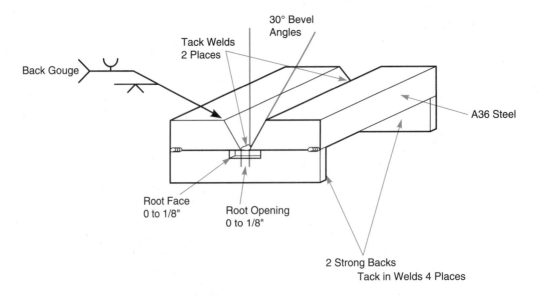

FIGURE 4–35   V-groove weld without backing.

UNIT 4: EXERCISE 8

Shielded Metal Arc Welding (SMAW): Padding Plate in Horizontal Position, Surfacing Weld

The electrode angle is important in completing this exercise. Student welders tend to move too quickly across the plate for fear that the weld pool will drip. You need not be overly concerned about this because the position of the electrode helps to counteract gravity. Move steadily, but not quickly, across the plate.

To hold the plate vertically for welding, a piece of steel should be welded on the back side. This will hold the plate in a fixture. Otherwise, the plate can be tack welded directly to the table or welded to another length of steel that is tack welded to the table. This will provide greater height if necessary to find a comfortable position for welding. Clamps can also be used. The setup for this exercise is shown in Figure 4–36.

Necessary Material and Equipment

Safety glasses	Helmet	Earplugs	Amperage: 130–150
Welding gloves	Slag hammer	Grinder	1 piece 3/8″ steel, 6″ × 8″
Protective clothing	Wire brush	E7018/E6010 electrodes, 1/8″	

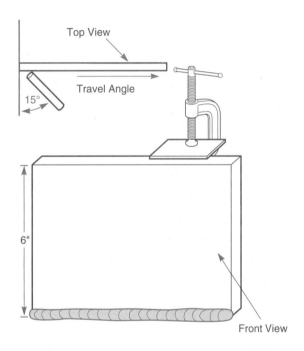
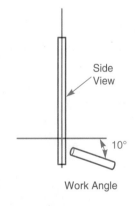

FIGURE 4–36   Padding plate in horizontal position.

Instructions

1. Beginning at the bottom, lay overlapping weld beads. Make one layer with the E7018 and another layer with the E6010.
2. Use the previous weld bead as a shelf to place each following weld bead.
3. Remember that the plate might have to be cooled occasionally between passes, or the amperage might have to be lowered on the welding machine. Otherwise, the plate will become too hot, causing the weld pool to drip.
4. The electrode must be tilted 10° to 15° in the direction of travel. This is in addition to the work angle of the electrode necessary to counteract gravity.
5. The ripples in each weld bead should be rounded. Ripples that come to a point indicate a travel that is too fast. High, narrow weld beads indicate an arc length that is too short, DCEN, or amperage too low.
6. Look for straight weld beads with even ripples and sufficient overlap.
7. Check with the instructor for evaluation, if necessary.
8. Practice as many padding plates as necessary to make one that meets the standards established for the course.

UNIT 4: EXERCISE 9

Shielded Metal Arc Welding (SMAW): Tee Joint in Horizontal Position, Fillet Weld

The **tee joint** in the horizontal position is a very common and challenging exercise (Figure 4–37). Two pitfalls to avoid are piling up weld metal and poor overlap. Piling up filler metal on the bottom piece can result in joint failure. A cross-section of the weld shows the problem when the welding instructions called for even distribution of filler metal between both members of the weldment. The first weld bead of each layer is critical. A good two-thirds of this weld bead should cover the bottom weld bead of the previous layer. Poor overlap within the completed weld results in valleys and ridges. Good overlap results in a smooth surface.

Note that the electrode angle is important in this exercise. Watch out for arc blow, and make any necessary adjustments.

Necessary Material and Equipment

Safety glasses	Helmet	E7018/7024 electrodes,	2 pieces 1/4″–3/8″ steel, 3″
Welding gloves	Slag hammer	1/8″	× 8″
Protective clothing	Wire brush	Amperage: 130–150	

Instructions

1. Begin each pass from the bottom. Lay 10 to 15 weld beads on each side, using the E7018 on one side and the E7024 on the other side. Cool to reduce the effects of heat.
2. Remove slag completely before laying the next weld bead.
3. Look for a smooth fillet weld with straight weld beads. Looking from the side, there should be as much fill on the top plate as on the bottom plate.
4. Check with the instructor for evaluation, if necessary.
5. Practice this exercise until you meet the standards established for the course.

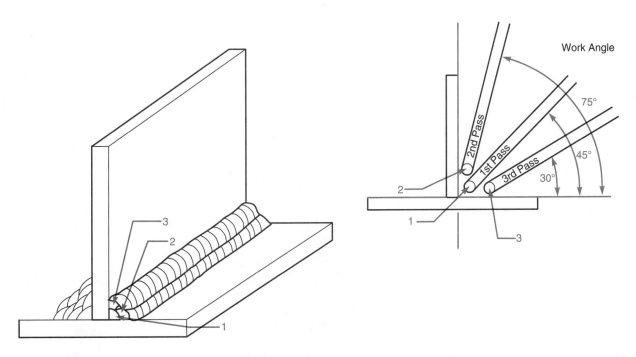

FIGURE 4–37   Tee joint in horizontal position.

UNIT 4: EXERCISE 10

Shielded Metal Arc Welding (SMAW): Butt Joint (Designed to Meet AWS *Structural Welding Code*) V-Groove Weld with Backing in Horizontal Position

The backing for this exercise is cut longer than for previous exercises to provide an area for beginning (arc strikes) and continuing each weld bead beyond the joint. Arc strikes on the base metal outside of the V-groove can bring about weld failure, so watch where you strike the arc. The joint root weld bead (first pass) should be done with a slight weave to fuse both plates to the backing, keeping slag pockets from forming during the joint root pass (Figure 4–38).

Necessary Material and Equipment

Safety glasses	Wire brush	2 pieces 1/8" to 1" A36	2 strong backs
Welding gloves	Earplugs	steel, 3" × 7" (6" if 1"	Grinder
Protective clothing	E7018 electrode, 1/8"	thick)	Cutting machine, 22.5°
Helmet	Amperage: 130–150	1 piece 1/4" bar, 1" × 8"	bevel
Slag hammer		(backing)	

Instructions

1. Follow Figure 4–38 in preparing the pieces for welding.
2. Grind the surfaces to be welded. Grind the top and bottom 1/4 inch beyond the weld area.
3. Fit up is important before tack welding the pieces together. The backing piece should have complete contact with the V-groove.
4. Do not weld over the top edges of the V-groove until the final cover layer. Use the edges as a guide for making the first weld bead of each layer. Chip off the slag and brush clean after each weld bead.
5. Air cool. Do not cool in water.
6. Prepare for the guided bend test, as described in Unit 10.
7. Check with the instructor for evaluation, if necessary.
8. Pass the guided bend test before moving on.

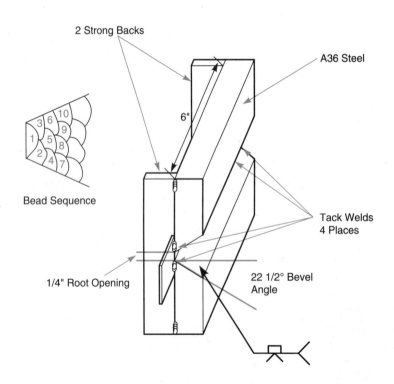

FIGURE 4–38   V-groove weld in horizontal position.

UNIT 4: EXERCISE 11

Shielded Metal Arc Welding (SMAW): Butt Joint (Designed to Meet AWS *Structural Welding Code*) V-Groove Weld without Backing in Horizontal Position

The keyhole is important in this exercise. **Poor root penetration** on the backside of this joint will lead to failure in the guided bend test. The amperage setting will control the size of the keyhole. Stop and increase the amperage if root penetration is incomplete. Stop and decrease the amperage if the keyhole widens beyond the size of the electrode. Stop welding if a problem arises with the size of the keyhole (Figure 4–39).

Necessary Material and Equipment

Safety glasses	Slag hammer	Amperage: 80–110	2 strong backs
Welding gloves	Wire brush	2 pieces 1/8″ to 1″ A36	Grinder
Protective clothing	Earplugs	steel, 3″ × 7″ (6″ if 1″	Cutting machine, 30° bevel
Helmet	E6010 electrode, 1/8″	thick)	

Instructions

1. Follow Figure 4–39 in preparing the pieces for welding.
2. Grind the surfaces to be welded. Grind the top and bottom 1/4 inch beyond the weld area.
3. Fit up, align, and tack weld the pieces before welding.
4. Use a whipping motion for the E6010 electrode. Whipping is a quick motion out of weld pool and back again, pushing the keyhole along. This is not a weaving motion.
5. Do not weld over the top edges of the V-groove until the final cover layer. Use the edges as a guide for laying the first weld bead of each layer. Chip off the slag and brush clean after each weld bead.
6. Watch the electrode angle. It is critical.
7. Prepare for the guided bend test, as described in Unit 10.
8. Check with the instructor for evaluation, if necessary.
9. Pass the guided bend test before moving on.

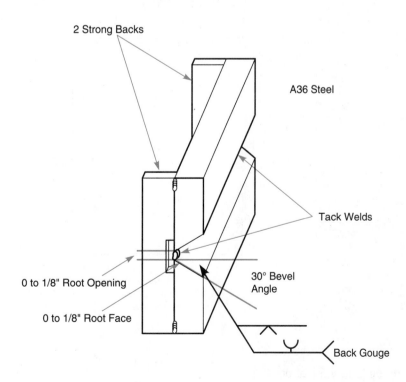

FIGURE 4–39 V-groove weld without backing.

UNIT 4: EXERCISE 12

Shielded Metal Arc Welding (SMAW): Padding Plate in Vertical Position, Surfacing Weld

The vertical position can be the most difficult position to learn. However, if you have completed the preceding exercises, you will find the vertical position much easier than the flat position. This is the one welding position where weaving is necessary. There are many different weaving methods; just one method is diagrammed. Use **uphill** welding for the following exercises unless otherwise indicated (Figure 4–40).

Remember to pause momentarily on each side of the joint to avoid undercut. Undercut is the unfilled groove on the base metal along the edge of the weld bead. A high spot in the middle of the weld indicates that travel has been too slow. Speed up, but always pause on the sides. In practicing this exercise, work on developing a rhythm for the motion of the electrode.

Necessary Material and Equipment

Safety glasses	Helmet	Earplugs	Amperage: 130–150
Welding gloves	Slag hammer	Grinder	1 piece 3/8″ steel, 6″ × 8″
Protective clothing	Wire brush	E7018, E6010 electrodes, 1/8″	

Instructions

1. Clean the surface with the grinder.
2. Tack weld a piece of scrap material to the plate. Clamp it in a fixture or weld it to the table at a comfortable height.

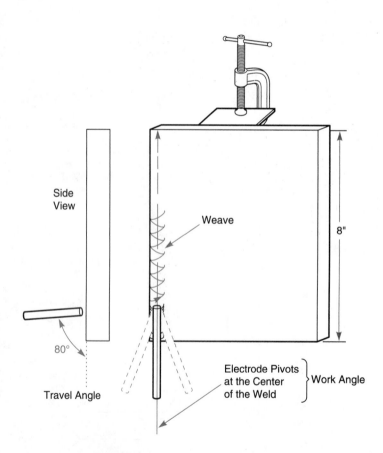

FIGURE 4–40   Padding plate in vertical position.

3. Weave one layer with E7018 and one layer with E6010. When weaving with E7018, the electrode should never leave the weld pool. Do not whip this electrode.
4. Begin from the bottom. Practice the weaving method. This is welding **uphill.**
5. Note the proper electrode angle, which is necessary to counteract gravity.
6. Avoid the tendency to move too fast. Use of the proper techniques (electrode angles, amperage, and so on) will prevent the molten metal from dripping out of the weld pool.
7. Look for straight weld beads with good overlap. You should find smooth ripples that are neither high in the middle nor undercut on the sides.
8. Check with the instructor for evaluation, if necessary.
9. Practice as many padding plates as necessary to meet the standards established in the course.

UNIT 4: EXERCISE 13

Shielded Metal Arc Welding (SMAW): Tee Joint in Vertical Position Fillet Weld

Having practiced padding plates, this exercise should be relatively easy. Once again, it is important to pause at each side to avoid undercut. Counting "one thousand one, one thousand two . . ." before moving across to the other side forces a pause. This may help to stop undercut.

The welding speed across the joint determines the final appearance of a fillet weld. Traveling too slowly causes excessive buildup in the middle (Figure 4–41).

Necessary Material and Equipment

Safety glasses	Helmet	Earplugs	Amperage: 130–150
Welding gloves	Slag hammer	Grinder	2 pieces 1/4"–3/8" steel, 6"
Protective clothing	Wire brush	E7018 electrode, 1/8"	× 3"

Instructions

1. Clean the surfaces with the grinder.
2. Tack weld two pieces together on each end.
3. Clamp the joint in a fixture or tack weld it to the table at a comfortable height.
4. Beginning at the bottom, practice the weaving method. The welding will be made uphill.
5. Avoid the tendency to move too fast. Use of the proper techniques (electrode angles, amperage, and so on) will keep the molten metal from dripping.
6. Make four or five layers on each side. Weld until the weave becomes too wide, cooling down the preceding layer. Chip off the slag and brush clean after each weld bead.
7. Look for consistent ripples with no gaps, a smooth **weld face,** and no undercut.
8. Check with the instructor for evaluation, if necessary.
9. Practice this exercise until you meet the standards established in the course.

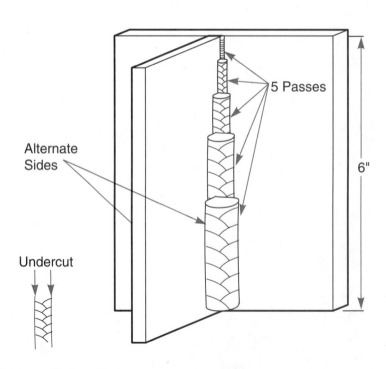

FIGURE 4–41   Tee joint in vertical position.

UNIT 4: EXERCISE 14

Shielded Metal Arc Welding (SMAW): Butt Joint (Designed to Meet AWS *Structural Welding Code*) V-Groove Weld with Backing in Vertical Position

Begin the arc below the joint on the backing strip. Arc strikes made on the surface of the joint outside the weld should be rejected upon a visual examination. Arc strikes outside the weld can crack during the guided bend test. Think ahead, and try not to end any weld beads, causing restarts, in the area where samples will be taken for testing. The joint root weld bead is very important. Be sure there is complete fusion all along the edge on the joint root weld bead. Gaps in the weld caused by incorrect weave technique can trap slag (Figure 4–42).

Necessary Material and Equipment

Safety glasses	Earplugs	1 piece 1/4" bar, 1" × 8"	Cutting machine, 22.5°
Welding gloves	E7018 electrode, 1/8"	backing	bevel
Protective clothing	Amperage: 130–150	2 strong backs	
Helmet	2 pieces 1/8" to 1" A36	Grinder	
Slag hammer	steel, 3" × 7" (6" if 1"		
Wire brush	thick)		

Instructions

1. Follow Figure 4–42 in preparing the pieces for welding
2. Grind the surfaces to be welded. Grind the top and bottom 1/4 inch beyond the weld area.
3. Fit up is important. The backing piece should have complete contact with the V-groove before you make the tack welds.

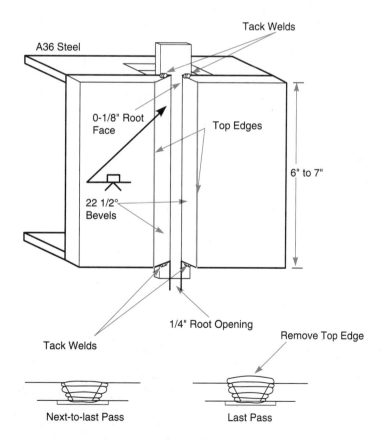

FIGURE 4–42   V-groove weld in vertical position.

4. Begin at the bottom. All weld beads will be completed uphill.
5. Do not weld over the top edges of the V-groove until the final cover layer. Use the edges as a guide in laying the first weld bead of each layer. Chip away the slag and brush clean after each weld bead.
6. Air cool. Do not cool in water.
7. Prepare for the guided bend test, as described in Unit 10.
8. Check with the instructor for evaluation, if necessary.
9. Pass the guided bend test before moving on.

UNIT 4: EXERCISE 15

Shielded Metal Arc Welding (SMAW): Butt Joint (Designed to Meet AWS *Structural Welding Code*) V-Groove Weld without Backing in Vertical Position

The keyhole is again important for this exercise. Poor root penetration on the backside of this joint will lead to test failure. Although the amperage is important, so is the space or gap between the plates. A joint root opening should be no larger than 1/8 inch as determined by the welding procedure. A joint root opening smaller than 1/8 inch must have complete root penetration (Figure 4–43).

Necessary Material and Equipment

Safety glasses	Slag hammer	Amperage: 80–110	Grinder
Welding gloves	Wire brush	2 pieces 1/8″ (1″) A36	Cutting machine, 30° bevel
Protective clothing	Earplugs	steel, 3″ × 7″ (6″)	
Helmet	E6010 electrode, 1/8″	2 strong backs	

Instructions

1. Follow Figure 4–43 in preparing the pieces for welding.
2. Grind the surfaces to be welded. Grind the top and bottom 1/4 inch beyond the weld area.
3. Fit up, align, and tack weld the pieces before welding.
4. Begin at the bottom. All weld beads should be completed uphill.
5. Use a whipping motion for the E6010 electrode. Use a quick motion out of weld pool and back again, pushing the keyhole along. This is not a weaving motion.
6. Do not weld over the top edges of the V-groove until the cover (final) layer.
7. Watch the electrode angle. It is critical.
8. Prepare for the guided bend test as described in Unit 10.
9. Check with the instructor for evaluation, if necessary.
10. Pass the guided bend test before moving on.

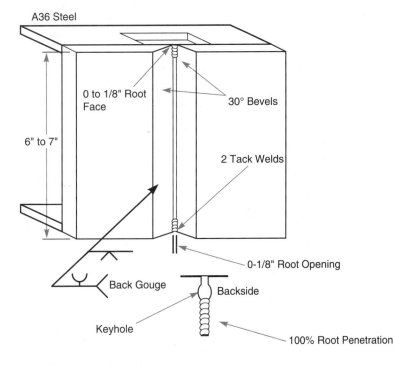

FIGURE 4–43 V-groove weld without backing.

UNIT 4: EXERCISE 16

Shielded Metal Arc Welding (SMAW): Tee Joint in Overhead Position Fillet Weld

Wearing leathers or extra protection on the upper body is worthwhile when welding in the overhead position. The padding plate has been omitted in this exercise. The joint root weld bead will involve a slight weaving motion to fuse both pieces. All other weld beads must be stringers. Find a comfortable position from which to view the weld pool. Position your helmet right under the work. Overhead welding will now take less time to master than welding in the flat position (Figure 4–44).

Necessary Material and Equipment

Safety glasses	Slag hammer	Grinder	2 pieces 1/4″–3/8″ steel, 6″
Welding gloves	Wire brush	E7018 electrode, 1/8″	× 3″
Protective clothing	Earplugs	Amperage: 130–150	Fixture for positioning
Helmet			

Instructions

1. Clean the surfaces with the grinder.
2. Tack weld two pieces together on each end.
3. Tack weld a piece of scrap material to the joint if needed. Clamp it in a fixture.
4. Begin on one side. Watch the electrode angle. It should be no more than 15°.
5. Avoid the tendency to move too fast. Use of the proper techniques (angles, amperage, and so on) will keep the weld from dripping.
6. Make four or five layers on each side.
7. Look for even weld beads with consistent ripples, good overlap, and no undercut.
8. Check with the instructor for evaluation, if necessary.
9. Practice this exercise until you meet the standards established for the course.

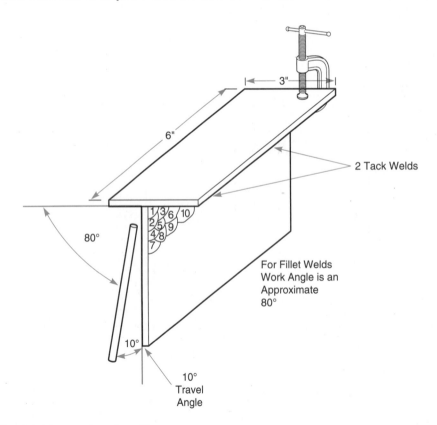

FIGURE 4–44   Tee joint in overhead position.

UNIT 4: EXERCISE 17

Shielded Metal Arc Welding (SMAW): Butt Joint (Designed to Meet AWS *Structural Welding Code*) V-Groove Weld with Backing in Overhead Position

Leathers or extra protection on the upper body is recommended for this exercise. Do not hurry the travel speed, especially on the joint root weld bead. Wash (touch) the edge of each piece with the weld pool on the joint root pass. This is considered a weave. All other weld beads should be stringers.

Watch the electrode angle and the arc length. Too great an angle will affect penetration and will produce a high weld face. Too large an arc length will cause unnecessary spatter (Figure 4–45).

Necessary Material and Equipment

Safety glasses	Wire brush	2 pieces 1/8″ to 1″ A36	2 strong backs
Welding gloves	Earplugs	steel, 3″ × 7″ (6″ if 1″	Grinder
Protective clothing	E7018 electrode,	thick)	Cutting machine, 22.5°
Helmet	1/8″	1 piece 1/4″ bar, 1″ × 8″	bevel
Slag hammer	Amperage: 130–150	(backing)	

Instructions

1. Follow Figure 4–45 in preparing the pieces for welding.
2. Grind the surfaces to be welded. Grind the top and bottom 1/4 inch beyond the weld area.
3. Fit up is important before tack welding. The backing piece should have complete contact with the V-groove.
4. Do not weld over the top edges of the V-groove until the final cover layer. Use the edges as a guide for making the first weld bead of each layer.
5. Chip off the slag and brush clean after each weld bead.

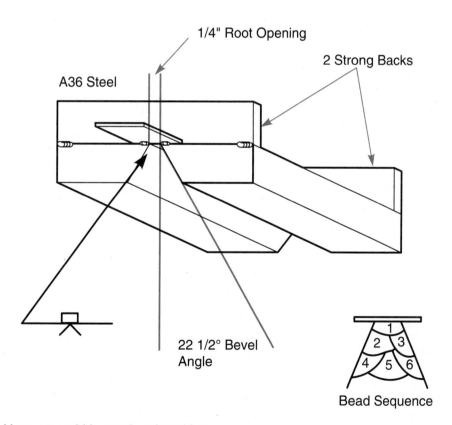

FIGURE 4–45   V-groove weld in overhead position.

6. Air cool. Do not cool in water.
7. Prepare for the guided bend test, as described in Unit 10.
8. Check with the instructor for evaluation, if necessary.
9. Pass the guided bend test before moving on.

UNIT 4: EXERCISE 18

Shielded Metal Arc Welding (SMAW): Butt Joint (Designed to Meet AWS *Structural Welding Code*) V-Groove Weld without Backing in Overhead Position

The keyhole is important during welding. Poor root penetration on the backside of this joint will lead to test failure. Although the amperage setting is important, so is the space or gap between the plates. Too large a joint root opening will melt away the edges. Too small a joint root opening will prevent complete root penetration (Figure 4–46).

Necessary Material and Equipment

Safety glasses	Slag hammer	Amperage: 80–110	2 strong backs
Welding gloves	Wire brush	2 pieces 1/8" to 1" A36	Grinder
Protective clothing	Earplugs	steel, 3" × 7" (6" if 1"	Cutting machine, 30° bevel
Helmet	E6010 electrode, 1/8"	thick)	

Instructions

1. Follow Figure 4–46 in preparing the pieces for welding.
2. Grind the surfaces to be welded. Grind the top and bottom 1/4 inch beyond the weld area.
3. Fit up, align, and tack weld.
4. Use a whipping motion for the E6010 electrode. This should be a quick motion out of weld pool and back again, pushing the keyhole along. This is not a weaving motion.
5. Chip off the slag and brush after each weld bead.
6. Do not weld over the top edges of the V-groove until the cover layer.
7. Watch the electrode angle. It is critical.
8. Prepare for the guided bend test, as described in Unit 10.
9. Check with the instructor for evaluation, if necessary.
10. Pass the guided bend test before moving on.

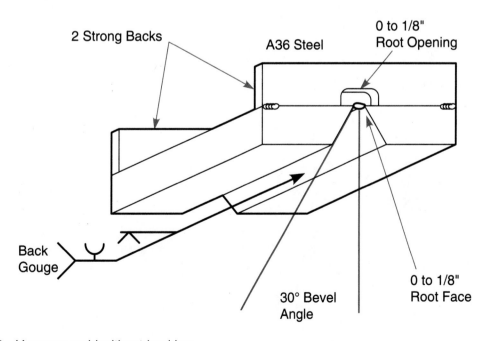

FIGURE 4–46   V-groove weld without backing.

GAS METAL ARC WELDING AND FLUX CORED ARC WELDING

"If you just stop and put tools away neatly you will both find the tool and also scale down your impatience without wasting time or endangering the work."

—Robert M. Pirsig, *Zen and the Art of Motorcycle Maintenance*

GOALS

- Develop an ability to do both gas metal arc welding and flux cored arc welding safely
- Pass destructive tests that meet the specifications set forth in the American Welding Society's *Structural Welding Code*

QUESTIONS

- What is the difference between gas metal arc welding and flux cored arc welding?
- What equipment is required?
- How does the equipment work?
- What are some of the problems in making quality welds?

SAFETY FIRST

The safety requirements for gas metal arc welding (GMAW) are generally the same as for any other electric **arc welding** process (Figure 5–1). Twenty-one safety reminders for electric arc welding are given toward the end of this unit. However, there are some things that make this **process** different from **shielded metal arc welding.**

FIGURE 5–1 Gas metal arc welding. *(Courtesy of R.H. Blake, Inc., Cleveland, Ohio.)*

Because gas metal arc welding uses a **shielding gas,** there is the concern associated with an **arc strike** on the cylinder. A cylinder under several thousand pounds of pressure can be dangerous if not handled safely. The **welder** should avoid unsafe conditions. Never do arc welding close to a gas cylinder in case of an accidental arc strike. Even if an arc strike does not rupture the cylinder, it could weaken the cylinder wall and cause trouble later.

Safe practices must also be followed when handling gas cylinders. Transport cylinders only when the cylinder **valve** is protected by the cylinder cap. **Secure** all cylinders in an upright position both in use and in storage. Use gas cylinders only for their intended purpose. And because gas metal arc welding means working for extended periods of time under the helmet, be alert for fires. Finally, dispose of the wasted **filler metal** properly. The **wire** is very sharp and can cause punctures.

GAS METAL ARC WELDING

Gas metal arc welding is a **semiautomatic** electric arc welding process in which wire is fed continuously into the **weld pool.** A gas is used to shield the weld pool from the atmosphere, to stabilize the **arc,** and to regulate penetration. Ultimately, **shielding gases** are used to produce **quality welds.**

Although gas metal arc welding is commonly referred to as **MIG** (metal inert gas), this is not an accurate description. Gas metal arc welding uses some **active gases,** such as **carbon dioxide** (CO_2), which is not **inert** and which breaks down at high temperatures. CO_2 is used extensively for shielding and is less expensive than gas mixtures using inert gases like **argon** and **helium.** Combinations such as 75% argon and 25% CO_2 are popular for some welding applications.

Gas metal arc welding has some advantages. It saves time and money because welding can be done without stopping to remove the **slag** from each **weld bead** or to replace an **electrode** stub with a fresh electrode. This welding process can produce quality welds on many **ferrous** and **nonferrous** metals. Unfortunately, gas metal arc welding has some disadvantages, too. Outdoor welding subjects the gas shield to the effects of the wind, and the cost of building a wind screen has to be considered. Furthermore, even the best **guns** (torches) for gas metal arc welding, especially the high-**amperage** guns, can cause fatigue during extended periods of welding. This process has limited applications because there are many places in which the gun will not fit or cannot be manipulated to complete the **weldment.** All in all, the advantages of gas metal arc welding do outweigh these disadvantages, and its popularity in manufacturing applications cannot be overlooked.

Constant Voltage

Gas metal arc welding uses a **constant-voltage power source** (Figure 5–2). This avoids the large drop in **voltage** that occurs during shielded metal arc welding, which uses **constant-current** electric **output.** With constant-current output, the welder can control the voltage to some extent and can affect the heat of welding by raising or lowering the electrode (changing the arc length). However, there is a limit to how fast the welder can **feed** the electrode into the **weld pool** because the welding **current** (amperage/heat) remains fairly constant.

Because gas metal arc welding uses constant voltage, it is not limited by how fast the welder can feed wire into the weld pool. With gas metal arc welding, voltage control is taken out of the welder's hands and handled by the voltage control on the **welding machine.** When the welder increases or decreases the distance of the **contact tube** from the base metal (by raising or lowering the gun), the resistance in the wire changes, and so too do the effects of the wire as it melts into the **weld.**

Increasing the distance of the gun from the base metal and holding that distance constant causes excessive buildup of weld metal. Decreasing the distance pushes the wire into the weld pool, causing a flatter-looking weld bead. With constant voltage, the **wire feed** system compensates instantaneously for changes in the distance of the gun from the base metal without sticking the wire or burning back to the contact tube as the welder moves along the **joint.** As shown in Figure 5–3, to keep the burn-off rate constant as the distance of the gun varies, the voltage change is small compared with a larger change in the welding current.

Equipment

The power source, **feeder system,** wire, shielding gas, and gun are the five primary pieces of equipment used in gas metal arc welding. Each is described in this section.

Power Source

The first major piece of equipment, the power source, is designed electrically for DC output. Two major controls may be attached to the power source or to a part of the feeder system. These are the voltage adjustment and the wire feed speed adjustment. These two controls work together in setting the proper amperage for a given welding job. Read through the **operator's manual** to learn the particulars of the voltage and wire feed adjustments and to identify any other control features

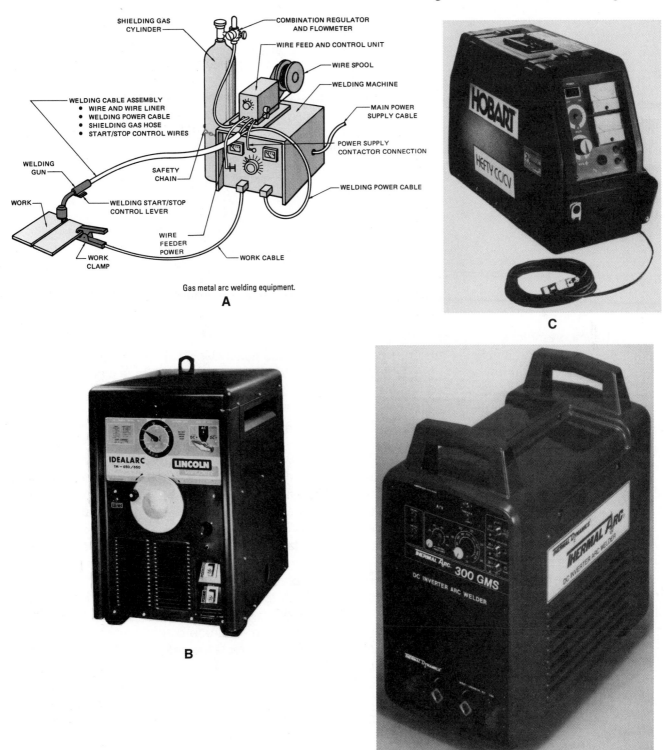

Gas metal arc welding equipment.

A

B

C

D

FIGURE 5–2   (A) Gas metal arc welding: Five primary pieces of equipment. (B) A movable, core-type welding machine. *(Courtesy of Lincoln Electric Co., Cleveland, Ohio.)* (C) Gas metal arc welding or shielded metal arc welding. *(Courtesy of Hobart Brothers Co., Troy, Ohio.)* (D) Inverter for gas metal, shielded metal, or gas tungsten arc welding. *(Courtesy of Thermal Dynamics, St. Louis, Missouri.)*

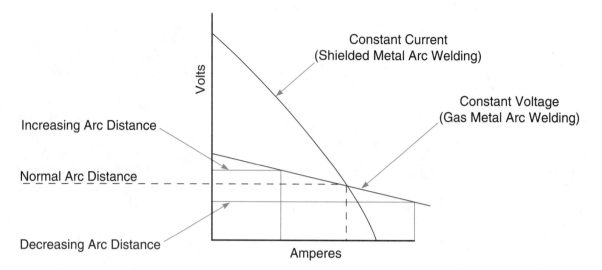

FIGURE 5–3   Graph that shows how changing the distance of the gun affects voltage and amperage.

that may be important for your power source. For example, some welding machines are equipped with a switch for the **manual** control of gas flow. Others have timers for setting **preweld** and **postflow** gas and a water flow switch for water-cooled guns.

Feeder System

The feeder system contains the drive rolls, which have to be adjusted or changed according to the **size** of wire being used. On some power sources, the wire is threaded through the feeder system and spring-loaded into position. On other systems, the pressure applied to the drive rolls is adjustable. Care must be taken not to tighten the drive rolls too much and distort (flatten) the wire, which can become tangled in the system.

On some power sources, the feeder system (Figure 5–4) can be conveniently separated from the welding machine and brought to the work to make adjustments on the spot; on other power sources, the feeder system cannot be separated (Figure 5–5). The feeder system contains the control for setting the wire feed adjustment. The major purpose of this system is to push the wire through to the gun at a constant rate.

Wire

Wire is the electrode or filler metal that is fed into the weld pool. Different types of wire are specially manufactured to meet the requirements for particular ferrous or nonferrous base metal applications. Wires may contain **deoxidizers,** which

react with the base metal and the shielding gas to produce quality welds. They are solid and may be coated with **copper** to improve electric contact and to prevent **rust.** Wires of several diameters are available. The 0.30, 0.35, and 0.45 diameter wires are among the most popular; they come in 2-, 10-, 30-, and 60-pound spools. Other sizes are also available, including 750-pound drums. These wires are used for amperage settings from 40 to 325 **amperes,** although a 3/32-inch solid wire is available for amperage settings approaching 700 amperes.

Electrode extension (wire **stickout**) is measured from the end of the copper contact tube (tip). Acceptable electrode extension can range from 1/4 inch to 1 inch, depending on the type of gas metal arc welding being done (Figure 5–6). The length of the electrode extension for a given rate of wire feed affects the characteristics of the weld. A longer length causes more resistance in the wire, taking heat away from the arc. The result is an excess of melted wire and lower penetration.

The **American Welding Society** provides a numbering classification system with specifications for the wires used in gas metal arc welding. It is important to match the wire with the base metal to be welded. Suppliers of welding products should have charts with information for matching the wire to the base metal. For example, the designation E70S-X states that the welding wire is a solid wire (indicated by S) for gas metal arc welding. It has a tensile strength of 70,000 **psi** (70) and certain other characteristics (Table 5–1).

A

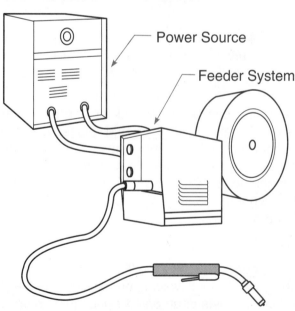

Power Source

Feeder System

B

FIGURE 5–4 (A) Feeder system. *(Courtesy of Lincoln Electric Co., Cleveland, Ohio.)* (B) Feeder system separate from power source.

TABLE 5–1:   PARTICULARS OF THE WIRE[a]

E70S-2 is a quality wire for use with 100% CO_2 or 98% argon/2% oxygen shielding gas on most carbon steel in all welding positions.

E70S-3 is a quality wire for use with 100% CO_2 or 98% argon/2% oxygen shielding gas on most carbon steels.

E70S-6 is a quality wire for use with CO_2 shielding gas on most carbon steels.

E70S-1B is a quality wire for use with 100% CO_2 or 98% argon/2% oxygen shielding gas on low-alloy steels. Stress-relieve the weldment at 1150 degrees for 8 hours.

E70T-1 is a quality flux-cored, tubular wire for use in single-pass or multipass welding in flat and horizontal positions where rust or mill scale must be tolerated, using CO_2 as the shielding gas.

E70T-4 is a quality flux-cored, tubular wire for use in single-pass or multipass welding in the flat position and the horizontal position without a shielding gas.

ER-1100 is an aluminum electrode wire that provides added ductility for use on commercially pure aluminum in decorative and architectural applications. The recommended shielding gases are 100% argon, 100% helium, or a mixture of the two.

ER-5356 is an aluminum wire with good ductility for use on base metals numbered 6061, 6063, 5454, 5456, 7005, and 7039. The recommended shielding gases are 100% argon, 100% helium, or a mixture of the two.

ER308 is the recommended stainless steel wire for base metals numbered 301, 302, 304, and 305. Argon shielding gas in combination with 1–5% oxygen can be used.

ER316 is the recommended stainless steel wire for the base metal numbered 316. Argon shielding gas in combination with 1–5% oxygen can be used.

E308T-X is the recommended flux-cored stainless steel wire for base metals 301, 302, 304, 305, 201, and 202. The shielding gas can be (1) 100% CO_2, (2) 98% argon, 2% oxygen, or (3) no gas at all.

---

[a]The AWS classification system also includes the chemical compositions and mechanical properties of these wires. Complete manufacturers' lists should be available from local welding supply companies.

Shielding Gas

The shielding gas consists of a gas cylinder and a **regulator** to control the gas flow. The inert gases helium and argon and the active gas carbon dioxide, which breaks apart into oxygen and carbon monoxide at elevated temperatures, are commonly used as shielding gases. The shielding gas replaces the **flux** by helping to stabilize the arc

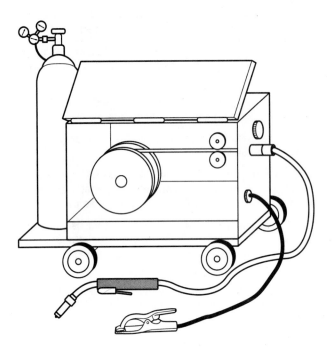

FIGURE 5–5   Feeder system inside power source.

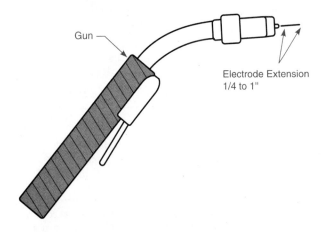

FIGURE 5–6   Electrode extension.

Gun

Electrode Extension
1/4 to 1"

A

B

FIGURE 5–7   (A) Regulator. *(Courtesy of Jeffus and Johnson, 1988.)* (B) Regulator/flowmeter. *(Courtesy of Concoa Controls Corp. of America.)*

Flowmeters can be calibrated for one particular gas or a combination of several gases. With the regulator-flowmeter combination, the pressure from the regulator to the flowmeter is preset by the manufacturer (Figure 5–7B). Some flowmeters are designed to handle CO_2, argon, helium, and **nitrogen.**

Gun

The gun, or torch, is the fifth major component of the gas metal arc welding process (Figure 5–8). The gun consists of an on/off switch that controls the wire coming out. It might also have a switch to control the welding current. The gun consists of a **nozzle** to direct the gas flow and an **insulator** to keep the nozzle from making electrical contact with the base metal. Electrical contact should occur only through the contact tube, which ener-

and to protect the weld pool. This eliminates the step of removing the slag.

The **pressure regulator** controls the gas shielding going into the gun (Figure 5–7A). A **flowmeter** might be used in addition to a regulator to adjust the flow of the shielding gas coming out of the gun, measured in cubic feet per hour **(cfh).** With a flowmeter, volume is indicated on a scale by a ball riding on top of the flow inside a tube.

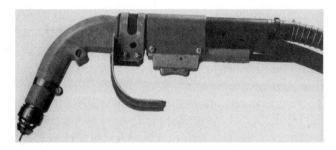

FIGURE 5–8 350-ampere gun. *(Courtesy of Lincoln Electric Co., Cleveland, Ohio.)*

gizes the wire. The gun also consists of a **liner,** which carries the wire from the spool. The liner should be matched to the size of the wire being used (Figure 5–9).

Many different styles of guns are being manufactured (Figure 5–10), but the system driving the wire through the gun is similar. It either pushes it through or pulls it through the gun. Some systems use a combination of these methods. When the drive rolls are located in the feeder system, the wire is pushed through the gun. When the drive rolls are located in the gun, the wire is pulled through. In a **spool gun,** the drive rolls and a spool of wire are located in the gun. A spool gun is used when the welder must cover a large area of the work space or when aluminum wire is being used. Aluminum wire, which is soft, can kink (twist up) while moving to the gun and out through the contact tube.

The gun is a most critical component in the gas metal arc welding process. Although welders may quickly notice that the power or the gas is off, they may be at a loss when there is trouble concerning the gun itself. If the nozzle is plugged

A

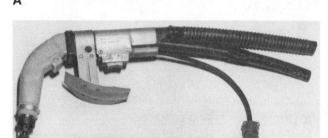

B

FIGURE 5–10 (A) Spool gun. (B) Water-cooled gun. *(Photos courtesy of Lincoln Electric Co., Cleveland, Ohio.)*

by **spatter,** the weld pool will not receive the necessary gas flow. If the nozzle is partially plugged, spatter can lodge between the contact tube and the nozzle. This can nullify the insulator, causing arcing off the nozzle whenever the nozzle touches the base metal. Spatter can be easily removed by tapping the nozzle against the palm of your hand. This will not damage the gun. If this does not work, the nozzle will have to be taken apart and the spatter cleaned out. If the gun's contact tube is worn, the wire will not receive the necessary electrical contact from the enlarged opening. Like the liner, the contact tube must be matched to the size of the wire being used (Figure 5–11).

Metal Transfer

Although many manufacturers have their own trade names, the American Welding Society describes two methods by which metal is melted across the arc in the gas metal arc welding process. **Short circuiting arc transfer** (GMAW-S) is one method. With this method, there is a transfer of metal across the arc, which is extinguished by the weld pool, only to reignite the arc and begin the cycle all over again. This cycle repeats up to 200 times per second. Short circuiting arc transfer

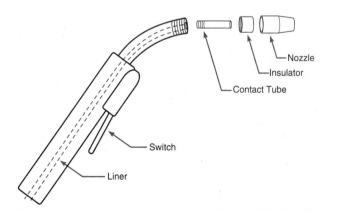

Nozzle
Insulator
Contact Tube
Switch
Liner

FIGURE 5–9 Exploded view of gun with parts labeled.

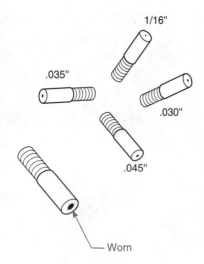

1/16"

.035"

.030"

.045"

— Worn

FIGURE 5–11 Contact tubes.

uses 0.030, 0.035, and 0.045 wires. Carbon dioxide and argon are the primary shielding gases in the welding of carbon steels. This process can be used in all welding positions; 225 amperes is the maximum current in the flat position. This process is popular on thin metals of up to 1/4 inch and can be an economical process for welding heavier metals in the **vertical** and **overhead** positions. Table 5–2 provides additional information about short circuiting arc transfer.

Drop transfer is the second method. Drops of metal cross the arc to the weld pool either by gravity or by the force of the welding circuit. Drop transfer is broken down into two types: **globular transfer** and **spray arc transfer** (GMAW-SP). Globular transfer produces a drop greater than the size of the wire. It is of use for low-current applications on thin metal. Although globular transfer uses CO_2 as a shielding gas at all amperage settings, it is not as effective as other welding processes, and spatter results.

The spray arc type of drop transfer happens when the drop changes to a spray by increasing the welding current. Spray does not occur when CO_2 is the shielding gas. A mixture of 98% argon, 2% oxygen is commonly used to initiate spray arc transfer. Table 5–3 lists the minimum amperage required to initiate spray transfer with various types and sizes of wire. For example, if a .045 mild-steel wire is selected for use with a 98% argon, 2% oxygen mixture, the minimum spray current is 220 amperes. This indicates that the higher amperage makes spray arc transfer suitable for use on thicker **steel** than would generally be used in the short circuiting arc transfer process.

Spray arc transfer is generally limited to use in the flat and horizontal welding positions because the heat that is generated in the weld pool cannot be controlled in the other positions. For this reason, some welding machines offer **pulsed spray. Pulsed gas metal arc welding** (GMAW-P) is a process in which the welding current is pulsed. A

TABLE 5–2:   SHORT CIRCUITING ARC TRANSFER: In One Pass Using 100% CO_2 or 75% Argon/25% CO_2 (approximate settings)

Steel Thickness	Wire Diameter	Shielding Gas	Voltage	Amperage
18–24 **gauge**	0.030	15–20 cfh	16–20	50–120
1/16 inch	0.035	20–25 cfh	19–22	110–130
1/8 inch	0.045	20–25 cfh	20–24	180–225

TABLE 5–3:   SPRAY ARC TRANSFER (minimum spray settings)

Wire Type	Wire Diameter	Shielding Gas	Amperage
Mild steel/low alloy	0.030 inch	98% argon/2% oxygen	150
Mild steel/low alloy	0.035 inch	98% argon/2% oxygen	165
Mild steel/low alloy	0.045 inch	98% argon/2% oxygen	220
Mild steel/low alloy	1/16 inch	98% argon/2% oxygen	275
Mild steel/low alloy	3/32 inch	98% argon/2% oxygen	350
Stainless steel	0.035 inch	99% argon/1% oxygen	180
Stainless steel	0.045 inch	99% argon/1% oxygen	225
Aluminum	0.045 inch	100% argon	135

pulsed arc produces two current levels, one at the spray level and one at a level too low for spray to occur. This pulsing feature permits welding in the vertical and overhead positions. Note that because the argon/oxygen gas mixture is more expensive, spray arc transfer may have limited applications on **steel** when other welding processes are more cost-effective.

Outstanding Features of Gas Metal Arc Welding

Gas metal arc welding is an economical welding process that can be carried out in all welding positions. It eliminates slag and some fumes and reduces **distortion** on thin metal without **melt-thru.** Short circuiting arc transfer is most popular for welding thin metal and can be adapted for making **spot welds.** Spray arc transfer has applications for welding thicker metal, and the pulsed arc feature makes it useful for vertical and overhead welding.

Gas Metal Arc Welding Problems

Gas metal arc welding has its own set of problems and solutions. If you become familiar with these problems and their causes, you can prevent them.

1. **Porosity** can be the result of dirty base metal, mill scale, poor or excessive gas flow, wind or draft, or a plugged nozzle.
2. Excessive spatter can be the result of excessive gas flow or the voltage is too high.
3. Excessive penetration (melt-thru) can result from wire speed and voltage adjustments that are set too high, travel speed that is too slow, or poor **joint design.**
4. Poor root penetration might be the result of wire speed and voltage adjustments that are set too low, travel speed that is too fast, or poor joint design.

Equipment failure can result in poor welding, so take good care of the equipment. Leaks in the shielding gas line will lead to inadequate shielding and an expensive waste of gas. Spatter in the nozzle or a worn insulator can cause arcing on the nozzle, which will affect welding and damage the nozzle. A dirty work environment can transfer dirt to the wire spool and then to liner, which can affect welding by interrupting the smooth flow of wire inside the liner. Repair all leaks, and remove spatter from the nozzle. Protect the electrode and workpiece leads from damage. Cover the equipment when not in use.

FLUX CORED ARC WELDING

Flux cored arc welding (FCAW) is different from gas metal arc welding. Flux cored arc welding is a semiautomatic welding process in which a **tubular wire** is fed continuously from a spool into the weld pool. The wire contains a flux inside that helps to shield the weld pool from contaminants on the base metal and in the atmosphere. The flux also helps stabilize the arc. Because the flux is inside the wire, it is protected and is unable to soak up moisture. The flux cored arc welding process allows a higher amperage than shielded metal arc welding. As a result, wire is deposited into the weld more quickly and with deeper penetration.

Welding machines that are set up for gas metal arc welding can be adapted for flux cored arc welding, although lower amperage machines limit the size of wire that can be used (Figure 5–12). Flux cored arc welding is most effective at higher amperages, allowing greater rates of deposition. Power sources designed for high-amperage welding feature water-cooled and gas-cooled guns with their own exhaust systems. At high amperages,

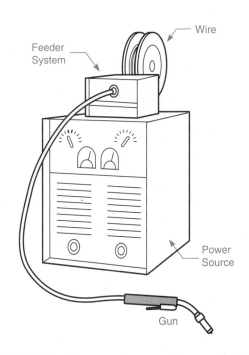

FIGURE 5–12   Flux cored arc welding equipment.

guns give off heat and produce a larger volume of fumes, requiring ventilation.

Flux cored arc welding is really shielded metal arc welding on a spool. Because these two processes are so much alike, more training is necessary to develop competency than gas metal arc welding processes. As in shielded metal arc welding, the gun angle is important in flux cored arc welding to counter the effects of gravity (Figure 5–13). Care must also be taken with this process when removing slag to avoid **slag inclusions** (slag trapped within the weld).

Flux cored arc welding has many applications, including the welding of storage vessels, railroad cars, earth-moving equipment, and steel structures. Although flux cored arc welding is a **self-shielding** welding process, a shielding gas can be added. Carbon dioxide is a popular choice. The composition of the wire determines whether the wire is self-shielding or requires a shielding gas.

Because the wire used in flux cored arc welding contains a flux, slag is created that must be removed. If all welding **parameters** are set correctly, the slag should chip off easily.

Although flux cored arc welding is appropriate for all welding positions, it is especially useful for welding plate (metal that is thicker than 3/16 inch) in the flat welding position. Flux cored arc welding is also a popular process for surfacing or building up equipment that is subject to wear.

DEVELOPMENTS IN TECHNOLOGY

This unit began with an illustration showing a microcomputer-based power source (see Figure 5–2B). Not only does computer technology make it possible to set welding parameters with the touch of a finger, but now this information can be stored in the computer's memory even after the power has been turned off. The use of computers will help welders do the job better (Figure 5–14).

Computer systems should not intimidate anyone. No matter what the packaging of the power source, welding skill is still going to be the main factor in getting the job done as welders flip down their helmets. New technology has replaced some dials and switches, offering welders more choices in completing the given task. Every development in technology is another opportunity for success.

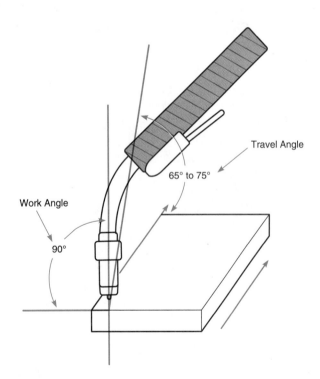

FIGURE 5–13 Work angle and travel angle for flux cored arc welding.

FIGURE 5–14 Digital solid-state power source for wire welding applications. *(Courtesy of Hobart Brothers Co., Troy, Ohio.)*

SAFETY REMINDERS

Review the safety procedures given below. Note that they are not identical to those given in the previous unit.

1. Wear **safety glasses** with side shields. Safety glasses protect your eyes from slag and from grinding particles. They also deflect **ultraviolet rays** which are given off by the welding arc and help prevent **arc flash.** Safety glasses are so important that no one should be in a welding shop without having them on. However, safety glasses are no protection against careless shop practices.

2. Wear earplugs in the shop to safeguard your hearing. Besides offering protection from nerve-damaging noise, earplugs help prevent hot slag and sparks from reaching the eardrum.

3. Use leather **welding gloves** to protect your hands and lower arms from burns and shock. Only gloves made for welding do an adequate job of handling heat without losing their shape. If you use tongs to handle hot metal, your welding gloves will last longer and not become stiff.

4. Wear cotton clothing, not synthetic fibers that will burn or melt to the skin, to shield exposed skin from arc rays which are more intense than sunlight. A leather jacket, leather sleeves, and chaps, although expensive, are preferred because they will save your clothing from the sparks and spatter of welding.

5. Be sure the shop is clear of any **flammable** or **volatile** materials. Gasoline and solvents have no place in a welding shop.

6. Keep cigarette lighters under pressure out of the welding shop.

7. Always wear a welding helmet with a filter lens of the correct shade to protect against arc flash. Arc flash is a painful, but usually a temporary, eye condition caused by the light of the **welding arc.** If you suspect arc flash is more serious, get proper medical attention. Replace cracked lenses or damaged helmet. Check the shade number on the lens. Gas metal arc welding and flux cored arc welding require a shade number of 10 to 14.

8. Welding shop should be equipped with fire extinguishers. Know their location and classification. See Figure 4–21.

9. Use the ventilation system to remove smoke and fumes.

10. Protect yourself against electric shock. Use gloves, and be sure your clothing and workplace are dry. If you stay dry while welding, you act as an insulator and not a conductor of electricity. Electricity always seeks the path of least resistance, so stay dry when welding to stay safe.

11. Steel-toe leather boots offer some protection against foot injury. Leave your athletic shoes at home.

12. Wear dark clothing, which absorbs light rays, to reduce the dangers of arc flash.

13. Before turning on the welding machine, be sure to stretch out the gun's **power lead** and the **workpiece lead.** Never wrap either lead around your body. Always place the **workpiece connection** on the work or as close to the work as possible for good electrical contact. The workpiece connection can be welded, clamped, or screwed into position.

14. Never weld on pressurized cylinders. A stray arc strike on a gas cylinder can cause an explosion.

15. As a common courtesy, always warn others in the area before striking an arc, and always begin an arc strike at the point of the welding.

16. Be alert to strange smells. Visibility is limited under a welding helmet, and your sense of smell becomes important. Unusual odors could indicate a fire or a toxic substance, alerting you to possible danger.

17. Pay attention to the work at hand. Never give in to daydreaming or distractions.

Safety for Gas Metal Arc Welding

18. Keep track of wire cut from the spool. Cut wire is a shop hazard. It is sharp enough to puncture the skin.

19. Turn off the welding machine if you must leave it unattended.

20. Replace parts of the gun when they wear out.

21. Secure the gas cylinder when setting up for welding, and take the necessary precautions to avoid an arc strike on a gas cylinder.

REVIEW

1. Name four points of safety for handling gas cylinders.
2. What is the purpose of the gas in gas metal arc welding?
3. List several advantages and disadvantages of gas metal arc welding.
4. Discuss the function of the voltage and wire feed adjustments in gas metal arc welding?
5. What are the five primary pieces of equipment used in gas metal arc welding? Describe each one.
6. How is a pressure regulator different from a flowmeter?
7. How do short circuiting arc transfer and spray arc transfer differ?
8. Name one gas that is used for spray arc transfer.
9. Describe four problems in the weld that might be encountered in gas metal arc welding.
10. What is flux cored arc welding?

Create Three Questions

1.
2.
3.

Related Math and English Questions

1. Which number is larger: 0.035 or 0.045? Write out these two numbers in words.
2. Is 0.045 larger than 1/16? Write out the decimal equivalent for 1/16.
3. Write a paragraph on the importance of wearing safety glasses.
4. Describe the relationship between travel speed and penetration during welding.

For Further Thought

1. Suppose that you are welding outside using the gas metal arc welding process, and the test weld is full of pinholes. What can you do?
2. Whenever the nozzle of the gun touches the base metal, the nozzle arcs. What can you do to eliminate this problem?
3. The wire is jamming up in the drive rolls. What could be the cause?
4. What advantages, when welding sheet metal, does gas metal arc welding have over shielded metal arc welding?
5. Can the same equipment be used for flux cored arc welding and gas metal arc welding? Explain.

SUGGESTED ACTIVITIES

1. Set up a demonstration of gas metal arc welding equipment. Name each part, and explain its function.
2. Compare some of the wires used in gas metal arc welding or flux cored arc welding. Become familiar with what the American Welding Society numbers mean.
3. Review an operator's manual and literature on welding machines provided by suppliers.
4. Learn how the feeder system works by changing wires and drive rolls threading the system.
5. Tour a business that uses gas metal arc welding or flux cored arc welding.

UNIT 5: WELDING EXERCISES

Setting Up to Weld

Follow the operator's manual when setting up the welding equipment for gas metal arc welding or flux cored arc welding. The successful completion of this unit partly depends on knowing what a quality weld looks like. If you can recognize a quality weld, you should be able to set up any welding machine to make quality welds even if the adjustments are not labeled.

1. When setting up for welding, remember that the voltage and wire feed adjustments work together. If the wire feed adjustment is too high for the voltage, the wire will not melt properly into the weld pool, and an excessive buildup of filler metal will result.

2. If the wire feed adjustment is too low for the voltage, the wire will burn back into the contact tube. To save the contact tube from becoming fused to the wire during welding, begin with a high wire feed and low voltage adjustments.

3. Some welding machines are equipped with fine adjustments for **slope** and **inductance.** Slope can be used in combination with the voltage adjustment. Increasing slope causes a decrease in current (amperage), which can cut back on spatter. Increasing inductance causes an increase in arc time (wetness of the weld pool) as the current rises. Inductance can also cut back on spatter by controlling the rate of current rise. Begin by setting these parameters in the middle range.

4. There is a relationship between travel speed across the joint and penetration. Penetration increases as travel speed decreases and vice versa. However, penetration should not be determined by travel speed alone. The adjustment of voltage in relation to wire feed speed is also involved.

5. The width of the weld bead is a primary consideration for setting the voltage, but travel speed and wire feed (amperage) are also important (Figure 5–15).

6. The height of the weld bead is a primary consideration for setting the voltage, and travel speed and wire feed (amperage) are also important.

7. The electrode extension, or stickout (length of wire measured from the contact tube), and the gun angle are two more secondary adjustments that can affect welding. Use a sidecutter and cut the wire to keep the stickout close when beginning the arc.

8. Check the nozzle to see that it is not plugged and that the opening of the contact tube is round. An **anti-spatter** spray or gel can be used to help prevent the spatter from sticking to the nozzle.

9. The sound of the arc during welding is an indication that the proper welding parameters are in place. Does the arc sound like eggs frying? It should.

10. Relax and find a comfortable position to weld. Good welding technique depends on being comfortable.

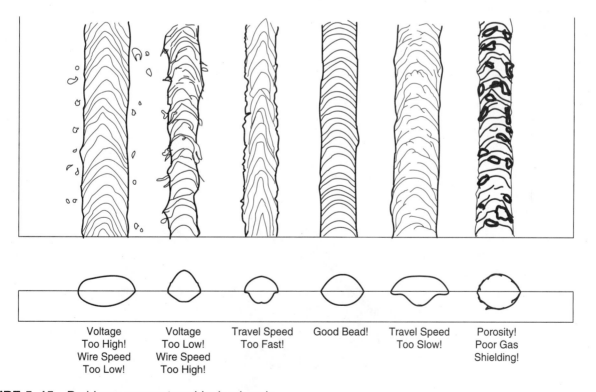

| Voltage Too High! Wire Speed Too Low! | Voltage Too Low! Wire Speed Too High! | Travel Speed Too Fast! | Good Bead! | Travel Speed Too Slow! | Porosity! Poor Gas Shielding! |

FIGURE 5–15   Problems encountered laying beads.

UNIT 5: EXERCISE 1

Short Circuiting Arc Transfer (GMAW-S): Padding Plate in Flat Position, Surfacing Weld

Practice using both the **forehand** and **backhand** methods (Figure 5–16). Work on making weld beads of consistent size, and concentrate on maintaining proper travel speed and gun angle. Several **passes** may be necessary to make proper adjustments on the voltage and wire feed controls. Do not be afraid to adjust the controls to examine the results of different settings. Figure 5–17 shows the setup for this exercise.

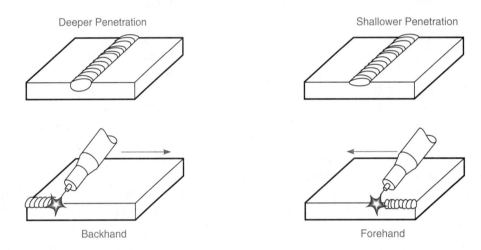

FIGURE 5–16   Backhand and forehand welding techniques.

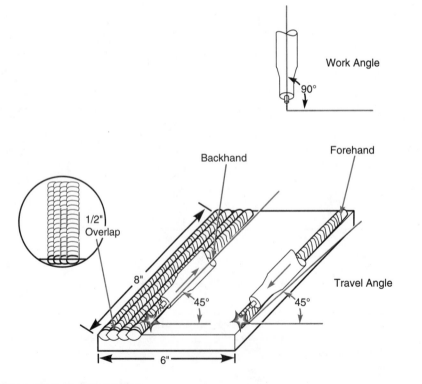

FIGURE 5–17   Padding plate in flat position.

Necessary Material and Equipment

Safety glasses	Helmet	Grinder	Spool wire (0.030, 0.035, or
Welding gloves	Earplugs	1 piece 3/8″ plate, 6″ × 8″	0.045)
Protective clothing	Sidecutter	(or scrap plate)	Shielding gas: carbon dioxide

Instructions

1. Examine the operator's manual, if necessary.
2. Be sure the power is off before plugging in the machine.
3. Screw out the pressure regulator before opening the gas cylinder valve, standing to one side. If attaching a cylinder or replacing a cylinder, follow safe practices. Move the cylinder only with the valve cap in place, and secure the cylinder so it cannot fall.
4. Turn on the machine. Adjust the gas flow from 15 to 25 cfh, with the voltage set between 15 and 25 volts.
5. Set the wire feed at 50%, and adjust it after running a test weld bead. What is the number of the wire being used?
6. Clean the metal surface by grinding. Run 1 to 3 layers of weld beads, practicing both forehand and backhand techniques.
7. Test the weld by **visual inspection.** Examine the weld beads for consistency and **overlap.**
8. Check with the instructor for evaluation, if necessary.
9. Practice this exercise until you meet the standards established for the course.

UNIT 5: EXERCISE 2

Short Circuiting Arc Transfer (GMAW-S): Corner Joint on Plate in Flat Position, Fillet Weld

This **corner joint** exercise requires consistent welding with complete **fusion** of a **fillet weld** into the base metal without **undercutting** the base metal. While practicing this exercise, occasionally examine the nozzle for spatter, which can affect the gas flow. Anti-spatter spray or gel can be used. The setup for this exercise is illustrated in Figure 5–18.

Necessary Material and Equipment

Safety glasses	Earplugs	2 pieces 3/8″ steel plate, 3″ × 8″	Spool wire (0.030, 0.035, or 0.045)
Welding gloves	Sidecutter		
Protective clothing	Grinder	Padding plate (scrap)	Shielding gas: carbon dioxide
Helmet			

Instructions

1. Set up the machine for welding. (Refer back to the beginning of these exercises, if necessary.) Be sure the power is off before plugging in the machine.
2. Turn on the machine. Adjust the gas flow from 15 to 25 cfh, with the voltage set between 15 and 25 volts.
3. Set the wire feed at 50%, and adjust it after running a test weld bead.
4. Clean the metal surfaces by grinding. **Tack weld** the two pieces of steel together, then tack weld them to the scrap padding plate.
6. Lay 10 to 15 weld beads with a slight weave. Practice both forehand and backhand techniques.
7. Check the overlap. There should be no valleys (grooves) between the weld beads.
8. Examine the weld for undercut along the base metal. If there is undercut but the wire feed and the voltage are set correctly, check the gun angle.

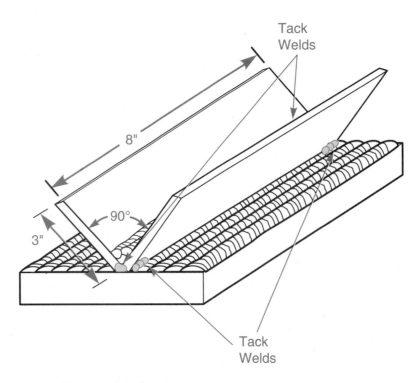

FIGURE 5–18 Corner joint, 10 to 15 weld beads.

9. Test the weld by visual inspection for bead consistency and overlap, avoiding undercut.
10. Check with the instructor for evaluation, if necessary.
11. Practice this exercise until you meet the standards established in the course.

UNIT 5: EXERCISE 3

Short Circuiting Arc Transfer (GMAW-S): Butt Joint on Plate in Flat Position, V-Groove Weld without Backing

This **butt joint** exercise provides an opportunity to test a welded joint by **destructive testing.** The first weld pass is critical because without complete **root penetration,** the weld will fail when tested. As you set up the machine for welding, run some test weld beads on scrap material. Figure 5–19 illustrates the setup for this exercise.

Necessary Material and Equipment

Safety glasses	Sidecutter	Spool wire (0.030, 0.035,	C-clamp fixture
Welding gloves	Grinder	0.045)	Cutting machine (to make
Protective clothing	2 pieces 3/8″ steel plate, 3″	Shielding gas: carbon	30° **bevel**)
Helmet	× 8″	dioxide	
Earplugs	2 strong backs		

Instructions

1. Set up for welding. Be sure the power is off before plugging in the machine.
2. Turn on the machine. Adjust the gas flow from 15 to 25 cfh, with the voltage set between 15 and 25 volts.
3. Set the wire feed at 50%, and adjust it after running a test weld bead.
4. Clean the metal surfaces by grinding. Grind from 0 to 1/8″ **root face.**
5. Tack weld the two pieces together with a 0 to 1/8″ **joint root** opening. Tack weld the strong backs (1-inch width) in four places.
6. Use the forehand technique on the joint root pass with a slight **weave,** concentrating on root penetration.
7. Four passes (if 3/8″ thick plate) should be enough to complete the welded joint.
8. Make a visual inspection. Look for poor root penetration, undercut, and **incomplete fill.**
9. Check with the instructor for evaluation, if necessary.
10. Pass the **guided bend test** before moving on.

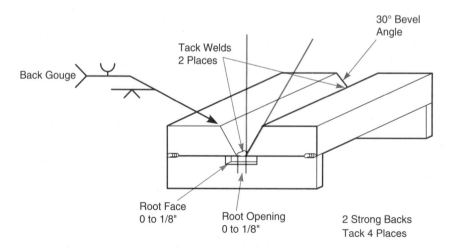

FIGURE 5–19   V-groove weld in the flat position without backing.

UNIT 5: EXERCISE 4

Short Circuiting Arc Transfer (GMAW-S): Lap Joint on Plate in Horizontal Position, Fillet Weld

The **lap joint** tests your ability to make weld beads of a consistent size. There must be sufficient overlap to keep the **legs** of the fillet weld within the thickness of the metal. Use the edge as a guide while making each pass. Direct the nozzle away from the top piece, which melts more easily than the bottom piece. The setup is shown in Figure 5–20.

Necessary Material and Equipment

Safety glasses	Earplugs	2 pieces 1/4″–3/8″ steel	Shielding gas: carbon
Welding gloves	Sidecutter	plate, 3″ × 8″	dioxide
Protective clothing	Grinder	Spool wire (0.030, 0.035,	
Helmet		or 0.045)	

Instructions

1. Set up the machine for welding. Be sure the power is off before plugging in the machine.
2. Turn on the machine. Adjust the gas flow from 15 to 25 cfh, with the voltage set between 15 and 25 volts.
3. Set the wire feed at 50%, and adjust it after running a test weld bead.
4. Clean the metal surfaces by grinding. Tack weld the two pieces together.
5. Use the backhand technique.
6. Test by visual inspection. The leg size of the weld should equal the thickness of the steel.
7. Check with the instructor for evaluation, if necessary.
8. Practice this exercise until you meet the standards established in the course.

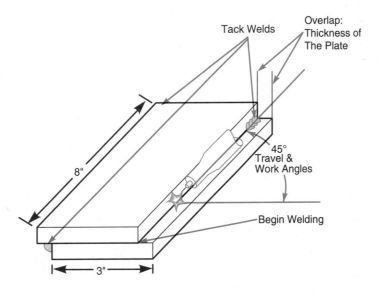

FIGURE 5–20   Lap joint in horizontal position.

UNIT 5: EXERCISE 5

Short Circuiting Arc Transfer (GMAW-S): Butt Joint on Plate in Horizontal Position, V-Groove Weld without Backing

The gun angle is important for this exercise. The joint root pass should be made with a 90° **work angle.** The gun nozzle should be angled downward when directing the bead toward the bottom piece. The gun nozzle should be angled upward when directing the bead toward the top piece. The angle of the gun will help prevent the weld pool from running down. Figure 5–21 shows the setup for this exercise.

Necessary Material and Equipment

Safety glasses	Sidecutter	Spool wire (0.030, 0.035,	C-clamp fixture
Welding gloves	Grinder	or 0.045)	Cutting machine (30°
Protective clothing	2 pieces 3/8″ steel plate, 3″	Shielding gas: carbon	bevel)
Helmet	× 8″	dioxide	
Earplugs	2 strong backs		

Instructions

1. Set up the machine for welding. Be sure the power is off before plugging in the machine.
2. Turn on the machine. Adjust the gas flow from 15 to 25 cfh, with the voltage set between 15 and 25 volts.
3. Set the wire feed at 50%, and adjust it after running a test weld bead.

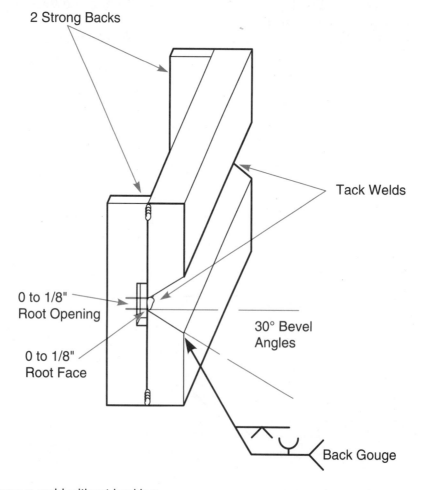

FIGURE 5–21   V-groove weld without backing.

4. Clean the metal surfaces by grinding. Grind from 0 to 1/8″ root face.
5. Tack weld the two pieces together with a 0 to 1/8″ joint root opening. Tack weld the strong backs (1-inch width) in four places.
6. Use the forehand technique on the joint root pass with a slight weave, concentrating on root penetration.
7. Make four passes. This should be enough to complete the welded joint.
8. Make a visual inspection. Look for poor root penetration, porosity, undercut, and incomplete fill.
9. Check with the instructor for evaluation, if necessary.
10. Pass the guided bend test before moving on.

UNIT 5: EXERCISE 6

Short Circuiting Arc Transfer (GMAW-S): Tee Joint on Plate in Vertical Position, Fillet Weld

This **tee joint** exercise in the **vertical position** requires both the **uphill** and **downhill** welding techniques. Remember to use the weaving technique for welding uphill. Use a slight weave on the first pass, a wider weave on the second pass, and a still wider weave on the third pass. Run **stringers** for welding downhill. Watch the weld pool moving downward, and move the gun just fast enough to keep the weld pool above the wire (to prevent weld pool from dropping away). Use the same angle for the nozzle of the gun when welding both downhill and uphill. Alternate welding on both sides of the joint. Figure 5–22 shows the setup for this exercise.

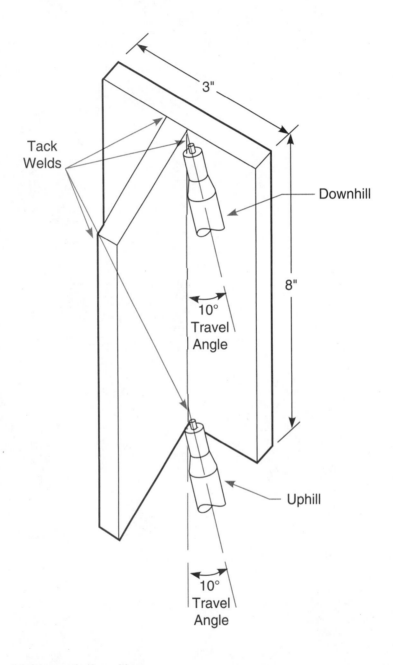

FIGURE 5–22   Tee joint in vertical position.

Necessary Material and Equipment

Safety glasses	Earplugs	2 pieces 1/4″–3/8″ steel	Shielding gas: carbon
Welding gloves	Sidecutter	plate, 3″ × 8″	dioxide
Protective clothing	Grinder	Spool wire (0.030, 0.035,	C-clamp fixture
Helmet		or 0.045)	

Instructions

1. Set up the machine for welding. Be sure the power is off before plugging in the machine.
2. Turn on the machine. Adjust the gas flow from 15 to 25 cfh, with the voltage set between 15 and 25 volts.
3. Set the wire feed at 50%, and adjust it after running a test weld bead.
4. Clean the metal surfaces by grinding. Tack weld the two pieces together.
5. Lay three passes uphill, using the weaving technique.
6. Make a visual inspection. Look for unequal leg size in the weldment, undercut, porosity, and incomplete fill. Correct these problems.
7. Check with the instructor for evaluation at any step in the sequence, if necessary.
8. Lay six weld beads downhill, making stringers.
9. Lay one to three more passes uphill.
10. Practice this exercise until you meet the standards established in the course.

UNIT 5: EXERCISE 7

Short Circuiting Arc Transfer (GMAW-S): Butt Joint on Plate in Vertical Position, V-Groove Weld without Backing

For this exercise, travel speed, in combination with the weaving technique, is important. Be sure to travel fast enough to penetrate each side of the joint equally well, using the uphill technique. Pause slightly on each side of the joint to avoid undercut. The setup is shown in Figure 5–23.

Necessary Material and Equipment

Safety glasses	Sidecutter	Spool wire (0.030, 0.035,	C-clamp fixture
Welding gloves	Grinder	or 0.045)	Cutting machine, 30° bevel
Protective clothing	2 pieces 3/8″ steel plate, 3″	Shielding gas: carbon	
Helmet	× 8″	dioxide	
Earplugs	2 strong backs		

Instructions

1. Set up the machine for welding. Be sure the power is off before plugging in the machine.
2. Turn on the machine. Adjust the gas flow from 15 to 25 cfh, with the voltage set between 15 and 25 volts.
3. Set the wire feed at 50%, and adjust it after running a test weld bead.
4. Clean the metal surfaces by grinding. Grind from 0 to 1/8″ root face.
5. Tack weld the two pieces together with a 0 to 1/8″ joint root opening. Tack weld the strong backs (1-inch width) in four places.
6. Use the forehand technique on the joint root pass with a slight weave, concentrating on root penetration.
7. Make four or five passes to finish the weld.
8. Make a visual inspection. Look for incomplete root penetration, undercut, porosity, and incomplete fill. Correct these problems.
9. Check with the instructor for evaluation, if necessary.
10. Pass the guided bend test before moving on.

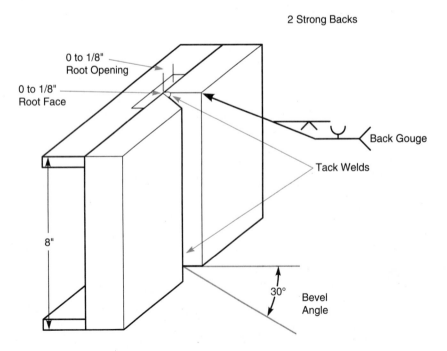

FIGURE 5–23   V-groove weld in vertical position without backing.

UNIT 5: EXERCISE 8

Short Circuiting Arc Transfer (GMAW-S): Lap Joint on Sheet Steel in Horizontal Position, Fillet Weld

Several tack welds can help minimize the problem of distortion when welding thin metal. **Intermittent welding** (welds that are not continuous over the entire joint) is another method for reducing the heat input. However, to develop skill welding thin metal, one continuous weld bead is recommended. The key to welding a joint of this kind is a close fit up. Figure 5–24 illustrates the setup for this exercise.

Necessary Material and Equipment

Safety glasses	Helmet	Grinder	Spool wire (0.030 or 0.035)
Welding gloves	Earplugs	2 pieces 16-gauge steel, 3″	Shielding gas: carbon
Protective clothing	Sidecutter	× 8″	dioxide

Instructions

1. Set up the machine for welding. Be sure the power is off before plugging in the machine.
2. Turn on the machine. Adjust the gas flow from 15 to 25 cfh, with the voltage set between 15 and 25 volts.
3. Set the wire feed at 50%, and adjust it after running a test weld bead. A lower heat input is required for welding thin metal.
4. Clean the metal surfaces by grinding. Tack weld the two pieces together. Two tack welds are not enough. Use a greater overlap than for previous exercises. Make the tack welds small enough to be remelted, using only one weld bead to complete each joint.
5. Use both forehand and backhand techniques. Which technique works best? Remember that the backhand technique concentrates greater heat at the joint root.
6. Make a visual inspection. Look for consistent size in the formation of the weld bead and an absence of both undercut and porosity.
7. Check with the instructor for evaluation, if necessary.
8. Practice this exercise until you meet the standards established in the course.

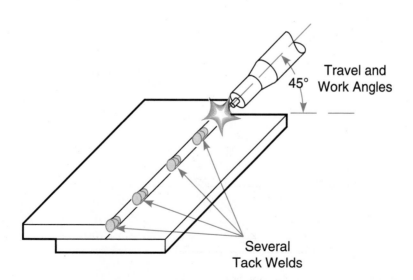

FIGURE 5–24   Lap joint on sheet steel in horizontal position.

UNIT 5: EXERCISE 9

Short Circuiting Arc Transfer (GMAW-S): Butt Joint on Sheet Steel in Flat Position, Square-Groove Weld

Distortion is a serious problem in a butt joint. If a box and pan brake is available, putting a small bend on the edge of very thin metal will result in less distortion in an **edge flange** joint. The bend in the flange melts with the filler metal into the weld. To achieve complete root penetration in this exercise (assuming correct adjustments on the welding machine), the travel speed is important. Note that these exercises can be more difficult than welding on actual projects. Weldments containing more mass experience less distortion. Figure 5–25 shows the setup for this exercise.

Necessary Material and Equipment

Safety glasses	Earplugs	Spool wire (0.030 or 0.035)	2 firebricks or C-clamp
Welding gloves	Sidecutter	Shielding gas: carbon	fixture
Protective clothing	Grinder	dioxide	
Helmet	2 pieces 16-gauge steel, 3″ × 8″		

Instructions

1. Set up the machine for welding. Be sure the power is off before plugging in the machine.
2. Turn on the machine. Adjust the gas flow from 15 to 25 cfh, with the voltage set between 15 and 25 volts.
3. Set the wire feed at 50%, and adjust it after running a test weld bead.
4. Clean the metal surfaces by grinding. Tack weld in several places. Fit up with the joint root opening equal to the diameter of the wire, or up to 1/16″.
5. Use the forehand technique, concentrating on complete root penetration.
6. Make a visual inspection. Look for complete root penetration and no porosity.
7. Check with the instructor for evaluation, if necessary.
8. Practice this exercise until you meet the standards established in the course.

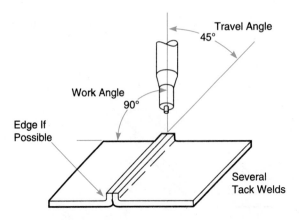

FIGURE 5–25 Butt joint on sheet steel in flat position.

UNIT 5: EXERCISE 10

Short Circuiting Arc Transfer (GMAW-S): Butt Joint on Sheet Steel in Horizontal Position, Square-Groove Weld

This exercise is similar to Exercise 9, but you will probably find it easier with the added practice. Remember to keep the gun at a 90° work angle. Keep moving along the joint to avoid melt-thru. The setup is shown in Figure 5–26.

Necessary Material and Equipment

Safety glasses	Earplugs	2 pieces 16-gauge steel, 3″	Shielding gas: carbon
Welding gloves	Sidecutter	× 8″	dioxide
Protective clothing	Grinder	Spool wire (0.030 or 0.035)	C-clamp fixture
Helmet			

Instructions

1. Set up the machine for welding. Be sure the power is off before plugging in the machine.
2. Turn on the machine. Adjust the gas flow from 15 to 25 cfh, with the voltage set between 15 and 25 volts.
3. Set the wire feed at 50%, and adjust it after running a test weld bead.
4. Clean the metal surfaces by grinding. Tack weld in several places. Fit up with the joint root opening equal to the diameter of the wire, or up to 1/16″.
5. Use the forehand technique, concentrating on complete root penetration.
6. Make a visual inspection. Look for complete root penetration and no porosity.
7. Check with the instructor for evaluation, if necessary.
8. Practice this exercise until you meet the standards established in the course.

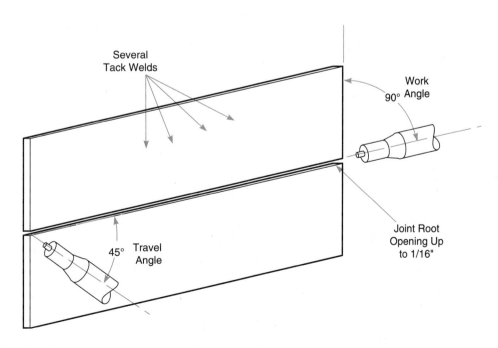

FIGURE 5–26   Butt joint on sheet steel in horizontal position.

UNIT 5: EXERCISE 11

Short Circuiting Arc Transfer (GMAW-S): Lap Joint on Sheet Steel in Vertical Position, Fillet Weld

This exercise is designed for the downhill welding technique. Downhill is very useful in welding very thin material when melt-thru can be a problem. Downhill is a useful skill, forcing the welder to maintain a faster travel speed than usual. The faster travel speed generates less heat. Figure 5–27 illustrates the setup for this exercise.

Necessary Material and Equipment

Safety glasses	Earplugs	2 pieces 16-gauge steel, 3″	Shielding gas: carbon
Welding gloves	Sidecutter	× 8″	dioxide
Protective clothing	Grinder	Spool wire (0.030 or 0.035)	C-clamp fixture
Helmet			

Instructions

1. Set up the machine for welding. Be sure the power is off before plugging in the machine.
2. Turn on the machine. Adjust the gas flow from 15 to 25 cfh, with the voltage set between 15 and 25 volts.
3. Set the wire speed at 50%, and adjust it after running a test weld bead.
4. Clean the metal surfaces by grinding. Tack weld in several places.
5. Keep the wire in the gap between the two pieces as you weld.
6. Make a visual inspection. Look for a weld bead that is of consistent size from one side of the weldment across to the other side.
7. Check with the instructor for evaluation, if necessary.
8. Practice this exercise until you meet the standards established in the course.

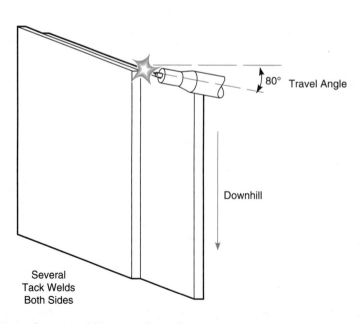

FIGURE 5–27   Lap joint on sheet steel in vertical position.

UNIT 5: EXERCISE 12

Short Circuiting Arc Transfer (GMAW-S): Butt Joint on Sheet Steel in Vertical Position, Square-Groove Weld

Controlling the size of the weld pool is the key to making weld beads of consistent size. Welding downhill makes this easier with the help of gravity. Keep in mind that penetration is not as deep when welding downhill as when welding uphill, so the downhill technique is not usually preferred when welding plate. The setup for this exercise is shown in Figure 5–28.

Necessary Material and Equipment

Safety glasses	Earplugs	2 pieces 16-gauge steel, 3″	Shielding gas: carbon
Welding gloves	Sidecutter	× 8″	dioxide
Protective clothing	Grinder	Spool wire (0.030 or 0.035)	C-clamp fixture
Helmet			

Instructions

1. Set up the machine for welding. Be sure the power is off before plugging in the machine.
2. Turn on the machine. Adjust the gas flow from 15 to 25 cfh, with the voltage set between 15 and 25 volts.
3. Set the wire feed at 50%, and adjust it after running a test weld bead.
4. Clean the metal surfaces by grinding. Tack weld in several places. Fit up with the joint root opening equal to the diameter of the wire, or up to 1/16″.
5. Maintain the proper **travel angle** to complete a weld with consistent appearance and 100% root penetration.
6. Make a visual inspection. Look for a straight weld bead of consistent size with 100% root penetration.
7. Check with the instructor for evaluation, if necessary.
8. Practice this exercise until you meet the standards established for the course.

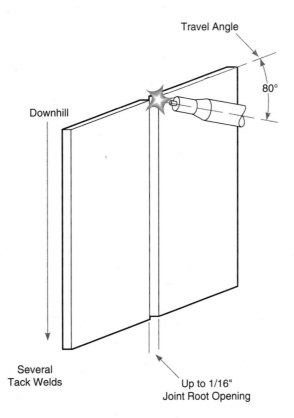

FIGURE 5–28   Butt joint on sheet steel in vertical position.

UNIT 5: EXERCISE 13

Spray Arc Transfer (GMAW-SP): Corner Joint on Plate in Flat Position, Fillet Weld

The higher heat input necessary for spray arc transfer requires that the weldment be cooled down on occasion. The use of a 98% argon/2% oxygen mixture allows a spray to develop in the arc. This method produces a high rate of deposition with deep penetration into the plate. If a pulsed spray capability is available on the welding machine, exercises designed for the vertical position can be added. Figure 5–29 illustrates the setup for this exercise.

Necessary Material and Equipment

Safety glasses	Earplugs	2 pieces 3/8″ steel plate, 3″	Spool wire (0.045)
Welding gloves	Sidecutter	× 8″	Shielding gas: 98% argon/
Protective clothing	Grinder	Padding plate (scrap)	2% oxygen
Helmet			

Instructions

1. Set up the machine for welding. Be sure the power is off before plugging in the machine.
2. Turn on the machine. Adjust the gas flow from 20 to 30 cfh, with the voltage set at 28 volts to begin with. The amperage should range from 225 to 275 amperes for .045 wire.
3. Set the wire feed at 50%, and adjust it after running a test weld bead. What is the number of the wire being used?
4. Clean the metal surfaces by grinding. Tack weld the two pieces together, then tack weld them to the scrap plate.
5. Examine the nozzle of the gun to be sure that it is not plugged. Porosity can be caused by poor gas flow.
6. Cool the steel as it overheats.

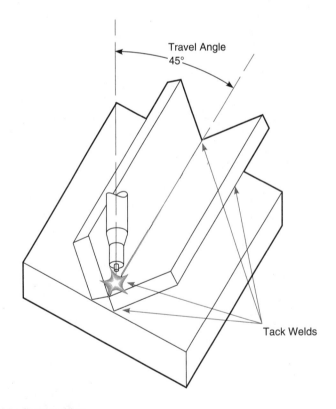

FIGURE 5–29   Fillet weld in flat position.

7. Lay 10 to 15 weld beads, and examine them for consistency of size. The beads should tie together.
8. Test by visual inspection. Look for porosity and undercut.
9. Check with the instructor for evaluation, if necessary.
10. Practice this exercise until you meet the standards established in the course.

UNIT 5: EXERCISE 14

Spray Arc Transfer (GMAW-SP): Tee Joint in Horizontal Position

The tee joint is popular in the design of weldments. For this exercise, the angle of the gun is an important consideration in addition to the voltage and wire feed settings. The first pass requires a 45° work angle. The work angles of the second and third passes are important, too. If there is not enough overlap on the second pass, the bottom plate receives too much filler metal. And if the gun angle (work angle) is targeted on the vertical plate, undercut will result. The setup for this exercise is shown in Figure 5–30.

Necessary Material and Equipment

Safety glasses	Helmet	Grinder	Spool wire (0.045)
Welding gloves	Earplugs	2 pieces 3/8″ steel plate, 3″	Shielding gas: 98% argon/
Protective clothing	Sidecutter	× 8″	2% oxygen

Instructions

1. Set up the machine for welding. Be sure the power is off before plugging in the machine.
2. Turn on the machine. Adjust the gas flow from 20 to 30 cfh, with the voltage set at 28 to begin with. What amperage is being used?
3. Set the wire feed at 50%, and adjust it after running a test weld bead.
4. Clean the metal surfaces by grinding. Tack weld the two pieces together.
5. Set the machine for welding on a piece of scrap steel.
6. Cool the joint between passes when it overheats.
7. Lay at least three to six weld beads on one side. Examine them to see whether the completed fillet weld has equal legs.
8. Make a visual inspection. Look for undercut and porosity.
9. Check with the instructor for evaluation, if necessary.
10. Practice this exercise until you meet the standards established in the course.

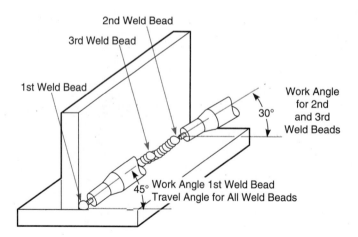

FIGURE 5–30   Tee joint in horizontal position.

UNIT 5: EXERCISE 15

Spray Arc Transfer (GMAW-SP): Butt Joint (Designed to Meet AWS *Structural Welding Code*) V-Groove Weld with Backing in Flat Position

A backing plate should be used for this exercise, which should be evaluated with a destructive test. Fit up is important, so be sure that the dimensions are accurate. Go over the tack welds when completing the first pass. Figure 5–31 illustrates the setup for this exercise.

Necessary Material and Equipment

Safety glasses	Sidecutter	1 piece 3/8″ A36 steel	Shielding gas: 98% argon/
Welding gloves	Grinder	plate, 2″ × 10″ (backing)	2% oxygen
Protective clothing	2 pieces 3/8″ steel plate, 3″	2 strong backs	Cutting machine, 22.5°
Helmet	× 7″	Spool wire (0.045)	bevel
Earplugs			

Instructions

1. Set up the machine for welding. Be sure the power is off before plugging in the machine.
2. Turn on the machine. Adjust the gas flow from 20 to 30 cfh, with the voltage set at 28 volts to begin with.
3. Set the wire feed at 50%, and adjust it after running a test weld bead.
4. Clean the metal surfaces by grinding. Bevel off 22.5° of material for a 60° groove angle.
5. Tack weld the two pieces to the backing plate. Set a 1/4″ joint root opening. Tack weld the strong backs in four places.
6. Use the forehand technique with a weave, making sure that there is complete fusion on the joint root pass.
7. Make a visual inspection. Look for complete fill within the joint.
8. Check with the instructor for evaluation, if necessary.
9. Pass the guided bend test before moving on.

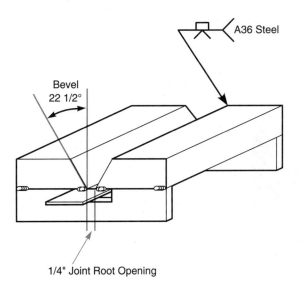

FIGURE 5–31 V-groove weld in flat position.

UNIT 5: EXERCISE 16

Flux Cored Arc Welding (FCAW): Corner Joint on Plate in Flat Position Fillet Weld

The purpose of this exercise is to give you some familiarity with flux cored arc welding. The greater the thickness of the steel, the higher the amperage setting must be. The larger the diameter of the wire, the higher the amperage setting must be. This exercise is designed for 1/16″ diameter wire and CO_2 shielding gas, but substitutions in wire size can be made. Some manufacturers make 1/16″ diameter self-shielding wires. If 3/32″ diameter self-shielding wires are used, plate of more than 3/8″ is also recommended. Figure 5–32 shows the setup for this exercise.

Necessary Material and Equipment

Safety glasses	Earplugs	2 pieces 3/8″ steel plate, 3″	Spool wire (1/16″)
Welding gloves	Sidecutter	× 8″	Shielding gas: carbon
Protective clothing	Grinder	Padding plate (scrap)	dioxide
Helmet			

Instructions

1. Set up the machine for welding. Be sure the power is off before plugging in the machine.
2. Turn on the machine. Adjust the gas flow from 25 to 35 cfh, with the voltage set at 27 volts to begin with.
3. Set the wire feed at 50%, and adjust it after running a test weld bead. Does the tubular wire used for this exercise require a shielding gas?
4. Clean the metal surfaces by grinding. Tack weld the two pieces of steel together, then tack weld them to the scrap steel.
5. Watch the angle of the gun to avoid having the weld pool run ahead. Use the backhand technique to control this problem.
6. Note the use of a greater wire stickout for flux cored arc welding. To eliminate stubbing (which occurs when the wire is driven through the weld pool, pushing the gun backward), increase the voltage or decrease the wire feed. Cool when necessary.
7. The higher amperage increases the size of the weld pool. Remove all slag and spatter before laying the next pass. Lay several passes.

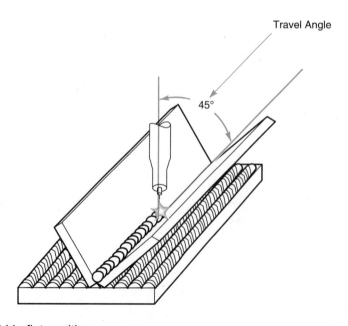

FIGURE 5–32   Fillet weld in flat position.

8. Make a visual inspection. Look for undercut and porosity.
9. Check with the instructor for evaluation, if necessary.
10. Practice this exercise until you meet the standards established in the course.

UNIT 5: EXERCISE 17

Flux Cored Arc Welding (FCAW): Butt Joint in Flat Position, Square-Groove Weld with Backing

As this exercise shows, a bevel is not always used when welding plate. The setup is shown in Figure 5–33.

Necessary Material and Equipment

Safety glasses	Earplugs	2 pieces 3/8″ A36 steel	2 strong backs
Welding gloves	Sidecutter	plate, 3″ × 7″	Shielding gas: carbon
Protective clothing	Grinder	1 piece 3/8″ steel plate, 2″	dioxide
Helmet		× 8″ (backing)	

Instructions

1. Set up the machine for welding. Be sure the power is off before plugging in the machine.
2. Turn on the machine. Adjust the gas flow to 25 to 35 cfh, with the voltage set at 27 volts to begin with.

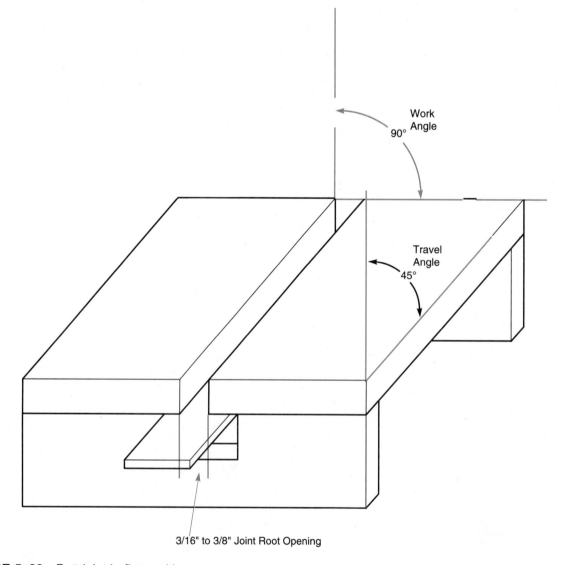

FIGURE 5–33 Butt joint in flat position.

3. Set the wire feed at 50%, and adjust it after running a test weld bead.
4. Clean the metal surfaces by grinding. Tack weld the two pieces together, and set a joint root opening equal to the thickness of the plate. Tack weld the joint to the strong backs.
5. Watch the angle of the gun to avoid having the weld pool run ahead. Use the backhand technique to control this problem.
6. Remove all slag before laying the next pass.
7. Make a visual inspection. Look for undercut and porosity.
8. Check with the instructor for evaluation, if necessary.
9. Pass the guided bend test before moving on.

GAS TUNGSTEN ARC WELDING

"The welder is in, an old man in his sixties or seventies, and he looks at me disdainfully—a complete reversal of the waitress."

—Robert M. Pirsig, *Zen and the Art of Motorcycle Maintenance*

GOAL

- Develop an ability to do gas tungsten arc welding safely

QUESTIONS

- What is gas tungsten arc welding?
- What equipment is required?
- How does the equipment work?
- What are the problems in making quality welds?

SAFETY FIRST

The safety requirements for **gas tungsten arc welding** (GTAW) are generally the same as for any other electric **arc welding** process. A list of 21 safety precautions for electric arc welding is given toward the end of this unit. The special features of gas tungsten arc welding present some unique safety concerns. Gas tungsten arc welding does not generate the sparks and **spatter** that other electric welding **processes** do, so heavy clothing is not necessary for protection. However, bare skin should not be exposed to the **arc,** which can cause burns similar to those produced by the other electric arc welding processes.

The **high-frequency** required for many gas tungsten arc welding applications causes interference in the transmission of radio and television communications. It may also affect pacemakers, which are used in the treatment of heart conditions, and battery-operated watches.

GAS TUNGSTEN ARC WELDING

Gas tungsten arc welding is a welding process in which an electric arc is produced off the end of a **nonconsumable tungsten electrode** (Figure 6–1). A bare wire (welding rod) is usually melted into the **weld pool,** giving added **strength** to the **weldment.** This process is sometimes referred to as **tig** (tungsten inert gas), but *gas tungsten arc welding* is the name preferred by the **American Welding Society.**

Gas tungsten arc welding is commonly used for welding **stainless steel, aluminum,** and **carbon steels.** It is also popular for joining exotic metals, among them **titanium** and **magnesium.** It is an important welding process for **critical welding** applications in which contaminants within the **weld** can cause weld failure. It has been found to be successful in welding very thin metals in which **melt-thru** can sometimes be a problem. For success with this welding process, the **joint** must be clean of **rust, mill scale, film,** and other contaminants, which will affect the **quality** of the **weld bead.**

Because gas tungsten arc welding can be an expensive welding process to learn, the skills developed for **oxyacetylene welding** can pay dividends now. The **manipulative** skills of oxyacetylene welding develop the coordination of both

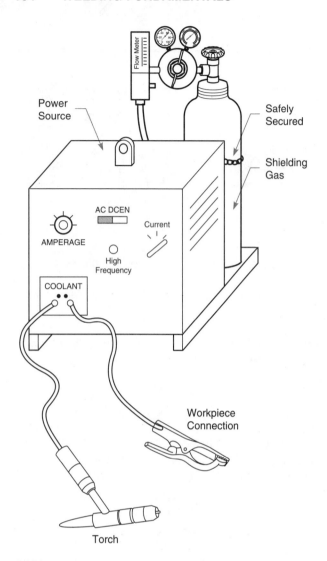

FIGURE 6–1   Gas tungsten arc welding equipment.

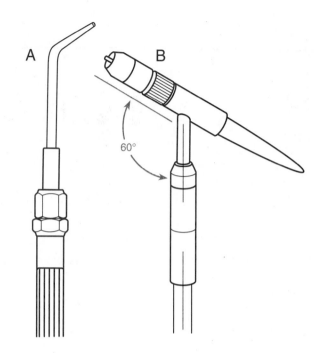

FIGURE 6–2   (A) Oxyacetylene welding torch. (B) Gas tungsten arc welding torch.

hands, which will shorten the time for training in gas tungsten arc welding. The equipment for these two processes differs somewhat (Figure 6–2). In gas tungsten arc welding, adjustments on the **welding machine** replace the knobs on the oxyacetylene welding torch to control the input of heat into the weld.

CHOICE OF CURRENT

Three choices of current are provided as **output** in gas tungsten arc welding. They are direct current electrode positive (DCEP), direct current electrode negative (DCEN), and alternating current with high-frequency (ACHF). DCEN and ACHF are recommended for most applications.

ACHF is preferred for welding aluminum and magnesium because **oxides** begin to form immediately, even after a chemical bath or mechanical cleaning (brushing or grinding) of the weldment before welding. The alternating current provides a cleansing action that breaks down the formation of these oxides. DCEN is preferred for welding all of the stainless and carbon steels, although ACHF is used in some applications. The choice of output current in gas tungsten arc welding should be specified as part of the **welding procedure.**

COMPONENTS OF GAS TUNGSTEN ARC WELDING

The **power source, shielding gas, welding rod,** and **torch** are four basic components of the gas tungsten arc welding process.

Power Source

Welding begins with the selection of the power source. Gas tungsten arc welding requires a **constant-current** power source so that a change in the **arc length** will not result in a large increase in arc current. Several power sources are easily converted to gas tungsten arc welding (Figure 6–3). These include the **transformer** electric welding machine, the transformer-**rectifier** electric welding machine, the **motor generator** electric welding machine, the **inverter,** and the **engine-driven generator.**

Power sources that are specially designed for gas tungsten arc welding contain several features that offer some advantages (Figure 6–4). These power sources have a greater range for **amperage** settings and are capable of fine-tuning current output that can greatly reduce heat down to one or two **amperes,** which is necessary on very thin metal. Welding an aluminum can is no problem for these welding machines.

The high-frequency units built into the power source provide off, start only (for **DC** welding), and continuous (for **AC** welding) (Figure 6–5). High-frequency is used to start the arc, which would otherwise require touching the tungsten electrode to the **base metal.** In critical applications, contamination of the tungsten electrode by contact with the base metal can lead to weld failure.

High-frequency is used for non-touch starts of the arc during DC welding and for keeping the arc going during AC welding. When welding with AC,

A

FIGURE 6–4 (A) Gas tungsten arc welding power source. *(Courtesy of Hobart Brothers Co., Troy, Ohio.)*

the sine wave cycle passes through zero every half cycle, putting out the arc (Figure 6–6). High-frequency superimposes a high **voltage** at radio frequencies to maintain the arc through the entire sine wave cycle to prevent the tungsten electrode from freezing to the base metal.

Some gas tungsten arc welding machines contain a **postflow** timer, which continues the flow of gas once the arc has been extinguished (Figure 6–7). This feature allows the tungsten electrode, the welding rod, and the weld pool to cool, preventing contamination. Like the welding rod and the weld pool, a red-hot tungsten electrode attracts contaminants, which can later be deposited into the weld. The shielding gas protects the welding rod from contamination while in the weld pool. This is one reason why moving

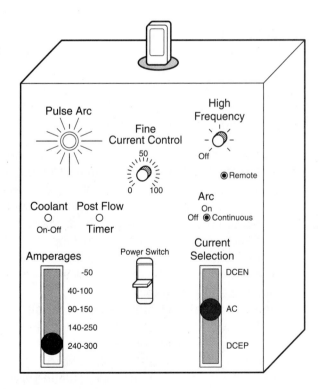

FIGURE 6–3 Conversion to gas tungsten arc welding.

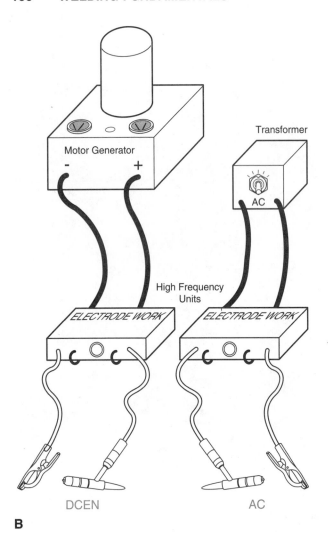

B

FIGURE 6–4 (B) Gas tungsten arc welding power sources.

FIGURE 6–5 High-frequency unit. *(Courtesy of Lincoln Electric Co., Cleveland, Ohio.)*

the welding rod in and out of the weld pool is not recommended.

On some gas tungsten arc welding machines, there is a pulse arc control, which interrupts the flow of current (Figure 6–8). This enables the **welder**

C

FIGURE 6–4 (C) Inverter power source for gas tungsten and shielded metal arc welding. *(Courtesy of Thermal Dynamics, St. Louis, Missouri.)*

to reduce heat input without pulling the torch away from the weld pool. **Pulsed gas tungsten arc welding** (GTAW-P) is a variation of gas tungsten arc welding, in which a pulsed change from high to low current takes place in the arc constantly. This give the welder some control over penetration and heat input.

Some gas tungsten arc welding machines provide a **remote-panel switch** so that current output can be adjusted by moving a button on the torch or by applying pressure to a foot control (Figure 6–9). This feature allows the welder to make current adjustments without having to stop work to reach the control panel of the welding machine. This feature makes it easier to fill the weld **crater** at the completion of welding.

Water-circulating units are used for both gas tungsten arc welding and gas metal arc welding high-amperage work to cool the torch or the gun while welding. These units consist of a reservoir, an electric motor, and a pump for circulating water

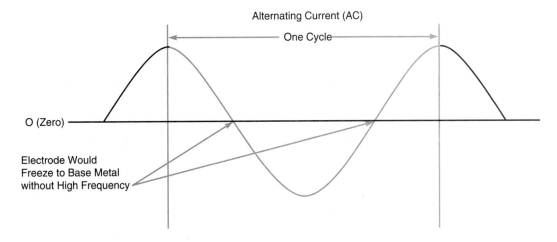

FIGURE 6–6 Alternating current sine wave cycle.

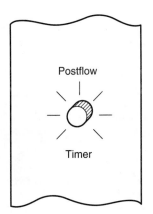

FIGURE 6–7 Postflow timer control.

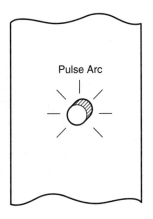

FIGURE 6–8 Pulse arc control.

through specially designed equipment (Figure 6–10).

Shielding Gas

The shielding gas is the second basic component of gas tungsten arc welding. **Argon** and **helium** are the two major shielding gases used in this type of welding (Figure 6–11). Although this process was originally developed using helium, helium is now less popular than argon, which is heavier and not as expensive. Argon provides a better gas shield for welding, especially when the wind or drafts can be a problem.

Because the shielding gas is contained in a **high-pressure gas cylinder,** of explosive force, never move the cylinder without the cap in place to protect the **valve. Secure** the cylinder in position so that it cannot fall, and always work at some distance from the cylinder to avoid **arc strikes** on the cylinder wall. A damaged or weakened cylinder compromises the safety of everyone in the shop.

The shielding gas requires a **regulator/flowmeter** (Figure 6–12). The regulator controls the pressure (set by the manufacturer) going to the flowmeter and has a gauge that indicates gas pressure remaining in the cylinder. The flowmeter sets the volume of gas measured in cubic feet per hour (cfh) released by the torch. The flowmeter is quite sensitive in measuring the cylinder pressure and setting gas flow, and it can be easily damaged if not handled properly. Always handle the regulator/flowmeter with care, attaching it immediately to the valve once the cylinder is in place.

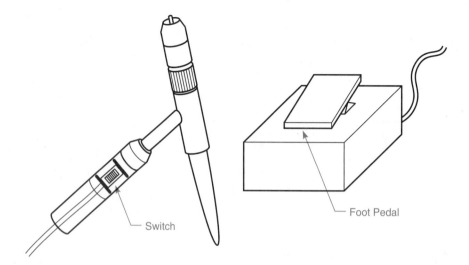

FIGURE 6–9   Two remote controls.

FIGURE 6–10   Water-circulating unit. *(Courtesy The Esab Group, Inc., Florence, South Carolina.)*

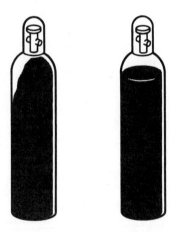

FIGURE 6–11   Argon and helium cylinders. *(Courtesy of Jeffus, 1993.)*

Welding Rod

The welding rod is the third basic component of gas tungsten arc welding. A variety of welding rods are available. These solid rods come in a standard length of 36 inches and have diameters of 1/16 inch to 3/8 inch. The selection of a welding rod depends on several factors, the most important of which is compatibility with the base metal. The size of the welding rod is also important. The larger-diameter rods take heat away from the weld pool. If the welding rod disappears so quickly into the weld pool that you have trouble keeping up, select a larger-diameter welding rod.

Welding rods designed primarily for oxyacetylene welding may not be suitable for gas tungsten arc welding, but the chemistry of wire used for gas

FIGURE 6–12 Regulator/flowmeter. *(Courtesy The Esab Group, Inc., Florence, South Carolina.)*

metal arc welding generally is suitable. In many situations, the selection of the correct welding rod for a given application is up to the welder. Steel, aluminum, **brass, copper,** stainless steel, titanium, and magnesium welding rods are among the varieties available (Figure 6–13).

Torch

The fourth basic component of gas tungsten arc welding is the torch (Figure 6–14). The torch

itself consists of a head, which is attached by its **power lead** to a **power adapter** (Figure 6–15). The power adapter connects the shielded metal arc welding power source and the shielding gas to the air-cooled torch (Figure 6–16). If the power source has been designed strictly for gas tungsten arc welding, the power adapter is removed, and the torch is connected directly to the gas tungsten arc welding power source. A water-cooled torch includes a **water coupling** for connecting the water hoses of the torch to the hoses of the water-circulating tank (Figure 6–17).

The selection of an air-cooled versus a water-cooled torch depends on the amperage setting and **duty cycle** of the welding machine. Depending on the manufacturer and the size of the torch, an air-cooled torch can handle from 70 to 200 amperes at 50% duty cycle and up to 220 amperes at 100% duty cycle. Depending on the manufacturer and the size of torch, a water-cooled torch can handle up to 500 amperes at 50% duty cycle and up to 600 amperes at 100% duty cycle. The key is to know the specifications of the torch being used.

Torch Head

The torch head is made up of five basic parts (Figure 6–18). The gas **nozzle,** made of heat-resistant material, is used to direct the flow of gas. It also acts as an **insulator** when rested against the base metal. Nozzles are available in various sizes and may become damaged if subjected to unnecessary force. A specially designed gas lens nozzle for use with a gas lens **collet body** is avail-

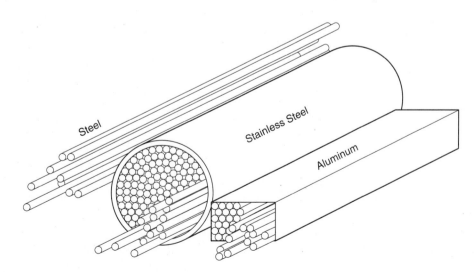

FIGURE 6–13 Welding rods.

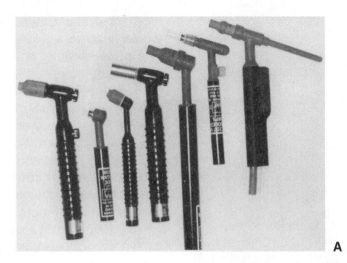

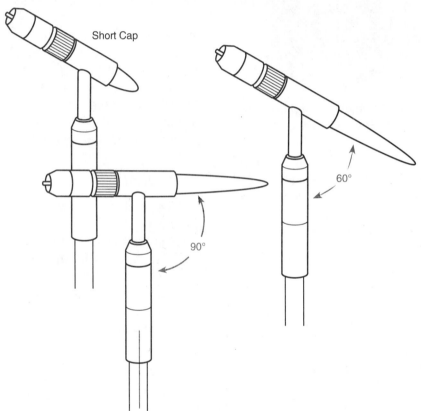

FIGURE 6–14   (A) Several torches used for gas tungsten arc welding. *(Courtesy of Larry Jeffus.)* (B) 60° and 90° torches with short cap.

able to supply a more stable gas stream coming from the torch.

The collet body screws into the torch head. It is designed to hold a tungsten electrode of a given diameter in position without slipping down out of the nozzle during welding. The collet body serves as a part of the electric circuit and comes in various sizes.

The **collet tube** fits into the collet body and should be matched to the size of the collet body (Figure 6–19). The collet tube also serves as part of the electric circuit. The collet tube and the collet body should be matched to the diameter of the tungsten electrode being used.

The tungsten electrode is the final point of the electric circuit in which the arc is created. The

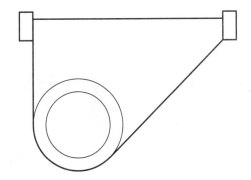

FIGURE 6–15 Power adapter.

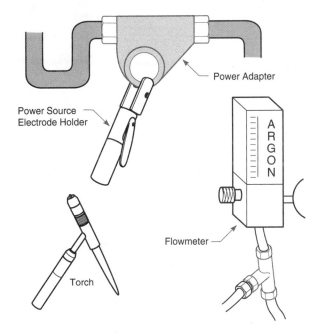

FIGURE 6–16 Setup for an air-cooled torch.

electrodes used in gas tungsten arc welding are nonconsumable and are not designed to become part of the weldment. Any pieces of the tungsten electrode that break off and become trapped in the weld are considered to be contaminants and should be removed. The tip of the tungsten electrode is the point of the electric circuit in which the arc forms. The purpose of the tungsten electrode is to concentrate a high-temperature but stable (does not wander) arc for quality welding.

The tungsten electrode is made of **tungsten**. Tungsten is a chemical **element** with the highest melting point of any known metal at 6,170° F and a boiling temperature of 10,700° F. The high boiling temperature makes tungsten suitable for gas tungsten arc welding because the molten ball that

forms on the tip will not drip under normal welding conditions. The several kinds of tungsten electrodes that are sold commercially are **color-coded** (Figure 6–20).

The **pure tungsten** electrode (EWP) is preferred for AC welding because it readily balls on the end for maximum arc stability. The 1% **thoriated** alloy tungsten electrode (EWTh-1) and the 2% thoriated alloy tungsten electrode (EWTh-2) can carry up to twice the current of the pure tungsten electrode, thereby increasing the operating amperage range for a tungsten electrode of a given diameter. Thoriated alloy tungsten electrodes are well suited for welding with DCEN. They can also be used for some AC applications. However, the low-level radioactivity of these thoriated alloy tungsten electrodes requires that welders take care not to inhale the particles given off by grinding. The **cerium** tungsten electrode (EWCe-2) does not have the current-carrying capacity of the thoriated tungsten electrode, but can be used as its replacement. The **zirconia** alloy tungsten electrode has the addition of zirconium oxide. It provides similar characteristics to those found in the thoriated tungsten electrode.

The five tungsten electrodes mentioned above are color-coded for identification. This marking should be left in place by using the opposite end of the tungsten electrode. Pure tungsten electrodes are green. The 1% thoriated tungsten electrodes are yellow. The 2% thoriated tungsten electrodes are red. The cerium tungsten electrodes are orange, and the zirconia tungsten electrodes are brown. These electrodes come in several popular sizes: .040, 1/16, 3/32, 1/8, 5/32, 3/16, and 1/4 inch diameters. There is a difference in the price, with the pure tungsten electrodes the least expensive.

The type of metal and the thickness of the metal are the primary factors in choosing a tungsten electrode. Once the tungsten electrode has been chosen, its diameter depends on the amperage setting. The higher the amperage, the greater the diameter required. The greater the diameter, the less the decomposition (breakup) of the tungsten.

Table 6–1 should be helpful in setting up for gas tungsten arc welding. It can be easily memorized. The table matches popular sizes of tungsten electrodes by increments of 50 amperes for ACHF

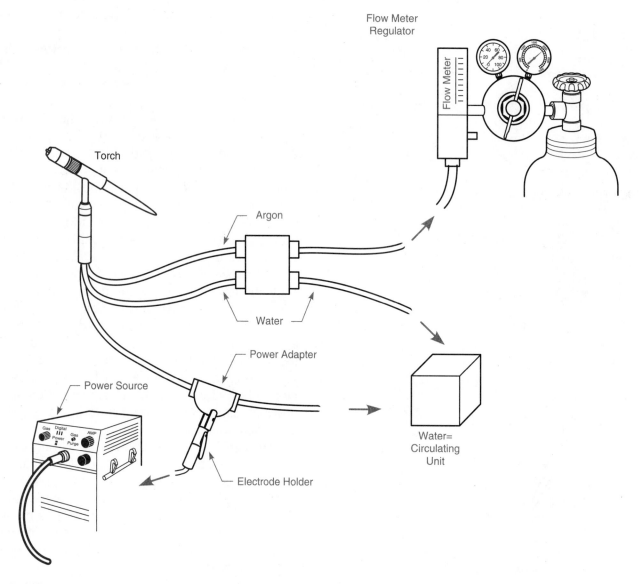

FIGURE 6–17   Setup for a water-cooled torch.

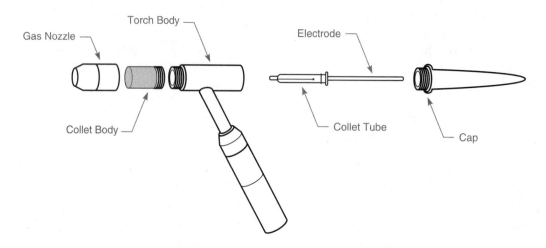

FIGURE 6–18   Exploded view of a torch head with the parts labeled.

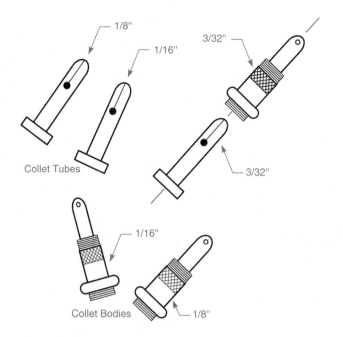

Collet Tubes

Collet Bodies

FIGURE 6–19 Collet bodies and collet tubes of various sizes.

and DCEN current outputs. If you know that the range of a .040-diameter tungsten electrode is 50 amperes maximum for pure tungsten, you can easily remember an approximation for every other tungsten electrode given in the table since all increments are of 50 amperes.

A ball will form on the end of a pure tungsten electrode during AC applications. The ball helps to stablize the arc during welding. If the ball becomes larger than one and a half times the diameter of the electrode, the amperage setting is too large for the size of the tungsten electrode, or the tungsten electrode has become contaminated (Figure 6–21).

For DC applications, a sharp point must be ground on the end of the tungsten electrode. The heat of the arc will cause a small ball to form on the end of the electrode. The point must be centered with a bevel extending twice the size of the diameter, and the grinding marks must be parallel to the length of the tungsten electrode to keep the arc from wandering off the side (Figure 6–22). For critical applications or for frequent gas tungsten arc welding, a pedestal grinder with a silicon wheel can be set aside only for use in sharpening tungsten electrodes.

The **cap** holds the tungsten electrode and the collet tube in position within the torch head. The cap also focuses the gas in one direction. A leak in the cap can affect the gas shielding during welding. Be sure the cap is tightly in position to seat the o-ring. Never use the cap to push in the tungsten electrode for a greater extension beyond the nozzle; this will damage the cap. Caps of different sizes are available (Figure 6–23). A smaller cap is necessary for welding in confined spaces.

SETTING UP THE EQUIPMENT FOR WELDING

1. Choose a type and size of tungsten electrode based on the kind of metal and its thickness.
2. Match the collet body and collet tube to the tungsten electrode, and then select the nozzle size based on the joint configuration.
3. Put the torch on its holder or in a place where the torch cannot fall and will not arc when the power source is turned on.
4. Check all gas connections. Be sure the cylinder is secured. Crack the cylinder valve before attaching the regulator/flowmeter, if required, to remove anything that might restrict the flow of gas.
5. Open the cylinder valve. Listen for leaks.
6. Inspect all water connections for leaks, if applicable. Open the system.

TABLE 6–1: MAXIMUM ACHF AND DCEN CURRENT OUTPUTS FOR VARIOUS TUNGSTEN ELECTRODES

| Size | AC (High Frequency) | | DCEN |
	Pure Tungsten	Thoriated	Pure and Thoriated
.040	50 amperes	100 amperes	100 amperes
1/16	100 amperes	150 amperes	150 amperes
3/32	150 amperes	200 amperes	200 amperes
1/8	200 amperes	250 amperes	250 amperes

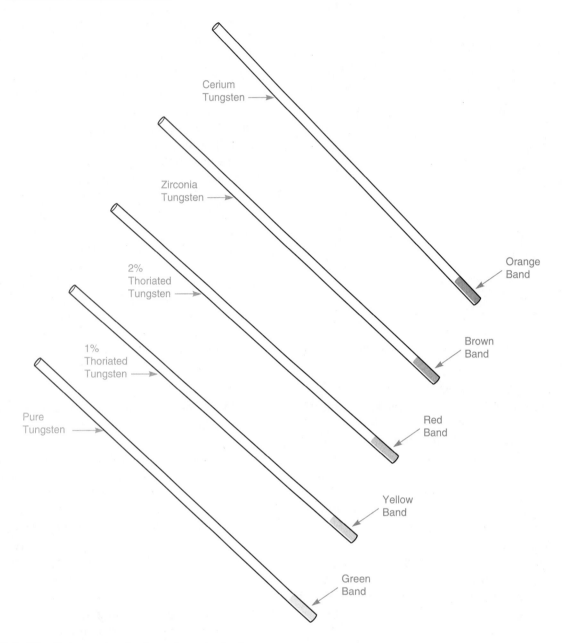

FIGURE 6–20   Color codes for tungsten electrodes.

7. Examine all electrical connections. Be sure the power source is off before plugging it in. Turn on the power switch or switches.

8. Check the gas flow at the torch. Feel the flow on your skin. Adjust the flowmeter.

9. Set the machine for the appropriate welding procedure. For example, turn on the high-frequency and set the machine for continuous welding, if required, or adjust the postflow timer.

10. Strike an arc to find out if the electric circuit is complete. Is the **workpiece connection** in position?

SHUTTING DOWN THE EQUIPMENT

1. Put the torch on its holder or in a place where the torch cannot fall and will not arc.

2. Close the valve on the gas cylinder.

3. Shut down the water system, if applicable.

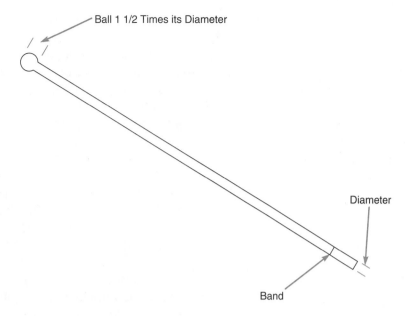

FIGURE 6–21   Pure tungsten electrode with ball.

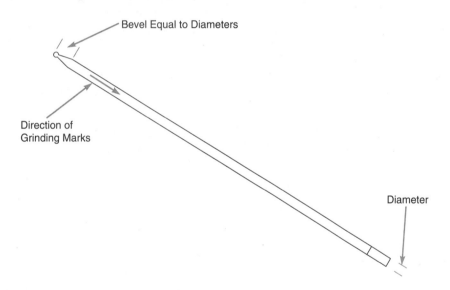

FIGURE 6–22   Sharpened point for DC applications.

4. Switch off the power source.
5. Follow shop procedures for cleanup.

POINTS TO REMEMBER

Your skill in gas tungsten arc welding will depend on practice and developing some coordination. Ultimately, you must learn to recognize and maintain the proper weld pool while welding. The size, or **wetness,** of the weld pool is an indication of whether adjustments have to be made. Four factors can affect the quality of the welding: arc length, amperage, travel speed, and angle of the torch.

Arc length is an important factor. If the arc length is too short, the tungsten electrode might touch or contaminate the base metal. On the other hand, if the arc length is too long, the heat generated by the arc might not be enough to melt the base metal, causing **poor root penetration,** or a concave (bowl-shaped) weld bead might be formed on thin metal, causing incomplete **root penetration.** In addition, if the arc length is too

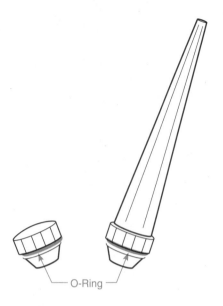

FIGURE 6–23 Two caps of different sizes.

long, the welding rod might not flow into the weld pool, causing a weld bead with pinholes as a result of poor gas shielding (Figure 6–24).

Amperage setting is a second factor that can affect the quality of the weld. The amperage setting might be too low if there is not enough heat to melt the base metal. The amperage setting can also be too low if it is necessary to have such a short arc length to be almost touching the base metal to form the weld pool. The amperage setting might be too low if the welding rod does not flow steadily into the

weld pool and seems to be setting on top of the base metal without melting in (Figure 6–25).

The amperage setting might be too high if unexpected melt-thru occurs on thin metal. Amperage setting can be too high when the tip of the tungsten electrode disintegrates, or if the odor of the torch becomes quite noticeable as the torch body overheats. The amperage setting also might be too high if the weld bead flattens out and becomes unusually wide (Figure 6–26).

Travel speed is the third factor. If the travel speed is too slow, melt-thru might occur on thin base metal, or the weld bead becomes unusually wide on thick base metal. If the travel speed is too fast, root penetration might be incomplete, or a narrow weld bead might seem to lay on the base metal without melting in (Figure 6–27).

The angle of the torch is the fourth factor. Hold the **work angle** of the torch at 90°. Hold the **travel angle** of the torch pointed in direction of travel with a slope of approximately 45° up from the base metal (Figure 6–28).

Outstanding Features of Gas Tungsten Arc Welding

Gas tungsten arc welding is most advantageous for welding thin metals (for metals of more than 1/4 inch, some other welding process may be more cost-effective) and for work on metals not commonly welded by other processes. It is useful

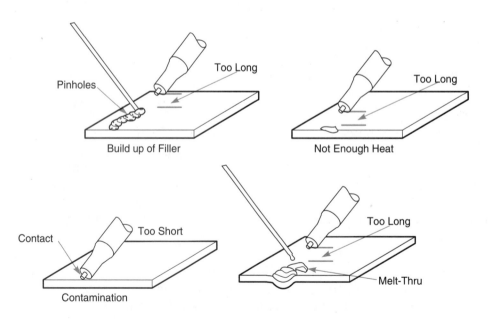

FIGURE 6–24 Incorrect arc lengths.

FIGURE 6–25   Amperage setting too low: The welding rod does not flow, and the weld pool that develops is small. *(Courtesy of Larry Jeffus.)*

in critical applications in which shielded metal arc welding or gas metal arc welding are too fast to ensure quality welding. **Hydrogen, oxygen, slag, and slag inclusions**—sometimes problems in shielded metal arc welding—weaken the weld and can be eliminated with gas tungsten arc welding. The smoke and spatter of gas metal arc welding also is not a problem for gas tungsten arc welding.

SAFETY REMINDERS

General Safety

1. **Safety glasses** with side shields are essential. They protect the eyes from grinding particles. Safety glasses worn under the welding helmet deflect some of the **ultraviolet rays** given off by the **welding arc,** helping to prevent **arc flash.** Safety glasses are so important that no one should be in a welding shop without wearing them. However, safety glasses are no protection against careless shop practices.
2. Wear earplugs in the shop to safeguard your hearing. Besides protecting you from nerve-damaging noise, earplugs can prevent hot slag and sparks from reaching the eardrum.
3. Use leather **welding gloves** to protect your hands and lower arms from burns and shock (Figure 6–29). Gas tungsten arc welding does not require the heavy leather gloves used for other electric welding processes.
4. Without the sparks or the spatter, gas tungsten arc welding allows you to wear

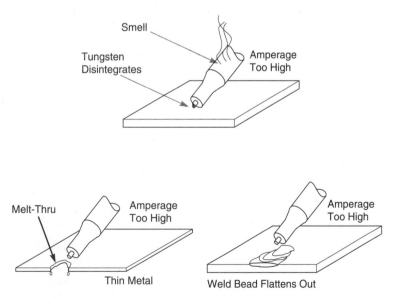

FIGURE 6–26   Amperage setting too high.

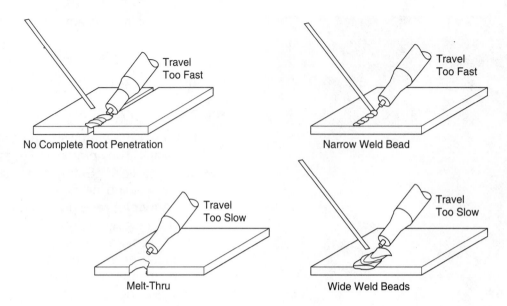

Travel Too Fast

No Complete Root Penetration

Travel Too Fast

Narrow Weld Bead

Travel Too Slow

Melt-Thru

Travel Too Slow

Wide Weld Beads

FIGURE 6–27 Incorrect travel speed.

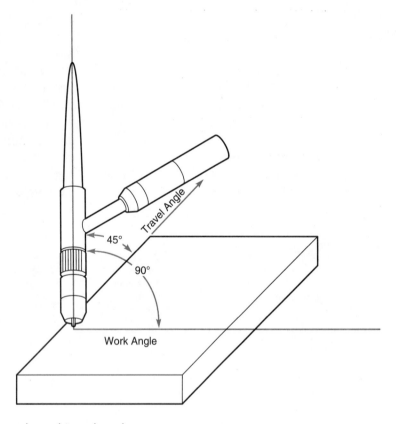

45°

Travel Angle

90°

Work Angle

FIGURE 6–28 Work angle and travel angle.

more comfortable clothing. However, the skin must be protected from the rays of light given off by the arc.

5. Be sure the shop is clear of any **flamma-ble** or **volatile** materials. Gasoline and solvents have no place in a welding shop.

6. Do not bring cigarette lighters under pressure into the welding shop.

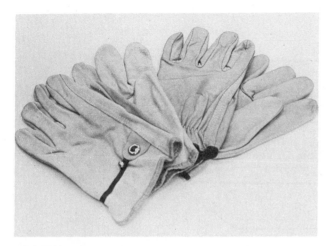

FIGURE 6–29 Lightweight welding gloves. *(Courtesy of Jeffus, 1993).*

7. Always wear a welding helmet with a filter lens of the correct shade to protect against arc flash (Figure 6–30). Arc flash is a painful, but usually a temporary, eye condition caused by the light of the weld-

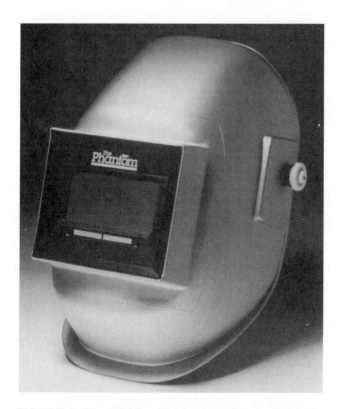

FIGURE 6–30 Welding helmet in which the light window darkens instantly on arc strike. *(Courtesy of Sellstrom Manufacturing Co.)*

ing arc. If you suspect an arc flash is more serious, get proper medical attention. Replace cracked lenses or damaged helmet. Check the shade number on the lens. Gas tungsten arc welding requires a shade number of 8 to 10 (Figure 6–31). In general, use a number 8 lens up to 10 amperes, number 9 up to 100 amperes, and number 10 up to 200 amperes.

8. Know the location of the fire extinguishers in the welding shop (Figure 6–32).

9. Use the welding shop's ventilation system to remove smoke and fumes.

10. Protect yourself against the danger of electric shock. Dry gloves and clothing and a dry workplace help prevent shocks. If you stay dry while welding, you act as an insulator and not a conductor of electricity. Electricity always seeks the path of least resistance, so stay dry when welding to stay safe.

11. Wear steel-toe leather boots to protect against foot injury. Athletic shoes do not provide adequate protection.

12. Dark clothing is preferred to white clothing because dark clothing absorbs light rays and white clothing deflects light rays. Arc flash can occur off aluminum and other sources of deflected light.

13. Before turning on the welding machine, be sure to stretch out the power lead of the torch and the **workpiece lead** (Figure 6–33). Never wrap either power lead or workpiece lead around your body. Always place the workpiece connection on the work or as close to the work as possible for good electrical contact. The workpiece connection can be welded, clamped, or screwed into position.

14. Never weld on pressurized cylinders. A stray arc strike on a gas cylinder can cause an explosion.

15. As a common courtesy, always warn others in the area before striking an arc. Get in the habit of always beginning an arc strike at the point of the welding.

16. Stay alert under the welding helmet to any unusual odors that might indicate a fire or a toxic substance.

17. Use common sense, and pay close atten-

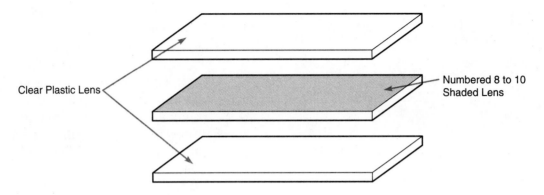

FIGURE 6–31 Shaded lens for gas tungsten arc welding.

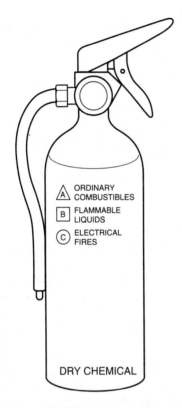

FIGURE 6–32 A-B-C fire extinguisher.

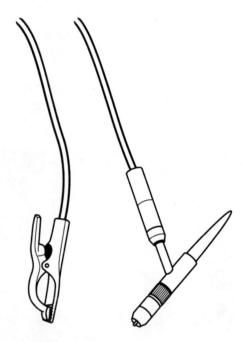

FIGURE 6–33 Leads stretched out.

tion to the work at hand. Never let day-dreaming or distraction have its way.

Safety for Gas Tungsten Arc Welding

18. Be aware that high-frequency can cause interference with television and radio communications. Pacemakers and battery-operated wristwatches may also be affected.

19. Be sure the gas cylinder is secured so it cannot fall.
20. Take care in handling the parts of the torch head, which can become very hot.
21. Take special precautions when welding specialty metals that produce toxic fumes.

REVIEW

1. Name two of the three metals commonly welded with the gas tungsten arc welding process.

2. Give the two current outputs that are most often used for gas tungsten arc welding.
3. What current output requires high-frequency, and what metal can be successfully welded?
4. What current output is preferred for welding stainless steel?
5. List the four basic components of gas tungsten arc welding.
6. What is the purpose of high-frequency?
7. Why is a postflow timer important?
8. Why is argon preferred as a shielding gas for gas tungsten arc welding?
9. When should an air-cooled torch be replaced by a water-cooled torch?
10. Name three of the five parts that make up the torch head, and give the function of each.

Create Three Questions

1.
2.
3.

Related Math and English Questions

1. If the gas flow is set for 15 cfh, how many cubic feet would be used up in eight hours of welding?
2. If an S cylinder holds 150 cubic feet of argon, for approximately how many hours of welding can you expect if the flowmeter is set at 20 cfh?
3. Write a paragraph using the following terms: *gas tungsten arc welding, nonconsumable, spatter, exotic metals, contaminants, mill scale.*
4. In a paragraph, explain the choice of gas tungsten arc welding over other welding processes.

For Further Thought

1. Why would gas tungsten arc welding be preferred over gas metal arc welding for welding 2-inch-diameter pipe?
2. Suppose a fabrication shop is making a spiral staircase. The steps are constructed out of 1/8-inch-thick square tubing. Would gas tungsten arc welding be the most cost-effective process for this job? Explain.
3. What methods can be used for cleaning the base metal before welding?
4. When welding outside, what can be done to protect against the wind?
5. If you have to repair a damaged aluminum bumper but are unable to generate enough heat to melt the aluminum, what are some of your options?

SUGGESTED ACTIVITIES

1. Write a paper on the different **types** of aluminum and their uses.
2. Write a paper on the different types of stainless steel and their uses.
3. Describe the different rods used for welding the various types of stainless steels and aluminums.
4. Gather metal samples of aluminum, magnesium, stainless steel, and so on. Be able to identify these and other metals.

UNIT 6: EXERCISES
Setting Up to Weld

There are a variety of power sources available for gas tungsten arc welding, and these have varying degrees of sophistication. You need only remember that the welding process does not change, no matter the complexity of the power source. As with any other piece of welding equipment, take a few minutes to become familiar with the control panel. Even with the best instructions, hands-on experience is necessary. Read the **operator's manual,** if available, as part of your learning.

1. Hold the torch so that your gloved hand and the equipment do not interfere with the welding (Figure 6–34).
2. Extend the tungsten electrode a distance of no more than two and a half times its diameter beyond the edge of the nozzle (Figure 6–35).
3. Set the argon shielding gas flow from 10 to 25 cfh. The type of metal and the welding conditions will determine the flow re-

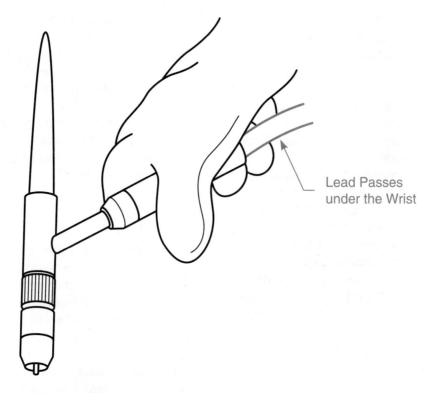

FIGURE 6–34 Holding the torch.

quired. Pinholes are an indication of poor gas shielding.

4. Set the amperage for 1/16-inch metal thickness to 140 amperes at most. **Joint design,** the type of base metal, the size and choice of tungsten electrode, and the electric output current (ACHF or DCEN) are important in selecting the amperage. Memorize Table 6–1. For all exercises using carbon steel as the base metal, is DCEN required?

5. Clean the metal before welding. This means both the base metal and the welding rod. Mill scale, rust, oil, dirt, and so on, will affect the quality of the weld.

6. To strike an arc, place the side of the gas nozzle on the workpiece. Raise the torch quickly to begin the arc (Figure 6–36).

7. **Forehand** welding is the preferred method. It offers the best gas shielding (Figure 6–37).

8. When the electrode becomes contaminated by the welding rod or the weld pool, stop work. Cut off the end of the tungsten electrode with a sidecutter, break off the

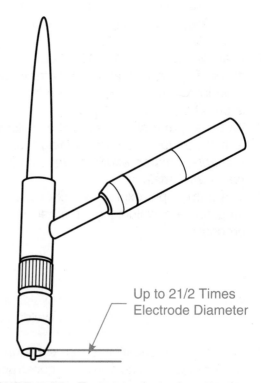

Up to 21/2 Times Electrode Diameter

FIGURE 6–35 Tungsten electrode extension.

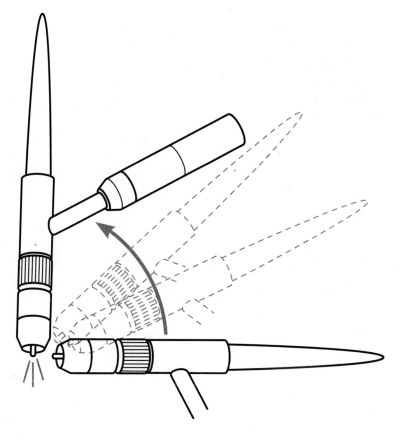

FIGURE 6–36 Striking the arc.

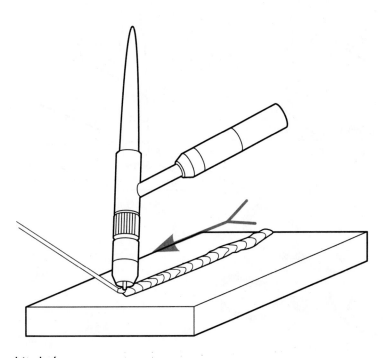

FIGURE 6–37 Forehand technique.

end of the electrode, or **short** the electrode, but not on the base metal (Figure 6–38). Remember to sharpen the electrode to a point for DCEN.

9. Welding conditions will determine the positioning of the torch. But for a reference, use a work angle of 90° and a travel angle at 45°.

10. Add the welding rod into the weld pool away from the electrode (Figure 6–39).

11. To avoid a crater forming at the end of a weld, slowly reduce the amperage (use the remote control attached to the torch) or slowly raise the tungsten electrode, reducing the size of the weld pool.

12. Put the torch on its holder when not welding.

Use the information in Table 6–2 as a reference for initially adjusting the amperage setting. The table provides data for three commonly welded metals but is limited since, for metals of more than 1/4 inch thickness, the cost of welding suggests an alternate welding process.

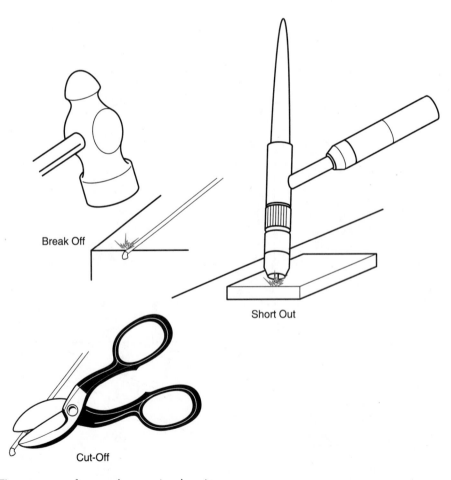

Break Off

Short Out

Cut-Off

FIGURE 6–38 Three ways of removing contaminants.

FIGURE 6–39   Positioning the welding rod. *(Courtesy of Jeffus, 1993.)*

TABLE 6–2:	AMPERAGE SETTING AND GAS FLOW		
Metal Thickness	Polarity	Max. Amperes	Argon (cfh)
CARBON/LOW-ALLOY STEELS			
18-GAUGE	DCEN	120	10–15
1/16 INCH	DCEN	140	10–15
1/8 INCH	DCEN	180	10–15
3/16 INCH	DCEN	280	10–15
1/4 INCH	DCEN	360	10–15
ALUMINUM			
18-GAUGE	ACHF	70	15–25
1/16 INCH	ACHF	90	15–25
1/8 INCH	ACHF	140	15–25
3/16 INCH	ACHF	190	15–25
1/4 INCH	ACHF	240	15–25
STAINLESS STEEL			
18-GAUGE	DCEN	80	10–15
1/16 INCH	DCEN	100	10–15
1/8 INCH	DCEN	160	10–15
3/16 INCH	DCEN	240	10–15
1/4 INCH	DCEN	320	10–15

UNIT 6: EXERCISE 1

Gas Tungsten Arc Welding (GTAW): Mild Steel in Flat Position, Padding Plate

This exercise in the **flat position** is an opportunity to practice striking the arc with and without a welding rod. Begin by traveling along the **plate,** completing one side, laying weld beads without a welding rod. Watch the size of the weld pool, using its size to control travel speed. Weld beads made without a welding rod should not overlap one another.

Turn the plate over, and complete a second side with a welding rod. Weld **stringers** with an overlap of one-half the bead's width. Observe the overlap, bead size, and bead straightness. Figure 6–40 shows the setup for this exercise.

Necessary Material and Equipment

Safety glasses	Earplugs	2% thoriated tungsten,	Shielding gas: argon
Welding gloves	Grinder	3/32″ diameter	
Protective clothing	1 piece 3/8″ steel, 6″ × 8″	Welding rod, 1/16″	
Helmet	(or scrap metal)		

Instructions

1. Clean the surfaces on both sides of the plate by grinding.
2. Beginning at the top of the plate, weld across the body using the forehand technique.
3. Use the size of weld pool to determine travel speed.
4. Do not move the torch from side to side.
5. Keep the nozzle off the plate.
6. Complete the exercise both without and with a welding rod.
7. Check with the instructor for evaluation, if necessary.
8. Practice this exercise until you meet the standards established in the course.

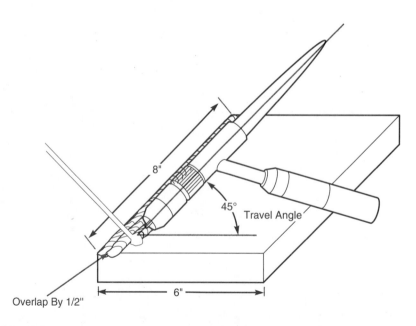

FIGURE 6–40 Padding plate.

UNIT 6: EXERCISE 2

Gas Tungsten Arc Welding (GTAW): Lap Joint (Mild Steel) in Horizontal Position, Fillet Weld

The two pieces of steel to be joined in the **lap joint** should be laid flat on the workbench or held by a C-clamp parallel to the floor. Even though the pieces are laid out flat, this is still considered the **horizontal position.** Melt both pieces of the joint, but concentrate the arc on the bottom piece. Metal always melts fastest along the edge. Note that in this exercise, root penetration does not mean melting through the bottom piece. Add enough welding rod to avoid **undercut** on the top piece. The setup for this exercise is illustrated in Figure 6–41.

Necessary Material and Equipment

Safety glasses	Earplugs	2% thoriated tungsten,	Shielding gas: argon
Welding gloves	Grinder	1/16″ diameter	
Protective clothing	2 pieces 1/16″ sheet steel,	Welding rod, 1/16″	
Helmet	2″ × 4″		

Instructions

1. Clean the surfaces to be welded by grinding.
2. **Tack weld** the two pieces together at each end.

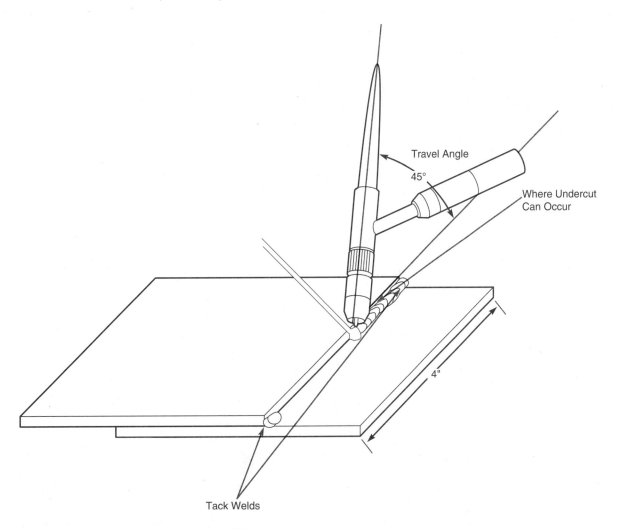

FIGURE 6–41   Lap joint in horizontal position.

3. Complete the weldment in one **pass.**
4. Examine the weld for unwanted undercut, and check the crater for incomplete fill. Is the weld bead consistent in size?
5. Cool, then test the weldment in a vice, positioning the weld just above the jaws. Hammer the top piece over.
6. Look for cracks and bending along any undercut after testing. Redo the weld if there is evidence of cracks or undercut.
7. Check with the instructor for evaluation, if necessary.
8. Practice this exercise until you meet the standards established for the course.

UNIT 6: EXERCISE 3

Gas Tungsten Arc Welding (GTAW): Tee Joint (Mild Steel) in Horizontal Position, Fillet Weld

An appropriate arc distance must be maintained to melt the base metal. With the **tee joint,** the nozzle size and the extension of the tungsten electrode beyond the nozzle are important. The electrode should extend into the joint root close enough to melt both pieces and still melt the welding rod. The tee joint requires a higher amperage than is needed for the lap joint. Figure 6–42 shows the setup for this exercise.

Necessary Material and Equipment

Safety glasses	Earplugs	2% thoriated tungsten,	Shielding gas: argon
Welding gloves	Grinder	1/16″ diameter	
Protective clothing	2 pieces 1/16″ sheet steel,	Welding rod, 1/16″	
Helmet	2″ × 4″		

Instructions

1. Clean the surfaces to be welded by grinding.
2. Adjust the extension of the tungsten electrode. Set a 45° work angle and a 45° travel angle.
3. Tack weld the two pieces together at each end.
4. Complete the weldment in one pass.
5. Examine the weld for undercut along the top piece. Was enough welding rod added? Is the weld bead of consistent size without pinholes?
6. Cool, then test the weldment by clamping the bottom piece in a vice. Hammer the top piece over against the weld, and squeeze the metal together.

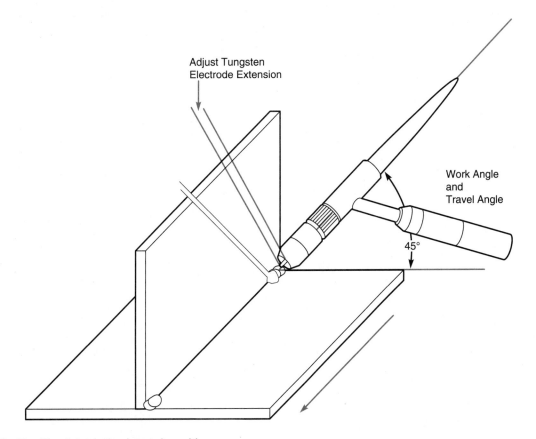

Adjust Tungsten
Electrode Extension

Work Angle
and
Travel Angle

45°

FIGURE 6–42 Tee joint in horizontal position.

7. Look for cracks, and check the amount of **fusion** after testing. Redo the weld if there is evidence of cracks or incomplete fusion.
8. Check with the instructor for evaluation, if necessary.
9. Practice this exercise until you meet the standards established in the course.

UNIT 6: EXERCISE 4

Gas Tungsten Arc Welding (GTAW): Butt Joint (Mild Steel) in Flat Position, Square-Groove Weld

A weld with **complete root penetration** requires preparation. A successful **butt joint** depends on grinding of the surfaces to be welded, **root opening,** tack welding, and amperage setting, to mention four things. The process of welding includes travel speed and mixing the welding rod into the weld pool.

Melt-thru is a common problem that has several possible causes, including too high an amperage setting, too slow a travel speed, and too wide a joint root opening. Poor root penetration is another problem. It is caused by too low an amperage setting, too fast a travel speed, or too narrow a joint root opening. The setup for this exercise is shown in Figure 6–43.

Necessary Material and Equipment

Safety glasses	Earplugs	2% thoriated tungsten,	Shielding gas: argon
Welding gloves	Grinder	1/16″ diameter	
Protective clothing	2 pieces 1/16″ sheet steel,	Welding rod, 1/16″	
Helmet	2″ × 4″		

Instructions

1. Clean the surfaces to be welded by grinding.
2. Set a 90° work angle and a 45° travel angle.
3. Tack weld the pieces together. Use the welding rod to set the joint root opening.
4. Complete the weldment in one pass.
5. Look for 100% root penetration.
6. Cool, then test the weldment in a vice, positioning the weld just above the jaws. Hammer the top piece over.

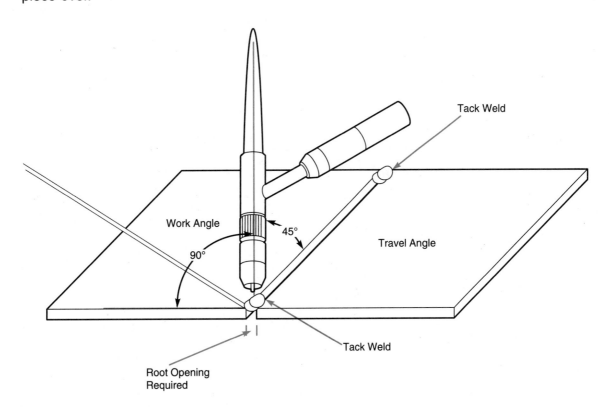

FIGURE 6–43   Butt joint in flat position.

7. Examine the weld for cracks or cold laps (incomplete fusion) after testing. Redo the weld if there is evidence of cracks, cold laps, or less than 100% root penetration.
8. Check with the instructor for evaluation, if necessary.
9. Practice this exercise until you meet the standards established in the course.

UNIT 6: EXERCISE 5

Gas Tungsten Arc Welding (GTAW): Butt Joint (Mild Steel) in Horizontal Position, Square-Groove Weld

Some of the same techniques used for welding in the flat position are necessary for this exercise. Keep in mind that the welding rod must be angled from above into the weld pool to counteract gravity. The addition of the welding rod should make the completed weld thicker than the base metal. This applies to all butt joints. Figure 6–44 illustrates the setup.

Necessary Material and Equipment

Safety glasses	Earplugs	Welding rod, 1/16″	Shielding gas: argon
Welding gloves	Grinder	2% thoriated tungsten,	
Protective clothing	2 pieces 1/16″ sheet steel,	1/16″ diameter	
Helmet	2″ × 4″		

Instructions

1. Clean the surfaces to be welded by grinding.
2. Set a 45° to 60° work angle and a 45° travel angle.
3. Tack weld the pieces together. Use the welding rod to set the joint root.
4. Complete the weldment in one pass.
5. Look for 100% root penetration.

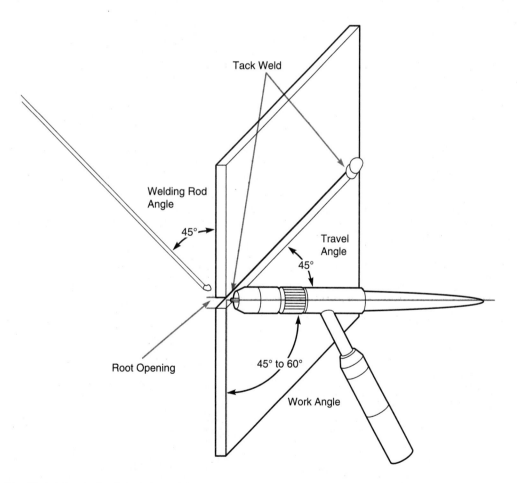

FIGURE 6–44   Butt joint in horizontal position.

6. Cool, then test the weldment in a vice, positioning the weld just above the jaws. Hammer the top piece over.
7. Examine the weld for cracks or cold laps. Redo the weld if there is evidence of cracks, cold laps, or less than 100% root penetration.
8. Check with the instructor for evaluation, if necessary.
9. Practice this exercise until you meet the standards established in the course.

UNIT 6: EXERCISE 6

Gas Tungsten Arc Welding (GTAW): Lap Joint (Mild Steel) in Vertical Position, Fillet Weld

Direct the tungsten electrode toward the bottom piece because less heat is required to melt the edge of the top piece. Apply the same technique used in the horizontal position for welding **uphill.** The relationship of the torch and welding rod to the lap joint does not change for the **vertical position.** The position of the welder changes, but not the technique.

Add the welding rod into the weld pool from above to avoid contamination of the electrode. If the electrode becomes contaminated, stop welding, remove the contaminants, and begin again. Figure 6–45 shows the setup.

Necessary Material and Equipment

Safety glasses	Earplugs	2% thoriated tungsten,	Shielding gas: argon
Welding gloves	Grinder	1/16″ diameter	
Protective clothing	2 pieces 1/16″ sheet steel,	Welding rod, 1/16″	
Helmet	2″ × 4″		

Instructions

1. Clean the surfaces to be welded by grinding.
2. Set a 90° work angle and a 45° travel angle with a 45° welding rod angle.
3. Tack weld the two pieces together at each end.
4. Complete the weldment in one pass with uphill welding.
5. Examine the weld for undercut, unfilled crater, and a weld bead of inconsistent size. Is the weld bead thicker than the base metal?

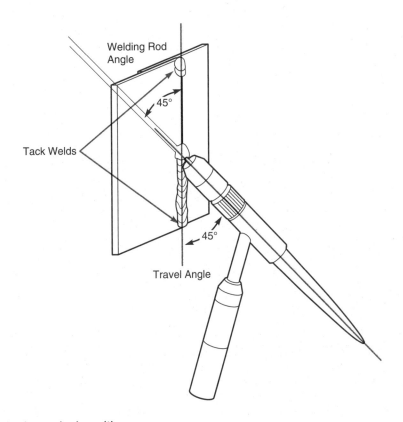

FIGURE 6–45   Lap joint in vertical position.

6. Cool, then test the weldment in a vice, positioning the weld just above the jaws. Hammer the top piece over.
7. Look for cracks and bending along any undercut. Redo the weld if there is evidence of cracks or undercut.
8. Check with the instructor for evaluation.
9. Practice this exercise until you meet the standards established in the course.

UNIT 6: EXERCISE 7

Gas Tungsten Arc Welding (GTAW): Butt Joint (Mild Steel) in Vertical Position, Square-Groove Weld

The travel angle of the torch and the angle of the welding rod are important. Watch the weld pool, directing heat equally to both pieces. Remember that preparation for welding is as important as welding. Pay attention to surface cleaning, joint root opening, amperage setting, and so on. The setup is shown in Figure 6–46.

Necessary Material and Equipment

Safety glasses	Earplugs	2% thoriated tungsten,	Shielding gas: argon
Welding gloves	Grinder	1/16″ diameter	
Protective clothing	2 pieces 1/16″ sheet steel,	Welding rod, 1/16″	
Helmet	2″ × 4″		

Instructions

1. Clean the surfaces to be welded by grinding.
2. Set a 45° travel angle and a 45° welding rod angle.
3. Tack weld the two pieces together. Use the welding rod to set the joint root opening.
4. Complete the weld.
5. Look for 100% root penetration.

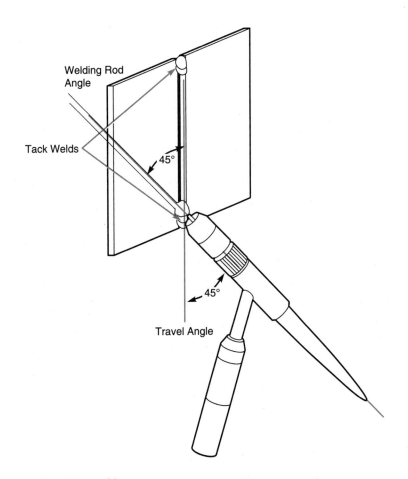

FIGURE 6–46 Butt joint in vertical position.

6. Cool, then test the weldment in a vice, positioning the weld just above the jaws. Hammer the top piece over.
7. Examine the weld for cracks or cold laps. Redo the weld if there is evidence of cracks, cold laps, or less than 100% root penetration.
8. Check with the instructor for evaluation, if necessary.
9. Practice this exercise until you meet the standards established for the course.

UNIT 6: EXERCISE 8

Gas Tungsten Arc Welding (GTAW): Thin and Thick Welding (Mild Steel)

Exercises 2 through 7 were designed for 16-**gauge sheet metal.** This exercise calls for welding thinner or thicker mild steel (perhaps both). Learn to make adjustments to the power source, the torch, and conditions within the weld pool. This exercise offers an opportunity to construct or repair some item using gas tungsten arc welding.

Necessary Material and Equipment

Safety glasses	Helmet	2% thoriated tungsten	Additional materials
Welding gloves	Earplugs	Shielding gas: argon	
Protective clothing	Grinder	Creative project	

Instructions

1. If possible, prepare the surfaces to be welded.
2. Clamp the pieces together. Tack weld them together, if necessary.
3. Complete the required welding.
4. Look for **defects** in the weld.
5. Test. If the weldment has been constructed or repaired, the real test is whether it holds up under application.
6. Examine for defects in the welding after testing.
7. Check with the instructor for evaluation, if necessary.
8. Practice this exercise until you meet the standards established for the course.

UNIT 6: EXERCISE 9

Gas Tungsten Arc Welding (GTAW): Lap Joint (Stainless Steel) in Horizontal Position, Fillet Weld

If you have successfully completed the first eight exercises in this unit, you should have no serious problems adapting to stainless steel. The previous exercises can be repeated with stainless steel if the metal is readily available. Because stainless steel is costly, exposure to this one stainless steel exercise will suffice.

Stainless steel is **high alloy** by definition because its **chromium** content is 12% or greater. Because of the chromium content, stainless steel cannot be cut by oxyacetylene without special equipment. The addition of chromium counteracts corrosion, which causes mild steel to rust more readily than other types of metal.

There are two major types of stainless steel: chromium and chromium-**nickel.** The 400 series stainless steels are the chromium types. The 200 series and 300 series are the chromium-nickel types. There are more than 30 types of stainless steel available for different applications. The 200 and 300 series are nonmagnetic; 304, 308, 309, 310, and 316 are some of the commonly welded stainless steels. It is necessary to match the welding rod to the type of stainless steel to be welded.

Preparing to weld stainless steel is similar to welding mild steel in some ways. Both the output current, DCEN, and the tungsten electrode (2% thoriated tungsten) can stay the same. However, the amperage setting required is less for welding stainless steel of the same thickness as mild steel. With stainless steel, the movement of heat through the metal is slower than through mild steel. This characteristic can cause overheating and distortion with stainless steel.

With stainless-steel oxidation, a black crust forms on the weld bead without adequate gas shielding. This should not be a problem with fillet welds, but it can be a problem with groove welds on the underside of butt joints. When welding stainless steel, a shielding gas can be pumped inside the pipe to eliminate the problem. A Y-fitting can be used for shielding both the inside and the outside of the joint during welding (Figure 6–47).

Necessary Material and Equipment

Safety glasses	Helmet	2% thoriated tungsten	Stainless-steel welding rod
Welding gloves	Earplugs	2 pieces of stainless steel	Shielding gas: argon
Protective clothing	Stainless-steel wire brush		

Instructions

1. Be sure there is adequate ventilation. Chromium and nickel, like other compounds found in welding fumes, should not be inhaled.
2. With the wire brush, clean the surfaces to be welded.
3. Tack weld the lap joint in several places to restrict distortion.
4. Clean with the wire brush before welding the first weld bead. Wire brush between each weld bead, if more than one.
5. Complete the weld.
6. Examine for consistent weld beads that are free of undercut.
7. Cool, then test the weldment in a vice, positioning the weld just above the jaws. Hammer the top piece over.
8. Look for cracks and bending along any undercut. Redo the weld if there is evidence of cracks or undercut.
9. Check with the instructor for evaluation, if necessary.
10. Practice this exercise until you meet the standards established in the course.

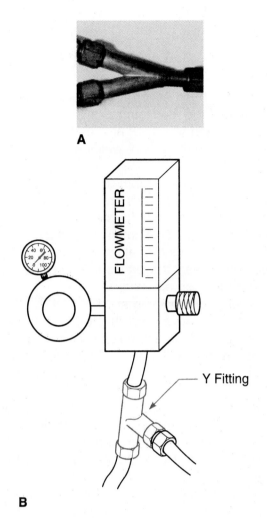

FIGURE 6–47   (A) Y fitting. *(Courtesy of Larry Jeffus.)* (B) Use of a Y fitting.

UNIT 6: EXERCISE 10

Gas Tungsten Arc Welding (GTAW): Lap Joint (Aluminum) in Horizontal Position, Fillet Weld

The aluminum welding exercise has been saved for last. All of the exercises designed for welding steel can be repeated using aluminum if the aluminum is available.

In one sense, the best has been saved for last. Of all metals, aluminum is one of the easiest to work with. This includes the steps leading up to the actual welding, such as lifting, cutting, grinding, and so on.

Aluminum is a popular product that has many applications. Aluminum is lightweight, weighing one-third an equal measure of steel. For example, one square foot of 1/4-inch-thick aluminum weighs about 3 1/2 pounds, as compared to about 10 1/2 pounds for steel. Aluminum is highly resistant to corrosion, but never let rusty steel bleed into aluminum. This causes oxidation of the aluminum.

Gas tungsten arc welding makes the welding of aluminum easy with some practice. The welding of aluminum requires different preparation than the previous exercises. Setting up to weld requires an electric current of ACHF and a pure tungsten electrode.

Welding aluminum is also different from welding steel in that the weld pool appears wet but does not change color during welding. The aluminum **oxides** that form and protect the aluminum from corrosion have a higher melting temperature than the aluminum itself. This means that a thick oxide surface must be removed before welding. As often happens when oxides are present, the aluminum melts away, leaving a mess before welding even starts. Paint, dirt, and other contaminants must be removed before welding.

There are eight types of aluminum (Table 6–3). 1100 is a commercially pure aluminum used in sheet metal work. 6061 aluminum has magnesium and silicon as its major alloying elements. It is commonly used in boats and transportation, where strength is required. This application emphasizes the importance of matching the aluminum with its welding rod to achieve integrity in the weldment. 4043 is the AWS number for a common filler rod used for welding most types of aluminum.

Keep in mind that the tungsten electrode is easily contaminated if it touches the weld pool, and oxides (black or gray coating) quickly form when the weld itself becomes contaminated.

Necessary Material and Equipment

Safety glasses	Helmet	Stainless-steel wire brush	Aluminum welding rod
Welding gloves	Earplugs	Pure tungsten	Shielding gas: argon
Protective clothing	Grinder	2 pieces of aluminum	

Instructions

1. Clean the surfaces to be welded by grinding.
2. Tack weld the lap joint in several places to restrict distortion.
3. Complete the weld. Wire brush the joint after starting and stopping the arc.

TABLE 6–3: DESIGNATIONS FOR ALUMINUM

Designation	Alloying Elements
1---	99% pure aluminum
2---	Copper
3---	Manganese
4---	Silicon
5---	Magnesium
6---	Magnesium and silicon
7---	**Zinc**
8---	Other elements

4. Examine for consistent weld beads free of undercut.
5. Cool, then test the weldment in a vice, positioning the weld just above the jaws. Hammer the top piece over.
6. Look for incomplete fusion and cracks along the weld bead resulting from undercut. Redo the weld if there is evidence of incomplete fusion or cracks.
7. Check with the instructor for evaluation, if necessary.
8. Practice this exercise until you meet the standards established in the course.

BRAZING, BRAZE WELDING, AND SOLDERING

"With the welding equipment you can build up worn surfaces with better than original metal and then machine it back to tolerance with carbide tools."

—Robert M. Pirsig, *Zen and the Art of Motorcycle Maintenance*

GOAL

- Develop an ability to do brazing, braze welding, and soldering safely

QUESTIONS

- What is brazing?
- What is braze welding?
- What is soldering?
- What equipment do these processes require?
- When are brazing, braze welding, and soldering used?

SAFETY FIRST

The **processes** covered in this unit require the heat of a **torch** or electricity for **soldering guns.** If necessary, review the safety procedures your shop will follow in case of an accident. Be sure to follow the manufacturers' recommendations for using their equipment, as explained in the **operator's manuals.** Ventilation is often overlooked when brazing, braze welding, and soldering, but breathing any toxic fumes given off by these processes can be dangerous to your health.

BRAZING

The **American Welding Society** recognizes 11 brazing processes. Brazing (B) is a group of processes in which **filler metal** heated to liquidus above 840° F but below the melting temperature of the **base metal** is pulled into the **joint** by **capillary action.** This term refers to the behavior at the surface of a liquid in contact with a solid. During brazing, capillary action occurs when the **cohesion** (attraction) of the molecules in a **liquidus** for each other is overcome by **adhesion** (attraction) to the base metal. In other words, the surface tension of the **molten pool** breaks down as molten metal flows on the base metal and into the joint.

The **strength** of brazing depends on a clean joint, spacing, and the **flux.** The pieces of metal to be joined together must be cleaned of dirt, petroleum products, and **oxides. Mill scale** is one form of oxide. It is produced in the metal manufacturing process. Paint, a second form of oxide, might be applied to protect the metal. **Rust** is a third form of oxide, produced during exposure of the iron in **steel** to the corrosive nature of moisture. If these oxides are not removed, the filler metal could attach itself to the oxides instead of to the base metal, weakening the joint (Figure 7–1).

Although many **joint designs** are used in brazing, two are most common. All brazing requires a

FIGURE 7–1   Clean metal before brazing.

joint design using the **lap joint** or the **butt joint** in a **scarf-groove weld** configuration (Figure 7–2). The spacing between the pieces making up the joint is important. A joint clamped tightly together will prevent capillary action. A joint with too much spacing will also prevent capillary action. The spacing of the pieces should fall between 0.002 inch (two-thousandths of an inch) and 1/64 inch for capillary action to occur.

Brazing depends on a flux, which protects the joint from oxides that form when **oxygen** in the air combines with the heat of the flame. The flux also cleans the base metal and helps the filler metal to flow. Fluxes are available in paste, powder, and liquid form and might be included on or within the filler metal (Figure 7–3). It is important to remove any flux residue left after brazing to prevent corrosion later.

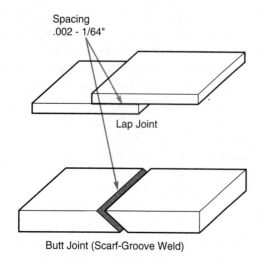

FIGURE 7–2   Lap joint and butt joint (scarf-groove weld).

FIGURE 7–3   Three types of flux.

Torch Brazing

This book limits its focus to **manual torch brazing** (TB) using the **oxyacetylene welding** torch (Figure 7–4). Torch brazing is popular for portable jobs when it is necessary to bring the brazing process to the work. Skillful brazing requires that enough heat be applied to the joint without **overheating,** which might cause the filler metal to flow out of the joint or the base metal to melt away.

In brazing, a filler **rod** is used to join base metals when **welding** is not necessary, would be too costly, would take too much time, might be next to impossible, could affect the inner structure of the base metal, or might cause

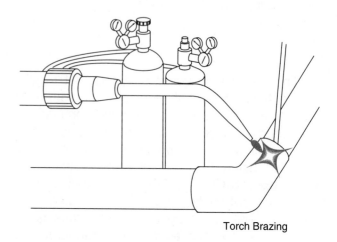

FIGURE 7–4   Torch brazing.

severe **distortion. Copper-zinc, aluminum-silicon,** and **silver alloys** are popular filler rods used for torch brazing.

Copper-zinc filler rods are used for joining **ferrous** and **nonferrous** metals when resistance to contamination by the flux is not a problem. Aluminum-silicon filler rods are used for joining **aluminums.** Silver alloy filler metals are used for joining ferrous and nonferrous metals, except aluminum and **magnesium.**

Brazing has many applications. Brazing with copper-zinc filler rods has application in the joining of steel **tubing.** Brazing with aluminum-silicon filler rods has application for aluminum castings. Brazing with silver alloy filler metals is used for **stainless steel, copper** tubing, and fastening machine cutting tools.

Once the base metal is prepared for brazing, the size of **tip** must be selected. The tip should produce just enough volume so as not to overheat or underheat the joint (Figure 7–5). The torch must be adjusted with the correct flame. Torch brazing requires a **neutral flame** to a slightly **carburizing flame** (Figure 7–6).

The surface of the joint, after being cleaned for brazing, may require a covering of flux. (If flux-filled or flux-covered rods are used, the surface might not require additional fluxing.) The heat of the tip should be directed away from the filler rod and away from the immediate area of the joint. The idea is to use the heat to draw the molten pool into the joint by capillary action (Figure 7–7) Molten metal will only flow into an area that has been heated sufficiently for capillary action to occur. The strength of the joint will ultimately be

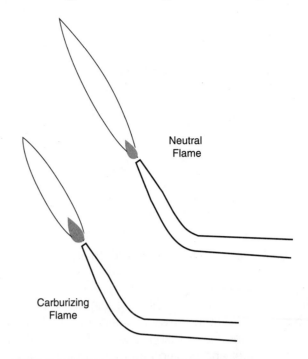

FIGURE 7–6 Carburizing and neutral flames.

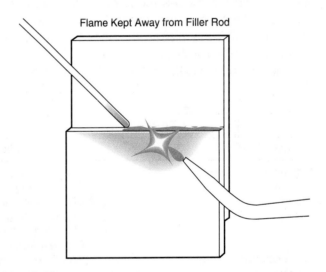

FIGURE 7–7 Correct position of the torch.

determined by filler metal that has been drawn into the joint and cannot be seen.

Filler metals come in various shapes. While rods are very common, coils and sheets are two other shapes in which filler metals can be purchased (Figure 7–8).

One method of adding filler metal is by letting heat draw the flowing flux-covered filler rod into the joint. Bare rods must be heated first and then dipped into a can of flux before being melted into the joint. A second method is allowing heat to draw

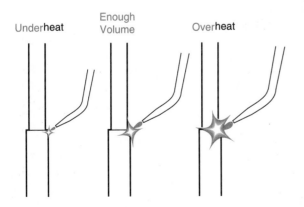

FIGURE 7–5 Correct and incorrect volumes of heat.

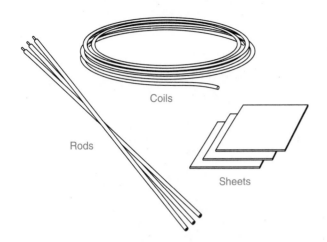

FIGURE 7–8   Common shapes for filler metals.

the filler metal from a coil. The joint must first be coated with flux in powder, paste, or liquid form. A third method is to take pieces of filler metal cut from a sheet and position them in the joint, which has been coated with flux. A vice grips can be used to hold the pieces together as the heat is applied and the pieces are drawn together. Spring pressure draws the pieces together as the filler metal melts (Figure 7–9).

The American Welding Society lists eight cate-gories of brazing filler metals. The filler metal and the flux must be matched with the base metal to meet the service requirements of a particular application. Table 7–1 lists the filler metals and fluxes that are appropriate for various types of base metals. Torch brazing may not be the recom-mended process for every application. Remember, there are 10 other brazing processes to consider.

How many different categories of filler metals are listed in the table?

The numbers 1, 2, 3A, and 3B are designations given to the fluxes by the American Welding Soci-ety. Number 1 is a flux in powder form for brazing temperatures between 700° and 1,200° F. Number 2 is a flux in powder form for brazing temperatures between 900° and 1,200° F. Number 3A is flux available in powder, liquid, and paste for brazing temperatures between 1,000° and 1,600° F. Fi-nally, number 3B is another flux available in pow-der, liquid, and paste. It is appropriate for brazing temperatures between 1,300° and 2,100° F.

Follow up the preparations for brazing laid out in this unit with practice. The results of practice brazing will be joints with a strength equal to or greater than that of many **fusion**-welded joints.

TABLE 7–1:   FILLER METALS AND FLUXES FOR BRAZING

Base Metal	Filler Metal	Flux (AWS No.)
Aluminum	BAlSi (brazing aluminum silicon)	1
Magnesium	BMg (brazing magnesium)	2
Copper	BAg (brazing silver)	3A
	BAu (brazing **gold**)	3B
	BCuP (brazing copper **phosphorus**)	3B
	RBCuZn (rod brazing copper zinc)	3B
Carbon steels and	BAg (brazing silver)	3A
low-alloy steels	BAu (brazing gold)	3B
	BCu (brazing copper)	3B
	RBCuZn (rod brazing copper zinc)	3B
	BNi (brazing **nickel**)	3B
Cast irons	BAg (brazing silver)	3A
	RBCuZn (rod brazing copper zinc)	3B
	BNi (brazing nickel)	3B
Stainless steels	BAg (brazing silver)	3A
and nickels	BAu (brazing gold)	3B
	BCu (brazing copper)	3B
	BNi (brazing nickel)	3B
Galvanized steels	BCuZn (brazing copper zinc)	3B
Tool steels	BAg (brazing silver)	3A
	BAu (brazing gold)	3B
	BCu (brazing copper)	3B
	BNi (brazing nickel)	3B

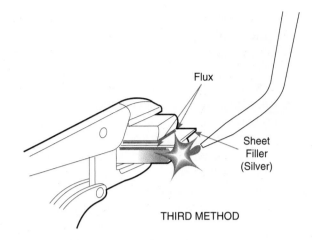

THIRD METHOD

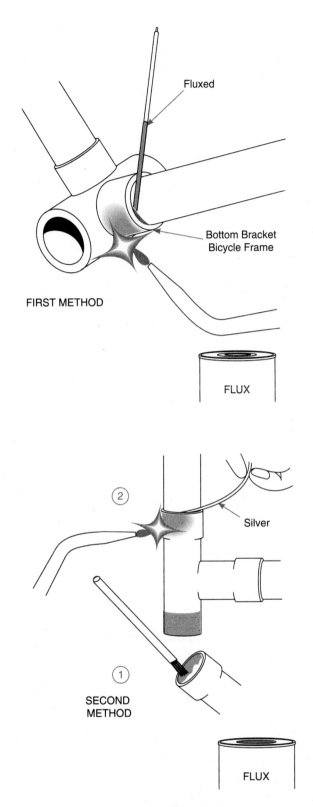

FIRST METHOD

FLUX

SECOND
METHOD

FLUX

FIGURE 7–9 Three methods of applying flux.

BRAZE WELDING

Braze welding (BW) is a metal-joining process that requires a torch and a filler rod. Oxyacetylene welding equipment is used for this process.

Braze welding is unlike brazing, in which a thin layer of filler metal is sandwiched between the two pieces of the base metal. Braze welding involves a **buildup** of filler metal in which capillary action does not take place. In braze welding, the base metal is heated above 840° F but below the melting temperature of the base metal.

Braze welding is generally used to repair steel and cast iron with copper-zinc filler rods. Braze welding is often overlooked in fabrication. It could replace fusion welding processes in many applications.

Braze welding has some advantages over other welding processes. Because less heat is required, less time is necessary, and no changes occur in the inner structure of the metal being welded. The end product is a **weld** that is readily **machinable;** the **stresses** of welding are contained within the joint itself.

Braze welding has some disadvantages, too. The strength of the joint is no greater than the strength of the filler metal, no matter how strong the base metal is. The color will not match. The operating temperature of a product in service is restricted to applications with an upper range of 500° F. Above 500°, the strength of braze welding is affected, even though copper-zinc **(brass)** alloys have a melting temperature approaching 1,600° F.

Unlike brazing, the process of braze welding can be applied to any of the five basic joints (**corner,** butt, lap, **tee,** and **edge**) used in fusion welding. And where **tensile testing** equipment is available, **v-groove weld** butt joints have been measured with **tensile strengths** exceeding 50,000 pounds per square inch.

Some of the techniques used for braze welding (and brazing, for that matter) are different from those used in oxyacetylene welding. Oxyacetylene welding takes more patience because heating the metal to melting temperatures requires more time. In oxyacetylene welding, the welder has the time for a close look at what is happening, since the act of welding is slower. Because braze welding never reaches the melting temperature of the base metal, less heat is necessary throughout the joining process. In braze welding, welders tend to dip the filler rod in and out of the joint. During oxyacetylene welding, dipping should not be done.

Braze welding is ideal for repairing cast iron castings when heating the base metal to melting temperature can cause cracking or a change in the inner structure of the metal. **Malleable cast iron** cannot be repaired by fusion welding without changing it into brittle **white cast iron.**

The process of braze welding begins with cleaning the joint before welding. Butt joints, which are common, should be **beveled** (**bevel-groove weld** or V-groove weld), depending on thickness. Cracks in castings might require drill holes at the ends of each crack. Use a grinder to clean the surface of the joint, if possible. On thick metal, a torch can be used for cutting or gouging out the crack. Use a grinder or chipping hammer to remove the **slag** before finishing the job with a wire brush (Figure 7–10).

Fit up before welding is important. If fit up is especially important, clamp the pieces into position, and **tack weld** before using the torch or grinder to make a bevel in the casting (Figure 7–11). Braze weld the joint, remove the clamps, and bevel the rest of the joint as required. Then complete the braze welding.

Preheat and examine the joint by melting a copper-zinc rod. If pieces of filler rod ball up and fall away, more preheating is required. When the base metal has turned a dark cherry red, the braze welding temperature has been reached, and filler metal should flow into the joint.

Precoating the joint with a thin layer of filler metal is the first step. This thin layer will take time. It might be necessary to work the rod so as to **bond** the molten metal with the base metal (Figure 7–12). Precoating is an important part of braze

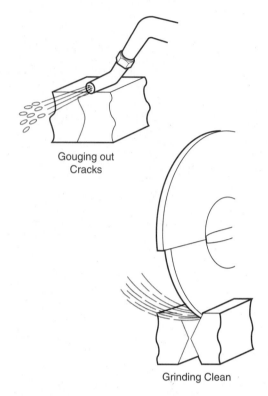

Gouging out Cracks

Grinding Clean

FIGURE 7–10   Gouging and cleaning the joint before braze welding.

welding. After precoating, the second and subsequent layers of filler metal should be easier to do because the filler rod will readily melt into the precoated layer. The final **pass** should extend beyond the **toe** of the weld (Figure 7–13).

SOLDERING

The AWS lists seven **soldering** (S) processes. This unit will touch on **torch soldering** (TS) and **iron soldering** (INS). Soldering is a group of processes in which the filler metal is heated to **liquidus** below 840° F and below the melting temperature of the base metal. Note the difference in melting temperatures between soldering and brazing.

For many soldering applications, an oxyacetylene welding torch will supply too much heat with even the smallest tip. The solution, then, is to select iron soldering or another form of torch soldering. A **propane** or air-acetylene torch are two methods of torch soldering that can be used (Figure 7–14).

In soldering, heat applied to a solder dissolves the surface of the base metal. This is the act of **wetting** the base metal to form a compound be-

Oxidizing flame

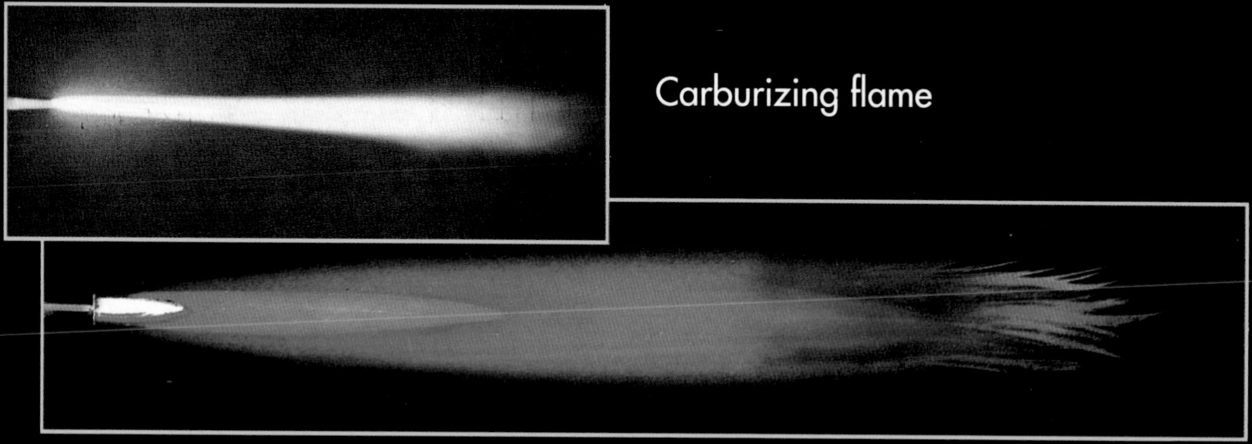

Carburizing flame

Neutral flame

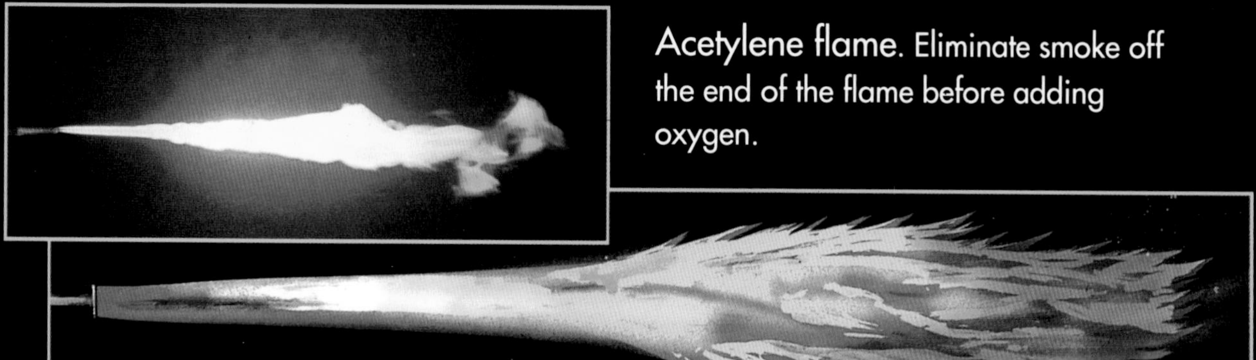

Acetylene flame. Eliminate smoke off the end of the flame before adding oxygen.

Heat base metal before adding filler metal.

Filler metal added to weld pool.

Filler metal melting off the front of the weld poo

ZING AND BRAZE WELDING

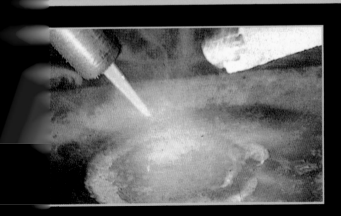

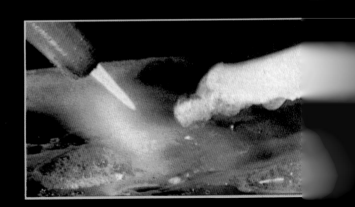

One: heat base metal to kindling
temperature

Two: push in lever on torch for jet of
oxygen.

Three: move the
torch along,
watching the
kerf.

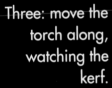

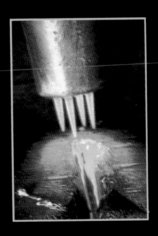

Four: move only
fast enough to
keep the kerf
forming

PLAMSA ARC CUTTING

Cutting attachment used for assistance

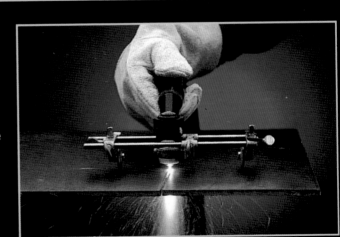

Please note the travel angle of th
electrode and direction of weldi

Note the weld pool.

Note the formation of slag.

Note the change
in travel angle

SHORT CIRCUITING ARC TRANSFER (GMAW-S)

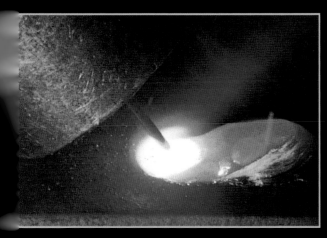

Backhand technique   Forehand technique

SPRAY ARC TRANSFER (GMAW-SP)

Use of high amperage for welding plate.

FLUX CORED ARC WELDING (FCAW)

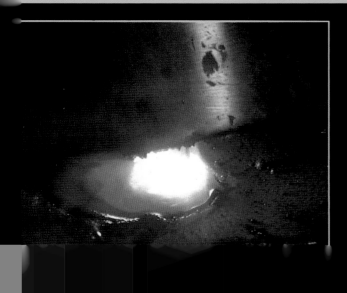
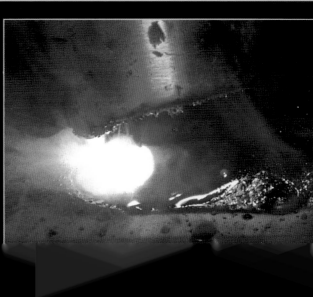

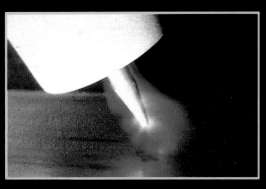
Welding on stainless steel.

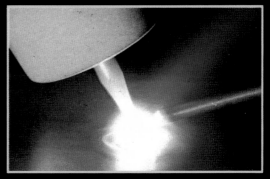
The filler rod is brought into the arc zone.

The end of the filler rod is dipped into the leading edge of the weld pool.

The rod is repeatedly dipped into the molten weld pool to form a bead.

SOME WELDING PROCESSES

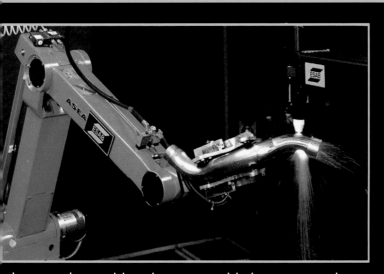

Above: Robot welds tailpipe assembly by Gas Metal Arc Welding (GMAW-S).
Right above: Laser Beam Welding (LBW). Joining a cap onto a thin walled tubing.
Right Below: Robot positions manifold for Plasma Arc Cutting (PAC).

Steel tapped from the furnace for a sample to be analyzed.

Molton steel is poured into ladle for transport to a casting machine.

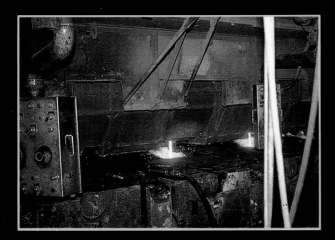

Molton steel flows into cast molds as bar stock.

Red hot bar stock comes out of casting.

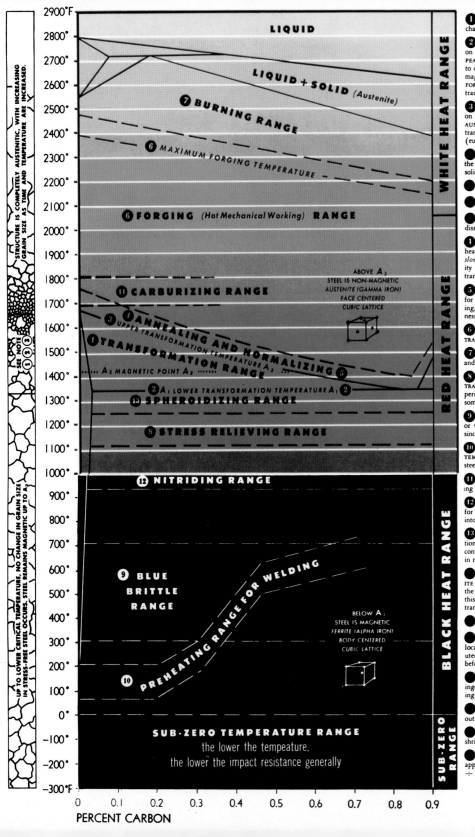

1 TRANSFORMATION RANGE. In this range steels undergo internal atomic changes which radically affect the properties of the material.

2 LOWER TRANSFORMATION TEMPERATURE (A_1). Termed Ac_1 on heating, Ar_1 on cooling. Below Ac_1 structure ordinarily consists of FERRITE and PEARLITE (see below). On heating through Ac_1 these constituents begin to dissolve in each other to form AUSTENITE (see below) which is non-magnetic. This dissolving action continues on heating through the TRANSFORMATION RANGE until the solid solution is complete at the upper transformation temperature.

3 UPPER TRANSFORMATION TEMPERATURE (A_3). Termed Ac_3 on heating, Ar_3 on cooling. Above this temperature the structure consists wholly of AUSTENITE which coarsens with increasing time and temperature. Upper transformation temperature is lowered as carbon increases to 0.85% (eutectoid point).

FERRITE is practically pure iron (in plain carbon steels) existing below the lower transformation temperature. It is magnetic and has very slight solid solubility for carbon.

PEARLITE is a mechanical mixture of FERRITE and CEMENTITE.

CEMENTITE or IRON CARBIDE is a compound of iron and carbon, Fe_3C.

AUSTENITE is the non-magnetic form of iron and has the power to dissolve carbon and alloying elements.

4 ANNEALING, frequently referred to as FULL ANNEALING, consists of heating steels to slightly above Ac_3, holding for AUSTENITE to form, then *slowly* cooling in order to produce small grain size, softness, good ductility and other desirable properties. On cooling slowly the AUSTENITE transforms to ferrite and pearlite.

5 NORMALIZING consists of heating steels to slightly above Ac_3, holding for AUSTENITE to form, then followed by cooling (in still air). On cooling, AUSTENITE transforms giving somewhat higher strength and hardness and slightly less ductility than in annealing.

6 FORGING RANGE extends to several hundred degrees above the UPPER TRANSFORMATION TEMPERATURE.

7 BURNING RANGE is above the FORGING RANGE. Burned steel is ruined and *cannot be cured* except by remelting.

8 STRESS RELIEVING consists of heating to a point below the LOWER TRANSFORMATION TEMPERATURE, A_1, holding for a sufficiently long period to relieve locked-up stresses, then slowly cooling. This process is sometimes called PROCESS ANNEALING.

9 BLUE BRITTLE RANGE occurs approximately from 300° to 700° F. Peening or working of steels should not be done between these temperatures, since they are more brittle in this range than above or below it.

10 PREHEATING FOR WELDING is carried out to prevent crack formation. See TEMPIL° PREHEATING CHART for recommended temperature for various steels and non-ferrous metals.

11 CARBURIZING consists of dissolving carbon into surface of steel by heating to above transformation range in presence of carburizing compounds.

12 NITRIDING consists of heating certain *special steels* to about 1000° F for long periods in the presence of ammonia gas. Nitrogen is absorbed into the surface to produce extremely hard "skins".

13 SPHEROIDIZING consists of heating to just below the lower transformation temperature, A_1, for a sufficient length of time to put the CEMENTITE constituent of PEARLITE into globular form. This produces softness and in many cases good machinability.

MARTENSITE is the hardest of the transformation products of AUSTENITE and is formed only on cooling below a certain temperature known as the M_s temperature (about 400° to 600° F for carbon steels). Cooling to this temperature must be sufficiently rapid to prevent AUSTENITE from transforming to softer constituents at higher temperatures.

EUTECTOID STEEL contains approximately 0.85% carbon.

FLAKING occurs in many alloy steels and is a defect characterized by localized micro-cracking and "flake-like" fracturing. It is usually attributed to hydrogen bursts. Cure consists of cycle cooling to at least 600° F before air-cooling.

OPEN OR RIMMING STEEL has not been completely deoxidized and the ingot solidifies with a sound surface ("rim") and a core portion containing blowholes which are welded in subsequent hot rolling.

KILLED STEEL has been deoxidized at least sufficiently to solidify without appreciable gas evolution.

SEMI-KILLED STEEL has been partially deoxidized to reduce solidification shrinkage in the ingot.

A SIMPLE RULE: Brinell Hardness divided by two, times 1000, equals approximate Tensile Strength in pounds per square inch. (200 Brinell ÷ 2 × 1000 = approx. 100,000 Tensile Strength, p.s.i.)

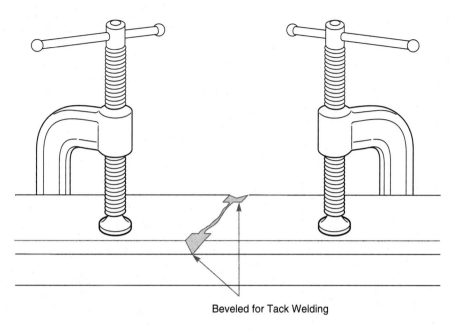

Beveled for Tack Welding

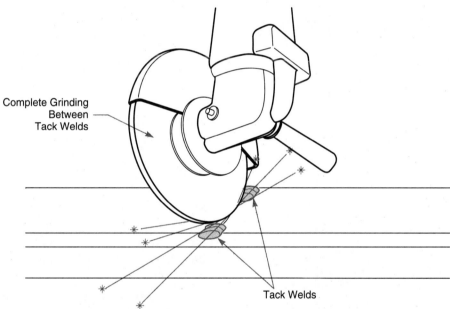

Complete Grinding Between Tack Welds

Tack Welds

FIGURE 7–11 Preparing the joint for braze welding.

tween the pieces that bonds them together. A soldered joint has nowhere near the strength of a brazed or welded joint. This is because there is a large difference between the strengths of solders and base metals.

Several types of solders are available. A complete list is available from the American Welding Society. The focus here is restricted to **tin-lead** solders. These are popular solders for joining most metals.

Solders are assigned a two-part designation.

The first number gives the percentage of tin content. The second number gives the percentage of lead content. For example, 5/95 solder is made of 5% tin and 95% lead. It has a high melting temperature and a small melting range. A small melting range means the solder is liquid for a small range of temperatures. When a small melting range is desired, a solder that freezes quickly can prevent unnecessary movement of the parts. At the other extreme, a 70/30 solder has a low melting temperature and also a small melting range. Pure

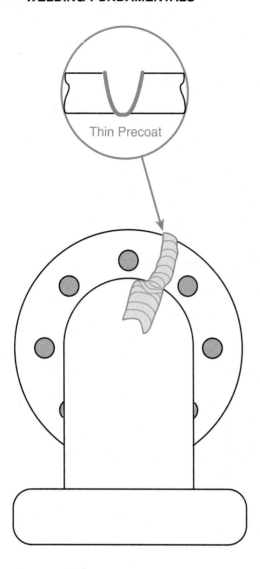

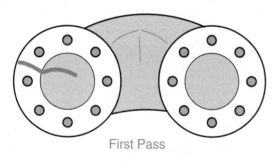

FIGURE 7–12   Precoating the joint.

TABLE 7–2:   MELTING TEMPERATURES OF COMMON METALS

Base Metal	Melting Temperature (° F)
Tin	450
Lead	620
Aluminum	1,218
Magnesium	1,240
Brass	1,660
Silver	1,762
Copper	1,891
Gold	1,945
Cast iron	2,300
Steel (0.4–0.7% carbon)	2,500
Stainless steel (nickel)	2,550
Steel (0.15–0.4% carbon)	2,600
Stainless steel (low carbon)	2,640
Steel (less than 0.15% carbon)	2,700
Iron (pure)	2,786
Tungsten	6,170

tin melts at 450° F. Pure **lead** melts at 620° F (Table 7–2). When tin and lead are combined, the lowest possible melting temperature is lower than 450°. When the tin-lead mixture reaches this lowest temperature, its **eutectic composition,** there is only one melting temperature.

A solder has a wider melting range the farther away the content is from its eutectic composition (Figure 7–15). A 30/70 solder has a wider melting range than a 50/50 solder. This is useful when it is necessary to shape the solder in a molten condition before the solder becomes solid.

Some soldering requires the use of a flux. The major purpose of a soldering flux is to remove any oxides or other compounds left on the base metal after cleaning and to prevent oxides from reforming during the soldering process. There are three basic types of soldering fluxes. They are categorized based on their ability to remove oxides.

Inorganic fluxes are made of very corrosive acids and salts. These fluxes work very well with a torch but must be removed to prevent corrosion around the joint. Brasses, coppers, stainless steels, and carbon steels are common metals appropriate for these fluxes.

Organic fluxes are not very corrosive acids and bases derived from living organisms. They have limited applications with a torch or flame because they are easily burned. Organic fluxes are commonly used with brasses, coppers, and stainless steels.

Rosin fluxes are the most nearly noncorrosive fluxes available. They do not have to be cleaned away after soldering. These fluxes are used in

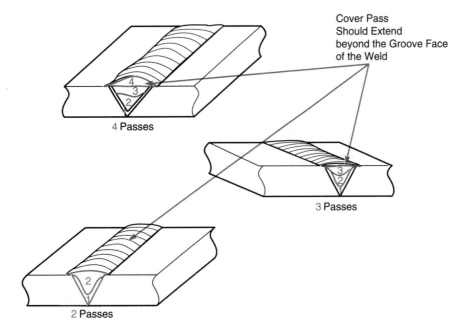

FIGURE 7-13   Three examples of layered braze welding.

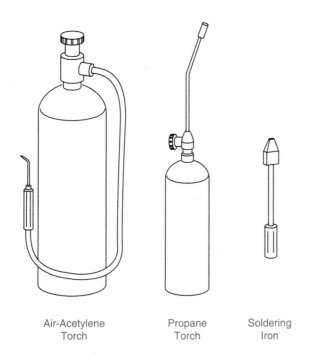

FIGURE 7-14   Soldering tools.

the electronics industry. Brasses and coppers are common metals that use rosin fluxes.

There is a special group of fluxes for soldering aluminum. Cast iron and magnesium should not be joined by soldering.

The butt joint and the lap joint are the two joint designs used in soldering (Figure 7–16). A number of different lap joint designs are used.

If a torch is to be used for soldering, keep the flame out of the joint (Figure 7–17). The traditional soldering iron, with a copper tip, should be given a precoating of solder before being used to solder a joint. First remove any oxides with a file, then dip the copper tip in a flux. The solder can be melted right off the tip of the soldering iron. The solder can also be applied to the pieces of the joint first before **sweat soldering** together (Figure 7–18). A soldering gun can be used like a soldering iron with electricity furnishing the heat instead of a torch (Figure 7–19).

For soldering, remember the following:

- Fit up the joint.
- Clean the base metal.
- Select the solder and flux.
- Do skillful soldering.
- Post-clean the joint.

SAFETY REMINDER

The importance of safety can never be emphasized enough. Ventilation is important when brazing, braze welding, or soldering. **Zinc,** a commonly

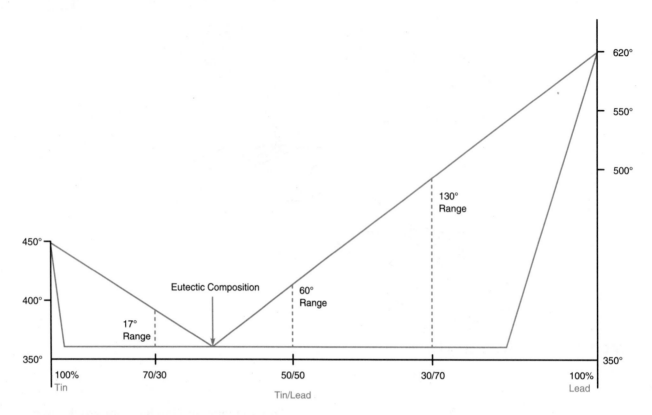

FIGURE 7–15 Melting range for tin/lead solders.

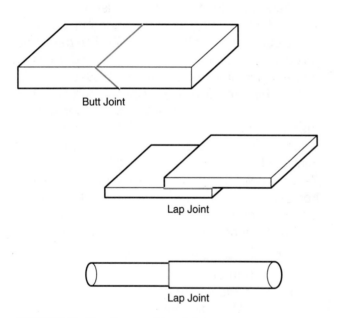

FIGURE 7–16 Butt and lap joints.

used metal that can become very toxic, is added to copper filler rods to add strength. Zinc is also found on carbon steels as a galvanized covering to prevent rust.

REVIEW

1. What things should be considered as a part of safety before doing any of the exercises in this unit?
2. Define brazing.
3. Describe capillary action.
4. Name three items that can lower the strength of brazing.
5. Explain the importance of spacing in a butt joint as a scarf-groove weld.
6. When might brazing be used?
7. What is braze welding?
8. List the advantages and disadvantages of braze welding.
9. How is braze welding different from brazing?
10. What is soldering, and how is soldering different from brazing?

Create Three Questions

1.
2.
3.

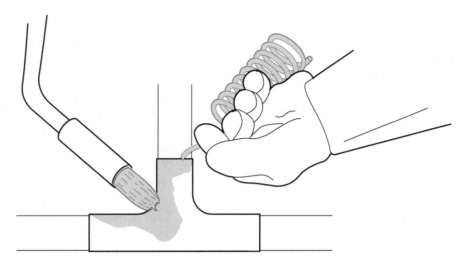

FIGURE 7–17 Propane torch sweat soldering two tubes.

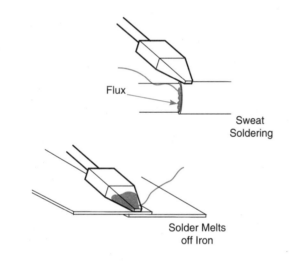

Flux

Sweat
Soldering

Solder Melts
off Iron

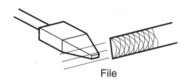

File

FIGURE 7–18 Using the soldering iron.

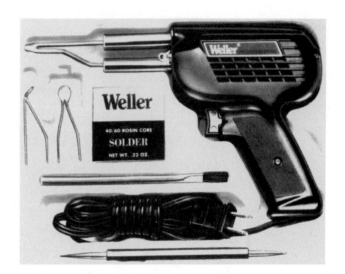

FIGURE 7–19 Soldering gun. *(Courtesy of Cooper Tools, Apex, North Carolina.)*

Related Math and English Questions

1. Which size braze welding rods require the most heat to melt into the joint?
   a. 0.0625″ or 1/8″
   b. 1/16″ or 0.094″
   c. 3/32″ or 0.125″
2. For capillary action to occur, the gap between pieces of the joint can be between 0.002″ and .016″. Which of the following fractions meets this gap requirement: 1/16″, 1/32″, or 1/64″?
3. Write a paragraph using the following terms: *brazing, filler rod, zinc, capillary action, torch, spacing, cleaning.*
4. In a paragraph, explain the reasons for using a flux.

For Further Thought

1. Why would braze welding not be recommended for repairing a crack in the firebox of a cast iron stove when it could be used

for repairing a broken leg on the same stove?

2. Is a tin-lead solder a good idea for use in joining copper tubing on a hot-water system? Explain.

3. Why is brazing used in the manufacture of some bicycle frames?

4. Suppose you had the job of repairing a cracked transmission housing. Would braze welding be a possibility? Explain.

5. There are several manufacturers of brazing and soldering materials, and each has its own brand names. What is the best way for trying to match filler metal and flux to the base metal?

SUGGESTED ACTIVITIES

1. Become familiar with the different pieces of equipment and supplies that are used in brazing, braze welding, and soldering.

2. Research the different kinds of cast iron. Find out which kinds are used for particular applications, and why.

3. Gather information about other brazing and soldering processes.

4. Learn about other types of solders besides tin/lead.

5. Tour a welding supply business.

UNIT 7: EXERCISE 1

Torch Brazing (TB): Lap Joint

This exercise is designed for use with copper-zinc filler rods. Keep the rod in the joint once the base metal is hot enough to melt the rod continuously. Move the flame over both pieces of metal, passing it over the end of the rod. If capillary action is to be complete, there should not be any gap along the edge of the top piece. Filler metal should be visible along the entire edge of the completed joint.

Size is important when using filler metal. For example, a 1/16″ diameter rod will melt more quickly, requiring less heat, than a 1/4″ diameter rod. The setup for this exercise is illustrated in Figure 7–20.

Necessary Material and Equipment

Safety glasses	Vice grips	Chisel	Copper-zinc filler rods,
Welding gloves	Sparklighter	Welding tip	1/16″
Protective clothing	Tip cleaners	2 pieces 16-gauge sheet	Flux (if necessary)
Goggles	Ball peen	steel, 2″ × 4″	Grinder
Earplugs			

Instructions

1. Clean the surfaces of the base metal by grinding one inch on each piece. Flux both surfaces.
2. Place the prepared surfaces together with enough overlap for the vice grips.
3. Clamp the pieces together at the middle with the vice grips. If enough heat is applied, molten filler metal will be drawn into the joint.

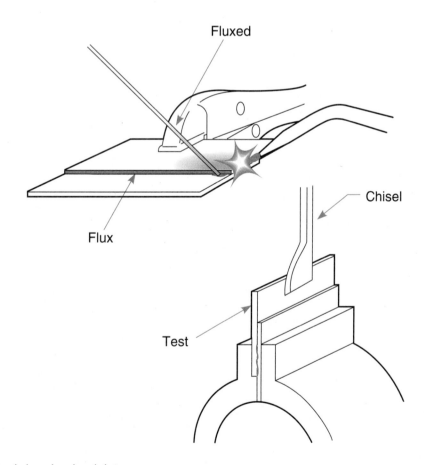

FIGURE 7–20   Torch brazing lap joint.

4. Adjust the torch to a slightly carburizing flame, and heat the base metal so that the filler rod will flow into the joint.
5. Complete the brazing process.
6. Examine for complete fusion along the joint. The edge of the top piece should not have any gaps where it meets the bottom piece.
7. Cool, then test in a vice by trying to separate the two pieces with a chisel. Note how far the filler was drawn into the joint. With success, the joint should not be easily separated.
8. Check with the instructor for evaluation, if necessary.
9. Practice this exercise until you meet the standards established in the course.

UNIT 7: EXERCISE 2

Torch Brazing (TB): Butt Joint on Steel Plate

This exercise is designed for the butt joint. In general, the butt joint is considered to be the joint configuration with the lowest strength. The amount of surface area for bonding is limited by the thickness of the metal. This exercise can be set up in any welding position (Figure 7–21).

Necessary Material and Equipment

Safety glasses	Earplugs	Ball peen	Copper-zinc filler rods,
Welding gloves	Vice grips	Welding tip	1/16″
Protective clothing	Sparklighter	2 pieces 1/4″ steel plate, 2″	Flux
Goggles	Tip cleaners	× 4″	Grinder

Instructions

1. Clean the edge of each piece by grinding.
2. Position the pieces as shown in Figure 7–21. Clamp them together with the vice grips so that the pieces will stand alone. Coat the edge of each piece of plate with a thin layer of flux.

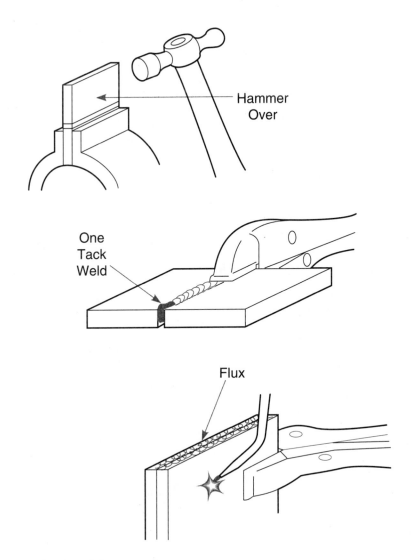

FIGURE 7–21   Torch brazing butt joint.

3. Without applying the flame directly to the flux, heat the flux to help it adhere.
4. Position the two pieces together without touching, using the vice grips to maintain the spacing.
5. Adjust the torch to a slightly carburizing flame, and heat the base metal to the point where the filler rod will flow into the joint.
6. Tack weld one side with small tack welds, and reposition the vice grips out of the joint. Complete the brazing process.
7. Cool, then place the **weldment** in a vice with the joint just above the jaws.
8. Test by hammering the joint over from the back. Examine to see whether the filler metal covered the entire edge.
9. Check with the instructor for evaluation, if necessary.
10. Practice this exercise until you meet the standards established in the course.

UNIT 7: EXERCISE 3

Torch Brazing (TB): Butt Joint on Steel Plate, Scarf-Groove Weld

This exercise is designed for the scarf-groove weld, which is used for brazed joints. Beveling the pieces to be joined increases the surface area of the joint for greater strength. To achieve effective strength for a given joint, increase the surface area by three times. Plate that is 1/4″ thick requires approximately 70° of bevel. To achieve the approximate bevel angle, measure the bevel in 5/8″ from the edge of the plate. Figure 7–22 shows the setup for this exercise.

Necessary Material and Equipment

Safety glasses	Earplugs	Ball peen	Copper-zinc filler rods,
Welding gloves	Vice grips	Welding tip	1/16″
Protective clothing	Sparklighter	2 pieces 1/4″ plate, 3″ ×	Flux
Goggles	Tip cleaners	3″	Grinder

Instructions

1. Grind a bevel into each piece of metal.
2. Cover each bevel with a thin coating of flux.
3. Without applying the flame directly to the flux, heat the flux to help it adhere.
4. Position the two pieces of metal together. Relying on the flux to maintain the spacing, clamp them together with vice grips.
5. Tack weld one side with small tack welds. Reposition the vice grips out of the joint. Complete the brazing process.
6. Cool, then place the weldment in a vice with the joint just above the jaws.
7. Test by hammering the weldment over from the back. Examine it to see whether the filler metal covered the entire scarf. If not, apply a greater volume of heat the next time.
8. Check with the instructor for evaluation, if necessary.
9. Practice this exercise until you meet the standards set for the course.

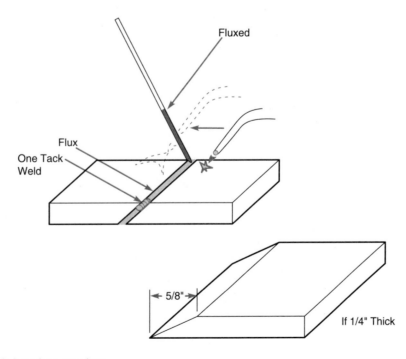

FIGURE 7–22   Torch brazing on plate.

UNIT 7: EXERCISE 4

Braze Welding (BW): Lap Joint

This exercise will help you develop skills that are very useful in repair welding. Braze welding can often shorten the welding time and be used to repair castings without causing concern about **cracks** from welding. The setup for this exercise is given in Figure 7–23.

Necessary Material and Equipment

Safety glasses	Earplugs	Ball peen	Copper-zinc filler rods,
Welding gloves	Vice grips	Welding tip	1/8″
Protective clothing	Sparklighter	2 pieces 3/8″ plate steel, 2″	Flux (if necessary)
Goggles	Tip cleaners	× 4″	Grinder

Instructions

1. Clean the surfaces to be joined by grinding.
2. Clamp the pieces together with a 1-inch overlap.
3. Adjust the torch to a slightly **oxidizing** flame, and heat the base metal to the point where the filler rod will flow into the joint.
4. Precoat the surfaces of the joint with a thin layer of filler metal.
5. Complete the braze welding process. The **legs** of the weld should equal the thickness of the base metal.
6. Examine the joint for appearance. The filler metal should extend beyond the edge of the top piece. That edge should not be showing anywhere along the entire length of the joint.
7. Cool, then place the weldment in a vice with the joint just above the jaws.
8. Test the weldment by hammering it over from the back. If only one or two solid blows from a ball peen is required, the joint fails.
9. Check with the instructor for evaluation, if necessary.
10. Practice this exercise until you meet the standards established for the course.

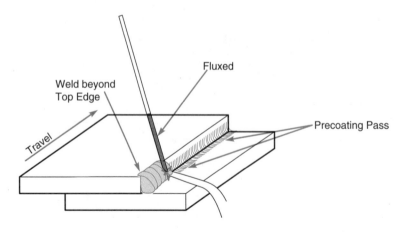

FIGURE 7–23   Braze welding lap joint.

UNIT 7: EXERCISE 5

Padding Plate: Surfacing Weld

This exercise provides practice in controlling heat. Weld 1/4 inch surface buildup without allowing the filler metal to overheat and flow off the plate. Figure 7–24 shows the setup.

Necessary Material and Equipment

Safety glasses	Earplugs	Welding tip	Copper-zinc filler rods,
Welding gloves	Vice grips	1 piece 1/4"–3/8" plate	1/8"
Protective clothing	Sparklighter	steel, 2" × 2"	Flux (if necessary)
Goggles	Tip cleaners		Grinder

Instructions

1. Clean the surface of the plate by grinding.
2. Adjust the torch to a slightly oxidizing flame, and heat the base metal to a point at which the filler metal flows.
3. Precoat the entire surface with a thin layer of filler metal.
4. Complete the braze welding process with a buildup of 1/4 inch over the entire surface of the plate.
5. Examine the plate for appearance. The 1/4-inch buildup should extend to all four edges.
6. Check with the instructor for evaluation, if necessary.
7. Practice this exercise until you meet the standards established in the course.

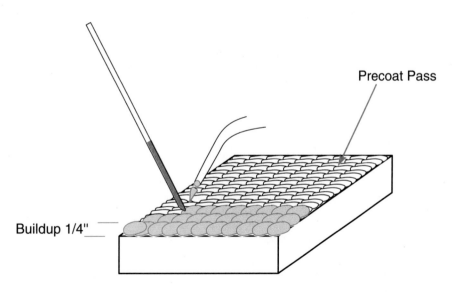

Precoat Pass

Buildup 1/4"

FIGURE 7–24   Padding plate.

UNIT 7: EXERCISE 6

Surfacing Weld Project

The goal of this exercise is to rebuild the surface of a broken or worn part. If no such part is available, a bolt head can be used. Grind or cut off part of the bolt head, and then rebuild the head. The setup for this exercise is illustrated in Figure 7–25.

Necessary Material and Equipment

Safety glasses	Earplugs	Welding tip	Flux (if necessary)
Welding gloves	Vice grips	Bolt, worn shaft, etc.	Grinder
Protective clothing	Sparklighter	Copper-zinc filler rods,	
Goggles	Tip cleaners	1/8″	

Instructions

1. Grind clean the part to be repaired.
2. Position the part, and precoat it.
3. Complete the braze welding, building it up beyond what is needed.
4. Grind or machine the part within tolerance.
5. Check with the instructor for evaluation, if necessary.
6. Place the part into service. This is the real test of whether the job has been completed successfully.

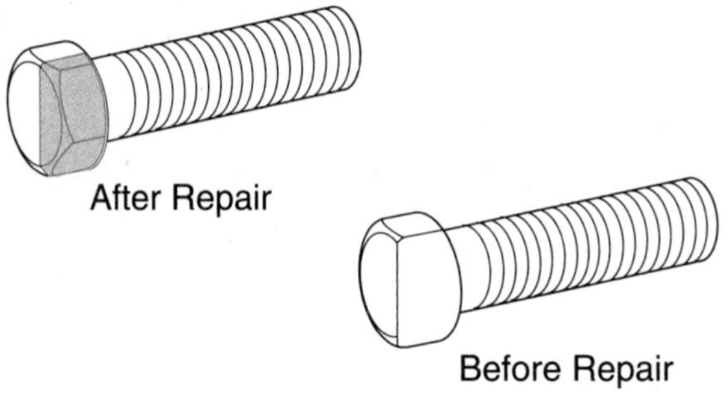

After Repair

Before Repair

FIGURE 7–25   Surfacing weld project.

UNIT 7: EXERCISE 7

Braze Welding Repair: Cast Iron

For cast iron, braze welding is easier than fusion welding. With braze welding, the entire object does not have to be welded at an elevated temperature, and braze welding does not require the time needed to slowly cool a cast iron part that has been welded.

Welding cold is an alternative to braze welding. Short **weld beads** are laid by electric welding. Each weld bead is cooled to the touch before adding the next bead. Welding cold avoids a heat imbalance within the weldment, which might result in stress within the cast iron which could lead to cracking.

If the piece to be repaired has been subjected to any petroleum products that have soaked in, heat the joint area to burn them off. Anything that might interfere with the flow of the copper-zinc filler rod should be removed. Figure 7–26 shows the setup for this figure.

Necessary Material and Equipment

Safety glasses	Goggles	Sparklighter	Copper-zinc filler rods
Welding gloves	Earplugs	Tip cleaners	Flux
Protective clothing	Vice grips	Cast iron project for repair	

Instructions

1. Grind clean the cast iron part to be repaired.
2. Position the part for repair. If required, clamp or tack weld the pieces together. Bevel all butt joints and any other joints in which the thickness exceeds 1/8 inch.
3. If alignment is critical, grind or cut a bevel through only part of the joint. Preheat the joint before adding tack welds. Tack weld solidly, then complete the bevel.
4. Precoat the joint, welding from each end toward the middle to avoid possible cracking.

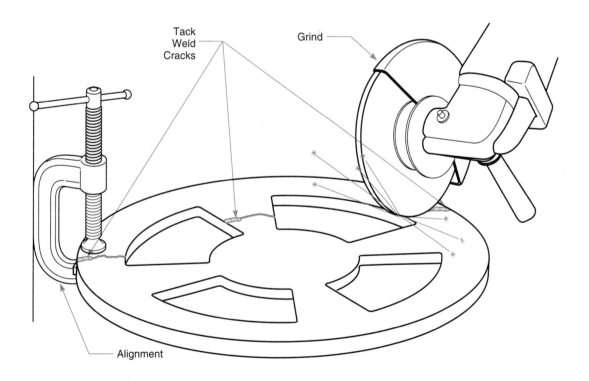

FIGURE 7–26   Braze welding repair.

5. Complete the braze welding process.
6. Check with the instructor for evaluation, if necessary.
7. Place in service for the ultimate test.

UNIT 7: EXERCISE 8

Torch Soldering (TS): Lap Joint

This exercise should be completed with an air-acetylene or air-propane torch. The heat of an oxyacety-lene welding torch (no matter what the tip size) may be too difficult to control without overheating the joint. A 50/50 tin-lead solder is recommended because the melting range of 60° F makes this solder easy to work with. Figure 7–27 shows the setup for this exercise.

The iron soldering process can be used instead if an air-acetylene or air-propane torch is not available. In that case, reread the section on iron soldering before beginning this project.

Necessary Material and Equipment

Safety glasses	Sparklighter	2 pieces 18- to 24-gauge	50/50 tin-lead solder
Protective clothing	Air-acetylene or air-	sheet steel, 2″ × 4″	Flux
Vice grips	propane torch		

Instructions

1. Clean the pieces of steel if required.
2. Flux the surfaces to be soldered.
3. Clamp the pieces together.
4. Tack weld both ends of the joint.
5. Begin the soldering process.
6. Test by opening up the joint from the back with a chisel. Examine it to see whether filler metal covers the length of the joint.
7. Check with the instructor for evaluation, if necessary.
8. Practice this exercise until you meet the standards established in the course.

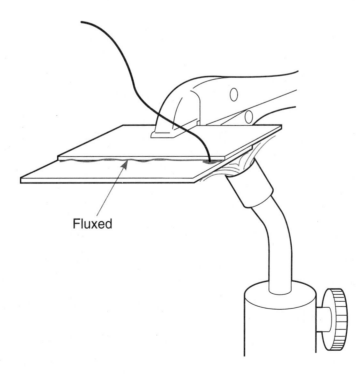

Fluxed

FIGURE 7–27   Torch soldering lap joint.

UNIT 7: EXERCISE 9

Torch Soldering (TS): Copper Tubing

The purpose of this exercise is to provide experience in a common soldering application. Another exercise can be used as a substitute if copper tubing is not available. The tubing should be laid out in a fixed horizontal position (the *5G position*). The setup is illustrated in Figure 7–28.

Necessary Material and Equipment

Safety glasses	Sparklighter	1 piece 1″ copper tubing, 2″	50/50 tin-lead solder
Protective clothing	Air-acetylene or air-		Flux
Vice grips	propane torch	1 piece 3/4″ copper tubing, 2″	Emery cloth

Instructions

1. Clean the joint with **emery cloth** where the two surfaces will touch.
2. Flux the joint.
3. Preheat the joint.
4. Begin soldering, concentrating the flame of the torch away from the solder and directing the heat where the solder is suppose to flow.
5. Test by opening up the joint. Examine it for a complete fill.
6. Check with the instructor for evaluation, if necessary.
7. Practice this exercise until you meet the standards established in the course.

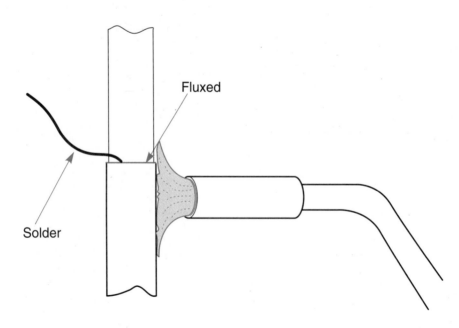

FIGURE 7–28 Torch soldering copper tubing.

Joint Design

"Steel has no more shape than this pile of dirt on the engine here. These shapes are all out of someone's mind."
—Robert M. Pirsig, *Zen and the Art of Motorcycle Maintenance*

GOAL

- Develop an awareness of the factors that figure into **joint design,** such as types of **weld,** structural shapes, and **distortion.**

QUESTIONS

- What is a groove weld, and what are its parts?
- What is a fillet weld, and what are its parts?
- What are some of the causes of weld failure?
- How are the welding positions defined?

GROWTH MEANS WELDING

Welding is everywhere. Welding is one of the key components for growth in the industrial sector. Many buildings—from skyscrapers to schools and from large factories to small businesses—are designed and constructed of **steel,** and this involves welding. Steel construction has moved light years from its humble beginnings in the 1870s, when preindustrial ironmaking was still in its glory. At that time and into the twentieth century, riveting was the common method for fastening both steel and iron together.

Imagine what construction was like earlier in this century. On bridge projects white hot rivets were tossed high above the ground and caught by workers standing on narrow beams. They positioned the rivets into holes and quickly hammered them tight before the metal cooled down. Fastening metal together by welding has advanced technology. The more massive structures that riveting-designed buildings and bridges re-quired were changed by the development of welding. The lighter-welded structures depended on joint designs that made use of new ways for fastening metal structures together.

WELDS FOR JOINING METAL

The **American Welding Society** lists 19 different welds that are used today for joining metal together. These different welds are used to join metal in the five basic joints—**butt joint, corner joint, tee joint, lap joint,** and **edge joint**—and in combinations of these different **joints.**

Groove Welds

For use in the butt joint, there are eight **groove welds.** The different parts of the groove weld are labeled in Figure 8–1. The **weld toe** runs the length of the weld, establishing the boundary of the weld face to the base metal. Each weld has two weld toes. The **weld face** is the surface of the weld that extends from the weld toe on one side to the weld toe on the other side. The **weld root** is where the weld has penetrated through the base metal into the **joint root** of both pieces in the joint. A **back weld** is made after welding on the opposite side of the joint, and a **backing weld** is made before welding on the opposite side of the joint.

Figure 8–2 provides a quick overview of the terminology encountered in preparing a butt joint for a **bevel-groove weld** or a **V-groove weld.** The bevel angle is made to aid penetration into the joint. The **groove angle** is formed when two bevels form a butt joint. The **groove face** is the surface after **beveling** has been completed. The **root opening** is the space between the pieces that form

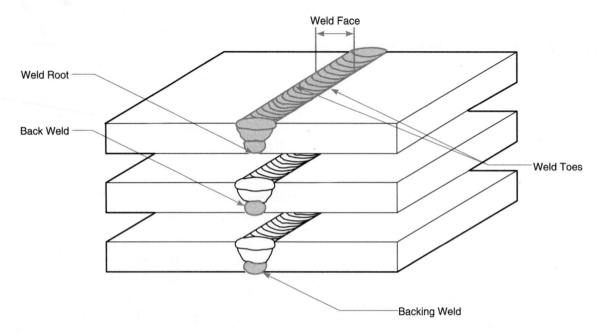

FIGURE 8–1 Groove weld with parts labeled.

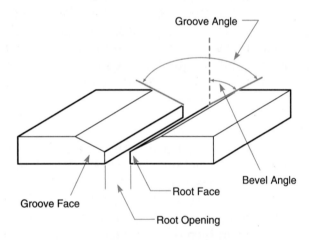

FIGURE 8–2 Butt joint V-groove weld with parts labeled.

the joint. The **root face** is the edge, not removed by beveling, that establishes the space of the root opening. Although not always used, a root face is sometimes required when the dimensions of the weldment are critical or when there is a possibility of **melt-thru** on the root **pass.** A grinder can be used on a bevel to form a root face.

The specifications for the first five groove welds that follow are from the **AWS *Structural Welding Code*** (ANSI/American Welding Society

D1.1). These specifications are used when developing tests to evaluate the skills of **welders.** The remaining groove welds are not listed in the AWS *Structural Welding Code* and are not described in as much detail here.

Square-Groove Weld

The **square-groove weld** is a use of the butt joint without edge preparation. The square-groove weld is effective for joining **sheet metal** and **plate** with limited edge preparation (Figure 8–3). For shielded metal arc welding, the maximum recommended **base metal** thickness is 1/4 inch for-complete **root penetration** when making a square-groove weld from both sides. Many factors go into welding plate with 1/4-inch thickness;

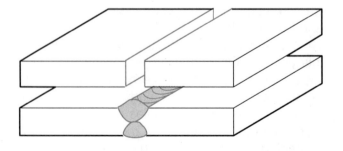

FIGURE 8–3 Square-groove weld.

among them are the **electrode,** welding position, **amperage** setting, and root opening. For **gas metal arc welding** or **flux cored arc welding,** the maximum recommended base metal thickness is 3/8 inch for square-groove welds.

Bevel-Groove Weld

The bevel-groove weld is effective when preparing only one piece of the joint (Figure 8–4). For shielded metal arc welding, the maximum recommended base metal thickness is unlimited for **complete root penetration** when welding from both sides. Preparation means making a 45° bevel angle with a root opening from 0 to 1/8 inch and a root face from 0 to 1/8 inch. For gas metal arc welding or flux cored arc welding with the same preparation, the maximum recommended base metal thickness is also unlimited. For shielded metal arc welding, when welding from one side with **backing material,** a 45° bevel angle with a root opening of 1/4 inch or a 30° bevel angle with a root opening of 3/8 inch is recommended.

For gas metal arc welding or flux cored arc welding when making bevel-groove welds from only one side with backing material, preparation means a root opening of 3/16 inch with a 30° bevel angle, a root opening of 1/4 inch with a 45° bevel angle, or a root opening of 3/8 inch with a 30° bevel angle.

V-Groove Weld

The V-groove weld is an effective weld when both pieces of the joint must be prepared (Figure 8–5). For shielded metal arc welding, the maximum recommended base metal thickness is unlimited for complete joint penetration. When welding from one side with backing material, the degree of the groove (two bevels) determines the size of the root opening: a 45° groove angle with a root

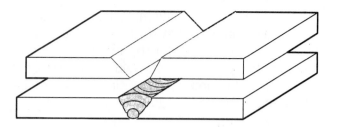

FIGURE 8–5   V-groove weld.

opening of 1/4 inch, a 30° groove angle with a root opening of 3/8 inch, or a 20° groove angle with a root opening of 1/2 inch. For shielded metal arc welding when welding from both top and bottom, the recommended preparation is a 60° groove angle with a root opening of 0 to 1/8 inch and a root face of 0 to 1/8 inch.

For gas metal arc welding or flux cored arc welding, the maximum recommended base metal thickness is unlimited for complete joint penetration. When making a V-groove weld from one side with backing material, the degree of the groove angle determines the size of the root opening: a 30° groove angle with a root opening of 3/16 inch, a 30° groove angle with a root opening of 3/8 inch, or a 45° groove angle with root opening of 1/4 inch. For gas metal arc welding or flux cored arc welding when welding from both top and bottom, use a 60° groove angle with a root opening of 0 to 1/8 inch and a root face of 0 to 1/8 inch.

J-Groove Weld

The **J-groove weld** is effective when one piece of the joint has been prepared by **machining** (Figure 8–6). For shielded metal arc welding, the maximum recommended base metal thickness is unlimited for complete joint penetration when weld-

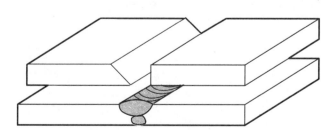

FIGURE 8–4   Bevel-groove weld.

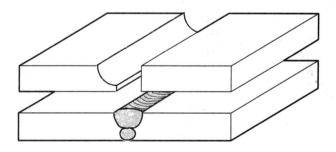

FIGURE 8–6   J-groove weld.

ing from both top and bottom. Preparation means a 45° bevel angle, a bevel radius of 3/8 inch, a root opening of 0 to 1/8 inch, and a root face of 1/8 inch. For gas metal arc welding or flux cored arc welding, the preparation is similar, except the bevel angle should be 30°.

U-Groove Weld

The U-groove weld is an effective weld when both pieces of the joint have been prepared by machining (Figure 8–7). For shielded metal arc welding, the maximum recommended base metal thickness is unlimited for complete joint penetration when welding from both top and bottom. Preparation means a 20° U-groove angle (for the **flat position** and the **overhead position**) and a 45° U-groove angle (for all positions), a bevel radius of 1/4 inch, a root opening of 0 to 1/8 inch, and a root face of 1/8 inch. For gas metal arc welding and flux cored arc welding, the preparation is the same, except the U-groove angle should be 20°.

Flare-Bevel-Groove Weld

A **flare-bevel-groove weld** is a welded joint in which one of the two members has a curved surface. Two pieces of sheet steel welded together with one of the pieces formed in a 90° angle is an example of the flare-bevel-groove weld (Figure 8–8).

Flare-V-Groove Weld

The **flare-V-groove weld** is a welded joint in which both members are curved surfaces. Two pieces of **pipe** welded together form an example of the flare-V-groove weld (Figure 8–9).

Scarf-Groove Weld

The **scarf-groove weld** was discussed in Unit 7. It is a special case of groove weld used in brazing (Figure 8–10).

Fillet Weld

The **fillet weld** is the most commonly used weld of all. It does not require the preparation necessary for the groove welds. A fillet weld joins the surfaces of two pieces of metal, usually at right angles to one other. In Figure 8–11, the eight parts of the fillet weld are labeled.

The **fillet weld legs** determine the size of the fillet weld. Each fillet weld has two legs. The weld toe runs the length of the weld, establishing the boundary of the weld face to the base metal. The weld face is the surface of the weld that extends from the weld toe on one side to the weld toe on

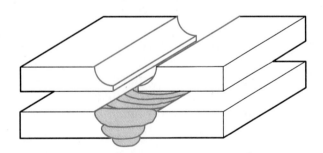

FIGURE 8–7   U-groove weld.

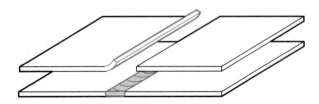

FIGURE 8–8   Flare-bevel-groove weld.

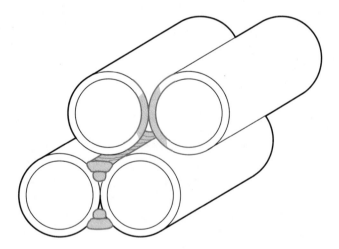

FIGURE 8–9   Flare-V-groove weld.

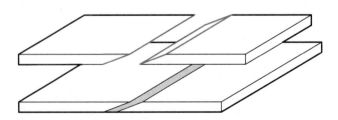

FIGURE 8–10   Scarf-groove weld.

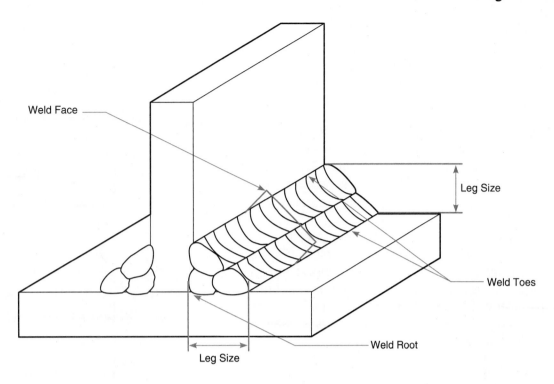

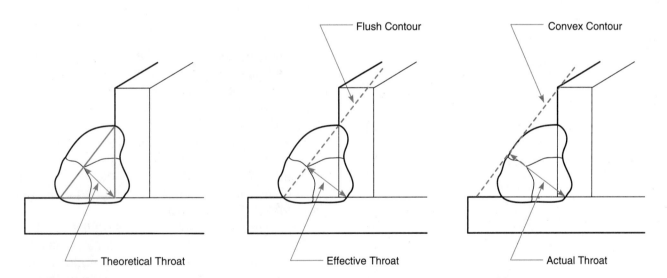

FIGURE 8–11 Fillet weld with parts labeled.

the other side. The contour is the shape of the weld face. The weld root is where the weld has penetrated through the base metal into the joint root of both pieces in the joint. The weld root is the farthest point from the weld face.

The **actual throat** is the shortest distance from the weld root to the weld face. The **effective throat** is the shortest distance from the weld root to the

weld face at the point of a **flush contour. A convex contour** or **concave contour** is not used in this measurement. The **theoretical throat** is the distance from the joint root (where the surfaces of the base metal are closest) to the hypotenuse of the largest triangle that would fit inside the fillet weld.

An understanding of the different terms that

describe sizing a weld is important. The ability to do more than just weld for the sake of welding separates a skilled welder from those who can just strike an arc. Putting too much weld into a joint can cause distortion, among other problems. Putting too little weld into a joint can cause the **weldment** to fail. It is important to do only as much welding as necessary to complete the job. Cost is the bottom line, and time is money. These are reasons why the **size** of the weld is so important.

Fillet Weld Gauge

A **fillet weld gauge** is a useful tool for the accurate measurement of fillet welds (Figure 8–12). Made of **cold-rolled stainless steel,** a gauge can be used to measure **convex** and **concave** fillet welds (Figure 8–13). A gauge is helpful in learning to tell the difference between, say, a 3/16-inch weld and a 1/4-inch weld.

Strength of Fillet Welds

For a fillet weld with legs of equal size (45° weld face), the **strength** of the weld is based on the leg size multiplied by the effective throat. In

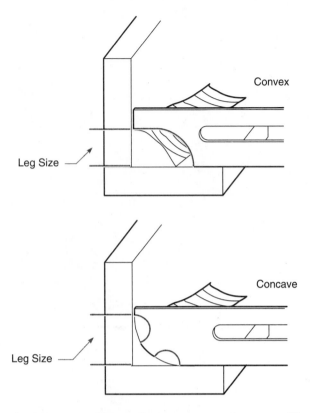

FIGURE 8–13 Measuring concave and convex fillet welds.

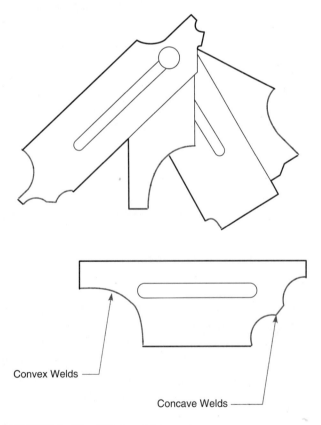

FIGURE 8–12 Fillet weld gauge.

determining the strength of a fillet weld with legs of equal size, 0.707 is the **coefficient** (constant number) that is always used for the effective throat. The theoretical throat must equal the effective throat (Figure 8–14).

When the legs are of unequal size, the strength of the weld is determined by the smaller of the two legs. Mathematical calculations are needed to determine the effective throat for unequal-legged fillet welds (Figure 8–15).

For greater strength, increasing the length of a fillet weld can be less costly than increasing its size. A 1-inch fillet weld is twice as strong as a 1/2-inch fillet weld, but requires four times the filler metal (Figure 8–16). On the other hand, doubling the size of one leg (1/2-inch leg by 1-inch leg) will double the filler metal required but add only one-tenth to the effective throat (Figure 8–17).

Plug Weld

The **plug weld** is considered to be a remnant of the riveting age. It is still used in applications in which welding along the edge of a joint is not

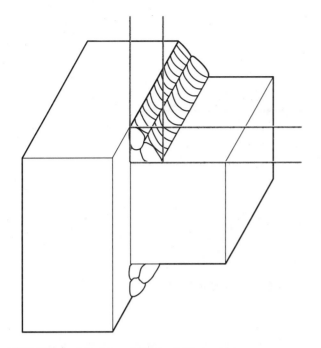

FIGURE 8–14   Equal-legged fillet weld.

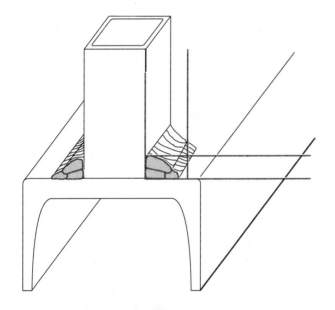

FIGURE 8–15   Unequal-legged fillet weld.

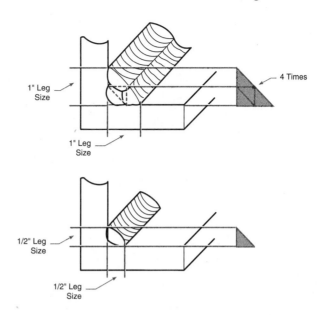

FIGURE 8–16   One-inch fillet weld versus half-inch fillet weld.

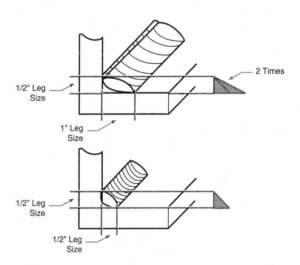

FIGURE 8–17   Unequal-legged half-inch by one-inch fillet weld versus equal-legged half-inch fillet weld.

enough or is not practical. A plug weld is a weld in a circular hole of one piece joined with a second piece in a lap joint (Figure 8–18). The size of a plug weld is given by the diameter of the hole. The hole is sometimes beveled for greater strength and is sometimes only partially filled in with weld.

Slot Weld

The **slot weld** is another remnant of the riveting age. It is really an elongated plug weld (Figure 8–19). The slot weld is also beveled for greater holding strength.

Spot Weld

The **spot weld** is popular for welding sheet steel. This weld is made on or between two overlapping pieces of metal (Figure 8–20). Spot welding between the two pieces of a lap joint is called **resistance spot welding.** Under pressure, the heat of an **arc** between the two pieces **fuses** them together.

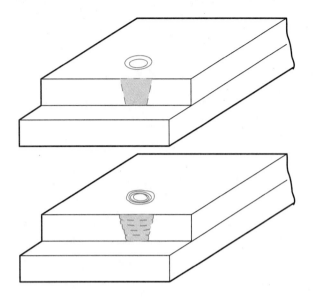

FIGURE 8–18 Plug weld.

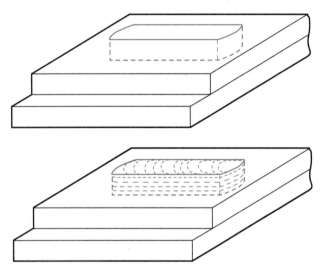

FIGURE 8–19 Slot weld.

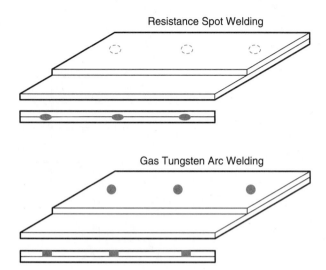

FIGURE 8–20 Spot weld.

Spot welding on the two pieces of a lap joint can be done with several arc welding **processes,** including gas metal arc welding. Some **power sources** for this type of welding have special **nozzles** and timers that control the time the arc is on to **feed** wire when the trigger of the **gun** is pressed.

Projection Weld

The **projection weld** is used in the **projection welding** process on lap joints (Figure 8–21). Projection welding is one form of resistance welding. Projection welds are predetermined raised surfaces that melt, resisting the flow of electrical **out-**

put. The metal fuses together at the points of the projections.

Seam Weld

The **seam weld** is a series of spot welds or a continuous weld between two overlapping pieces of metal (Figure 8–22). There are several **resistance seam welding** processes. By definition, the **arc seam weld** can be performed by any of the arc welding processes described in this book. The resistance seam welding process can involve the melting of nuggets (weld metal) or the melting of the base metal itself by resistance to electric output in an overlapping joint.

Backing Weld and Back Weld

Whether a weld is a backing weld or a back weld is entirely a matter of sequence (Figure 8–23). A backing weld is a weld that is made before a groove weld. A back weld is a weld that is made after a groove weld. A backing weld is made on the outside of the joint before making the groove weld. The inside of the backing weld might have to be cleaned before the groove weld is made. A back weld is made on the outside of a joint after the completion of the groove weld. The joint root bead of the groove weld might have to be cleaned by gouging, grinding, and so on, before the back weld is made.

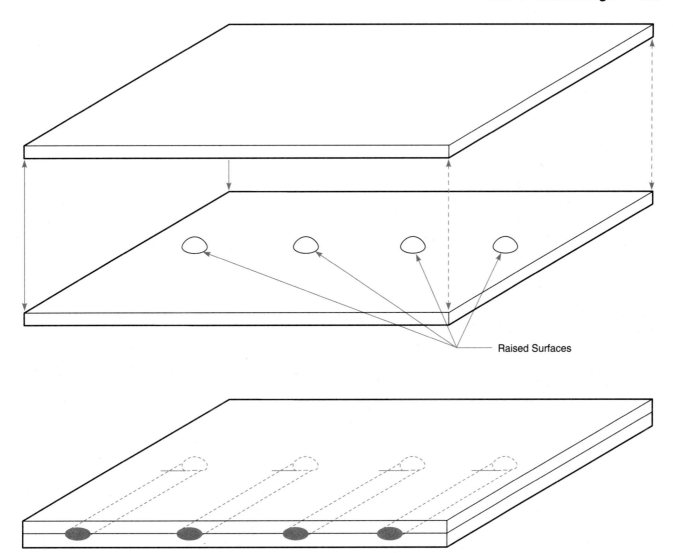

Raised Surfaces

FIGURE 8–21   Projection weld.

Surfacing Weld

A surfacing weld is applied to the surface of the base metal instead of becoming part of any joint (Figure 8–24). Surfacing welds are commonly used to build up surfaces subject to wear. **Filler metals** that are resistant to wear are available for welding.

Edge-Flange Weld

Usually made in sheet metal, a **flange** is a bend that forms a curve. An **edge-flange weld** is made with two flanged pieces forming the joint (Figure 8–25).

Corner-Flange Weld

A **corner-flange weld** is made when only one flanged piece is used to form the joint (Figure 8–26).

STRUCTURAL SHAPES

There are many structural shapes to choose from in the fabrication of any project. Steel construction sometimes involves making a choice between **hot-rolled steels** or the more expensive cold-rolled steels. Hot-rolled steels are deformed into shape at temperatures that allow coarse **grains** to **crystallize** within the metal. Cold-rolled

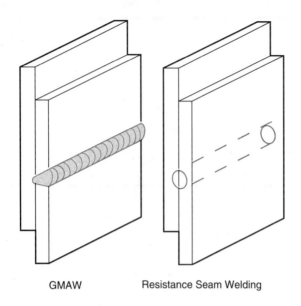

GMAW   Resistance Seam Welding

FIGURE 8–22   Two types of seam welds.

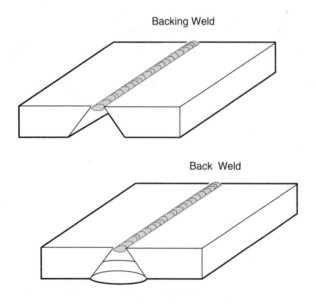

Backing Weld

Back Weld

FIGURE 8–23   Backing and back welds.

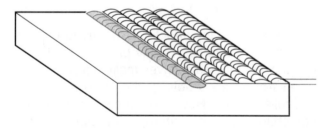

FIGURE 8–24   Surfacing weld.

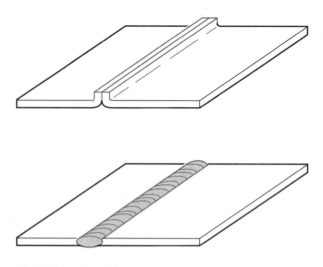

FIGURE 8–25   Edge-flange weld.

steels are deformed into shape below crystallizing temperatures. As a result, cold-rolled are finer grained steels with more accurate dimensions, greater strength, and lower **ductility.**

Sheet (up to 3/16 inch thick) and plate (more than 3/16 inch thick) are available in varying widths and lengths. One popular size is 48 inches by 96 inches. **Strip** (up to 3/16 inch thick and 12 inches wide) and **flat bar** (more than 3/16 inch thick and 6 inches wide) are available in 20-foot lengths (Figure 8–27).

Round bar, hexagon bar, and square bar are three more popular structural shapes. They come up to 20 inches thick in random 10- to 12-foot lengths. Pipe and **tubing** have wide applications. Pipe is measured from the inside diameter if it is 12 inches or less in diameter. Tubing is measured from the outside diameter (Figure 8–28).

Angle, channel, S-beam, square, and rectangular tubing are five more popular structural shapes (Figure 8–29). Before welding begins, the welder may be required to order necessary materials. A **bill of materials** lists the materials for a weldment in an acceptable format that can be easily understood. Note how dimensions are listed for the different structural shapes.

JOINT DESIGN

Selecting a joint design for welding depends on several factors. The first objective is to choose

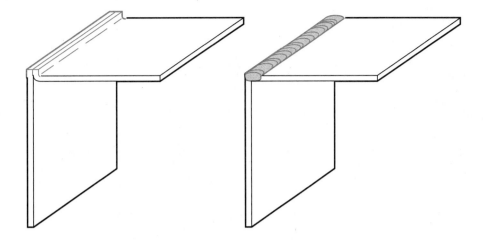

FIGURE 8–26   Corner-flange weld.

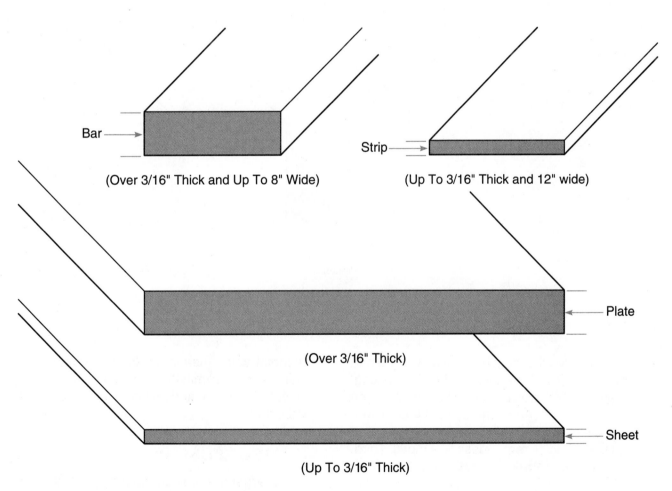

FIGURE 8–27   Structural shapes: bar, strip, plate, and sheet.

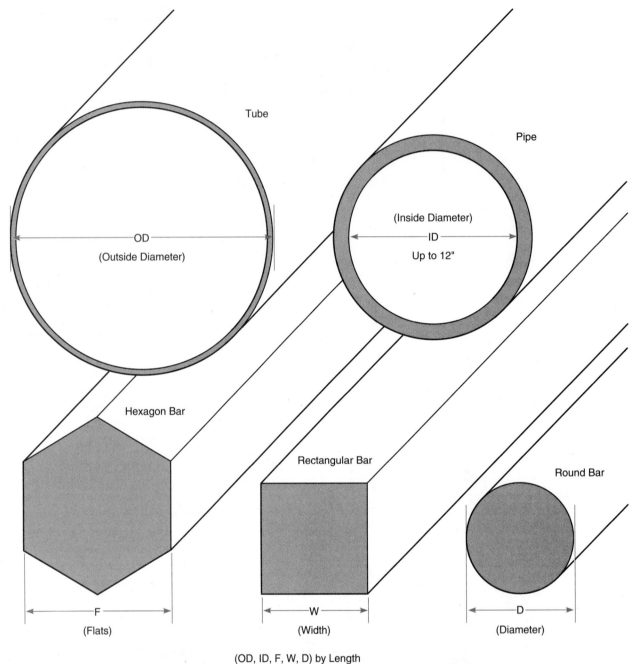

FIGURE 8–28 Structural shapes: tube, pipe, and bar.

joints that involve the least amount of welding and the least amount of preparation before welding. For example, fillet welds require less preparation than groove welds. A second point is that thinner metals, the use of smaller root openings, and smaller bevel angles require less filler metal. Third, for thick plate, **double-groove welds** are preferred over single-groove welds, when necessary (Figure 8–30). Double groove welds require less

filler metal, which means less time and less money. Other ways of eliminating unnecessary welding can cut expenses and make a company more competitive.

WELDING POSITIONS

The larger the project, the more important the design becomes. The design must include the

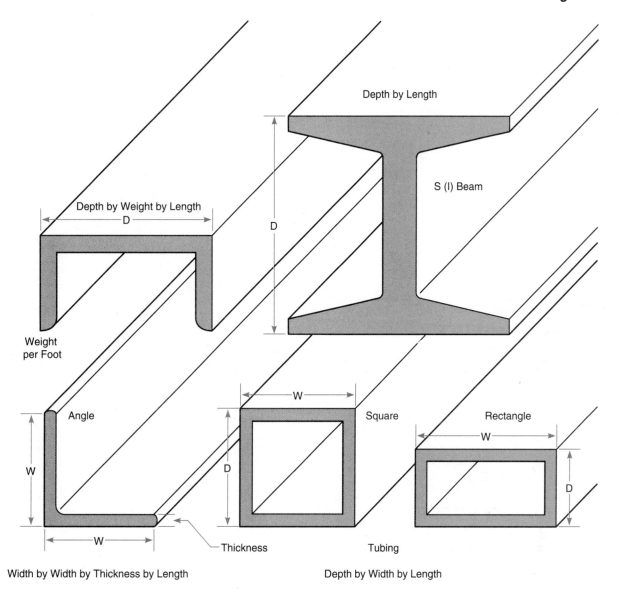

FIGURE 8–29 Structural shapes: angle, beam, and tubing.

placement of the welds. If it is possible, having all welding done in the flat position cuts down on welding costs. Welding in the flat position produces **quality welds** that can be done faster with a greater deposition of filler metal because high amperages can be maintained without the problems caused by gravity. Furthermore, the skill of the welders does not have to be as high so the cost of training should be less.

Although **positioners** are helpful for turning objects over to aid in welding, not all welding can be done in the flat position, especially in the field far from the manufacturing plant. Welders with the skills necessary to weld in all positions are always in demand.

The four welding positions are flat, horizontal, vertical, and overhead. The American Welding Society has laid out the boundaries for these basic positions. A position is considered flat if the plane of welding is from 0° to 15° (Figure 8–31). Most welding is completed in the flat position, and this position is the easiest to learn. Welding in the flat position can lay the most filler metal over a given period of time. Why would that be? (Think about gravity for a moment before answering.)

A position is considered horizontal if the plane

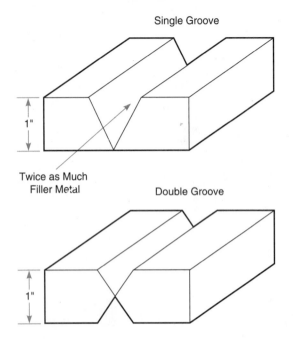

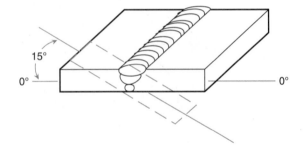

FIGURE 8–30 Double-groove weld versus single-groove weld.

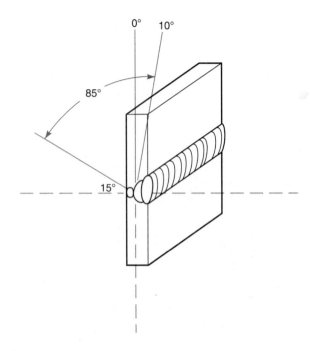

FIGURE 8–32 Plane for the horizontal position.

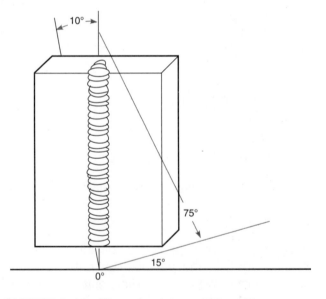

FIGURE 8–31 Plane for the flat position.

FIGURE 8–33 Plane for the vertical position.

of welding is from 15° to 100° (Figure 8–32). Welding in the **horizontal position** should be practiced from both directions.

A position is considered vertical if the plane of welding is from 15° to 100° with 10° of pivot (Figure 8–33). Welding in the **vertical position** can be done by the **uphill** or the **downhill** method. Downhill welding should be limited to noncritical welding—for example, welding on sheet metal when the concentration of heat can cause excessive **melt-thru.**

Finally, a position is considered overhead if the plane of welding is from 0° to 80° (Figure 8–34). The overhead position is more time-consuming than the other three positions. The effects of gravity limit the size of the **weld pool,** the amount of

heat, and consequently the amount of filler metal that can be deposited in a single pass.

Most of the time invested in learning how to weld focuses on the flat welding position. Beginners should not become discouraged about the time required to learn welding. Each subsequent welding position should take less time.

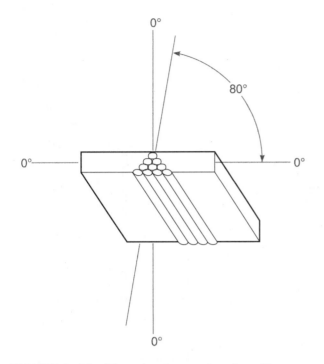

FIGURE 8–34   Plane for the overhead position.

DISTORTION

If you have spent time around metal barrels on a hot sunny day, you might remember those popping sounds. If so, you have heard the effects of the **coefficient of linear expansion.** Metals expand when heated, and the coefficient of linear expansion is used to measure this expansion of metal when heated.

The coefficient of linear expansion is a number given to materials to measure their expansion per inch per degree of rise in temperature. This principle is applied every time the flame of a torch expands a tight nut on a rusted bolt. Using the flame of a torch is controlled expansion by heat. For example, in the morning a length of aluminum is measured at 16 feet. If the temperature is raised 53° by noon, what is the new length? 16 multiplied by 12 (inches) multiplied by 53 multiplied by .0000123 equals 0.1251648 or 1/8 inch. New length at noon is 16 feet 1/8 inch. Table 8–1 gives the expansion numbers for some common metals used in welding.

Welders quickly learn what welding can do to metal. Distortion is a change in the shape of metal that can result from welding. The heat generated by welding is largely responsible for distortion. When **stresses** created during heating begin to

TABLE 8–1:   EXPANSION COEFFICIENTS FOR COMMON METALS

Metal	Coefficient of Expansion per Inch
Aluminum	.0000123
Copper	.0000088
Steel	.0000063
Cast iron	.0000056

act on a metal, distortion is the result. When heated, metal expands in all directions. When cooled, metal contracts (shrinks) in all directions. If metal is heated evenly and allowed to cool evenly, it will return to its original shape. In welding, unfortunately, metal never returns to its original shape because heating is uneven. With the temperatures used in welding, internal restraints within the metal itself break down, and the metal begins to twist.

Controlling Distortion

Welders must anticipate how welding will affect a given weldment and must be prepared to handle any distortion. Upon cooling, the molten weld pool will contract, pulling on the base metal. Keep the following points in mind when welding several pieces together on any fabrication project:

1. **Tack weld** all of the pieces together (Figure 8–35).
2. Try to spread out the welding to limit the concentration of heat at any one joint. If the joint is allowed to cool, the effects of heat can be minimized.
3. Do less welding as quickly as possible to reduce distortion. Keep the filler metal to a mimimum (Figure 8–36).
4. Become familiar with the **neutral axis.** This is an imaginary line in metal where there is no distortion. The closer welding takes place to the neutral axis, the less distortion there will be (Figure 8–37).

The use of **fixtures** is sometimes a solution for handling distortion, though fixtures can also build stresses within the weldment. Fixtures are used to help restrain an assembly from movement. C-clamps and **strong backs** are used in preassembly (Figure 8–38). **Intermittent welding** and

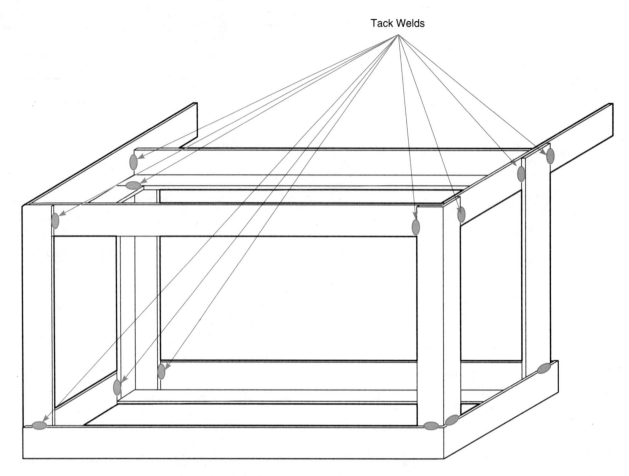

Tack Welds

Angle Iron Motor Mount

FIGURE 8–35   Assembly that has been tack welded together.

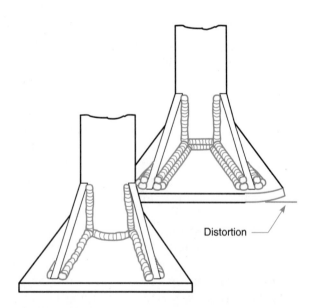

Distortion

FIGURE 8–36   Comparison of two welded joints.

prepositioning the joint are two more methods used to control distortion (Figure 8–39).

Peening the weld can sometimes help as a final measure to relieve stress, expanding the weld to counteract any shrinkage (Figure 8–40). If all of these measures fail to prevent distortion, jacks, cleats, wedges, clamps, and the heat of a torch may have to be used in the time-consuming activity of realignment (Figure 8–41).

WELD FAILURE

All of the effort put into welding will be lost if the part fails. Not all failures are the result of welding, however. For example, faulty design could put more stress on a part than any weld is capable of handling. There are numerous reasons why welds fail, and several might be involved at the same time. Learning the fundamentals of welding and

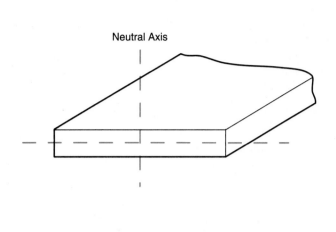

Neutral Axis

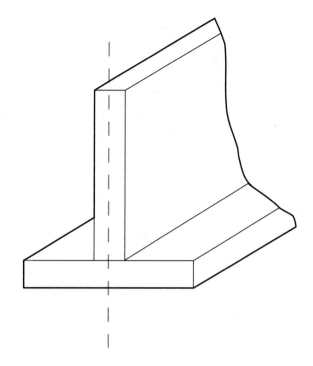

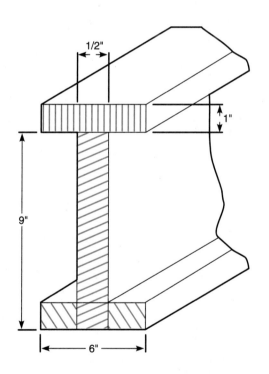

1/2"

1"

9"

6"

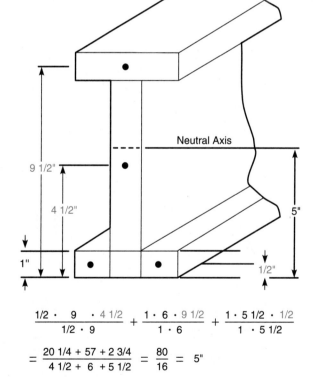

Neutral Axis

9 1/2"

4 1/2"

1"

5"

1/2"

$$\frac{1/2 \cdot 9 \cdot 4\,1/2}{1/2 \cdot 9} + \frac{1 \cdot 6 \cdot 9\,1/2}{1 \cdot 6} + \frac{1 \cdot 5\,1/2 \cdot 1/2}{1 \cdot 5\,1/2}$$

$$= \frac{20\,1/4 + 57 + 2\,3/4}{4\,1/2 + 6 + 5\,1/2} = \frac{80}{16} = 5"$$

FIGURE 8–37   Examples of the neutral axis.

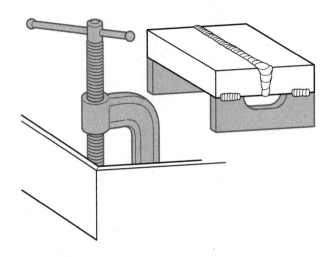

FIGURE 8–38   C-clamp and strong backs.

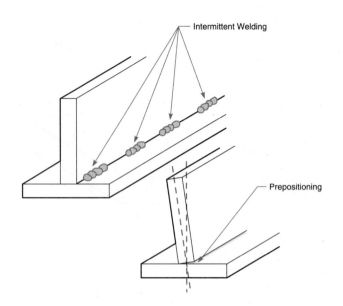

FIGURE 8–39   Intermittent welding and prepositioning the joint to control distortion.

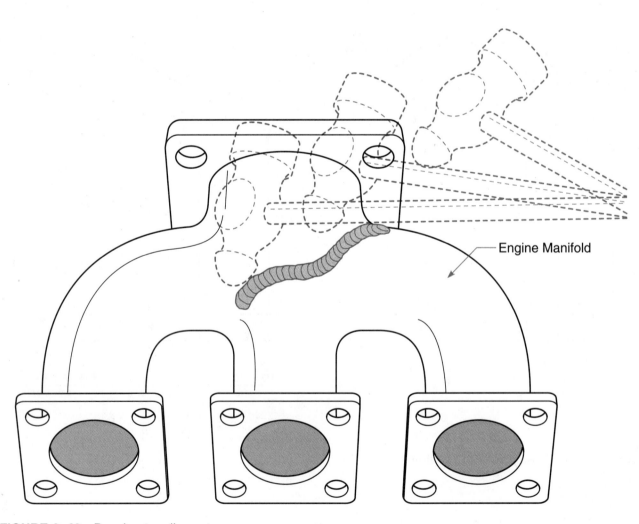

FIGURE 8–40   Peening to relieve stress.

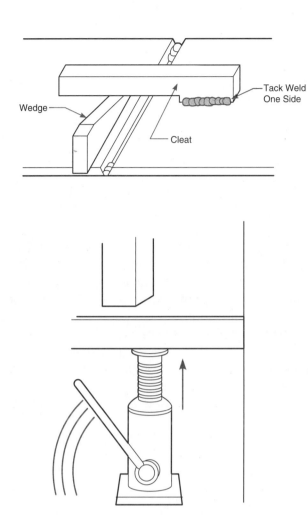

FIGURE 8–41 Jack, cleat, and wedge used in realignment.

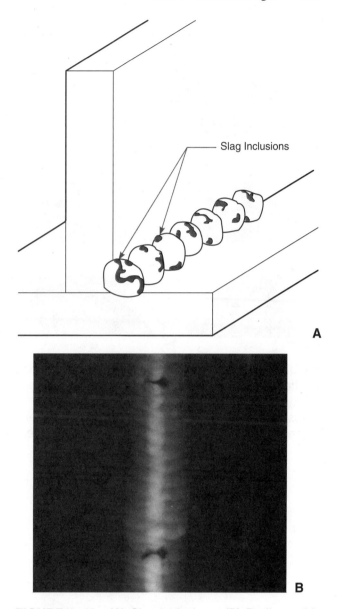

FIGURE 8–42 (A) Slag inclusions. (B) Radiographic image of slag inclusions—usually nonmetallic impurities that were not removed between weld beads. *(Courtesy of E.I. DuPont de Nemours and Co., Inc.)*

practicing these techniques will help to prevent weld failure. In some situations, though, faulty welding is the cause. There are several types of weld failure. A few of the **discontinuities** that can lead to failure are described in this section.

Slag inclusions occur when nonmetallic materials become trapped in the weld metal (Figure 8–42). Slag inclusions can result from a failure to remove the slag before making another pass or from improper welding techniques, including the wrong choice of electrode or incorrect **electrode angle.**

Porosity is the pinholes that appear in the weld metal as the result of gas pockets (Figure 8–43). Porosity can occur when there is improper cleaning of the weld between passes. Porosity can occur when the weld pool is not molten long enough for the gases to escape. Porosity can also occur when the amount of gas shielding is too low or too high.

Undercut is the groove that appears in the base metal along the toe of the weld (Figure 8–44). An undercut can result from too high an amperage setting, a failure to pause on the edges, or an incorrect electrode, gun, or torch angle.

Excessive **spatter** is the outcome when too many particles of molten metal are thrown out of the weld pool (Figure 8–45). Spatter can be caused by **arc blow,** too high an amperage setting, or too long an **arc length.**

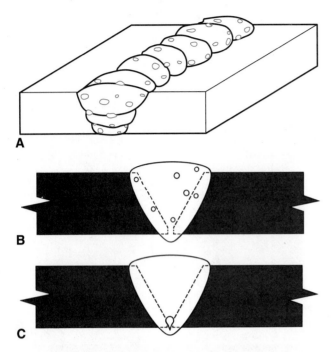

FIGURE 8–43   (A) Porosity. (B) Welding defect: Scattered porosity. Rounded voids random in size and location. (C) Welding defect: Root pass aligned porosity. Rounded and elongated voids in the bottom of the weld aligned along the weld centerline. *(Courtesy of E.I. DuPont de Nemours and Co., Inc.)*

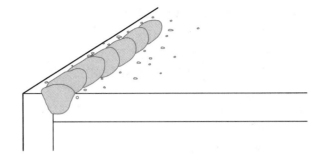

FIGURE 8–45   Excessive spatter.

Incomplete fusion is a void between the filler metal and the base metal (Figure 8–46). Incomplete fusion can be the result of welding too fast, too low a welding amperage, poor joint fit up, or improper joint design.

Crater cracks are depressions left at the point where a weld is stopped (Figure 8–47). Crater cracks occur when the arc is broken without filling in the **crater** (depression).

Transverse cracking is a **defect** that occurs

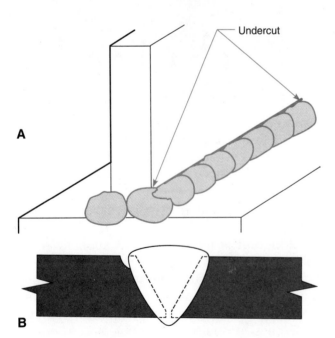

FIGURE 8–44   (A) Undercut. (B) Welding defect: External undercut. A gouging out of the piece to be welded, alongside the edge of the top or "external" surface of the weld. *(Courtesy of E.I. DuPont de Nemours and Co., Inc.)*

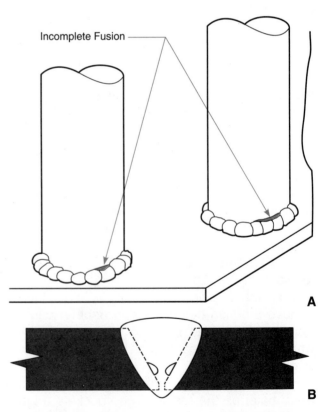

FIGURE 8–46   (A) Incomplete fusion. (B) Welding defect: Lack of sidewall fusion (LOF). Elongated voids between the weld beads and the joint surfaces to be welded. *(Courtesy of E.I. DuPont de Nemours and Co., Inc.)*

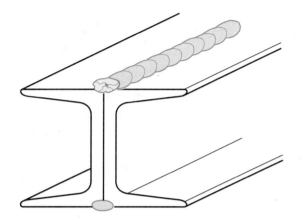

FIGURE 8–47 Crater cracks.

in the weld metal from toe to toe (Figure 8–48). Transverse cracking can be the result of using the wrong filler metal, of making too small a weld for the size of the joint, or of allowing the weld metal to cool too quickly.

Longitudinal cracking is a defect that occurs in the weld face or the base metal parallel to the toes of the weld (Figure 8–49). Longitudinal cracking can happen when the joint is unable to move or when it moves too slowly as the weld cools. Choosing a more ductile filler metal, selecting a

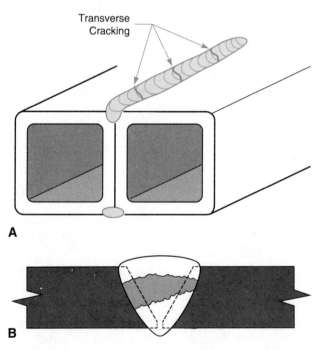

A

B

FIGURE 8–48 (A) Transverse cracking. (B) Welding defect: Tranverse crack. A fracture in the weld metal running across the weld. *(Courtesy of E.I. DuPont de Nemours and Co., Inc.)*

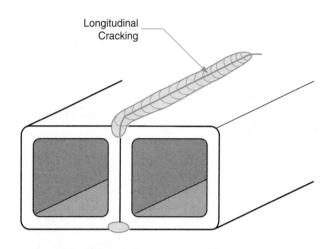

FIGURE 8–49 Longitudinal cracking.

less rigid joint design, or **preheating** can sometimes be helpful.

Underbead cracking is a defect that can occur in **high-carbon** and **alloy steels** when cracks appear in the base metal (Figure 8–50). **Hydrogen,** either from wet surfaces or absorbed within the **flux,** can become trapped in the **heat-affected zone** (HAZ) and result in cracking. **Low-hydrogen electrodes** that have been properly stored can help to prevent underbead cracking.

CALCULATING THE STRENGTH OF WELDS

Someone always has the job in product design of figuring out how much welding is enough. Knowing the strengths of welds should be of interest to those who do the welding.

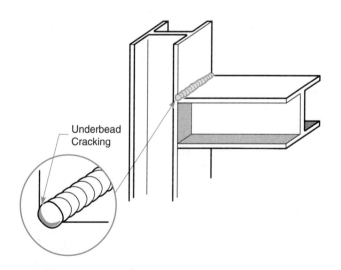

FIGURE 8–50 Underbead cracking.

Groove welds in which the effective throat is equal to or greater than the thickness of the base metal should provide greater strength than the base metal in **low-carbon** and alloy steels. For applications in which partial fill of the joint is required, the strength of groove welds can be calculated just like fillet welds.

Fillet Welds

In general, the accepted practice is to make fillet welds with a leg size equal to the thickness of the base metal. Sometimes, however, continuous welding along the entire length of the joint is not necessary. At other times, it would be beneficial to have some idea of how much strength a bracket needs to hold up a given weight. The allowable **shear force** on a fillet weld that is 1 inch long is determined by the following formula:

$$F = 0.707 \times L \times S$$

where: F = allowable shear force that a 1-inch-long, equal-legged fillet weld will support
0.707 = effective throat for an equal-legged fillet weld
L = leg size of fillet weld
S = allowable shear stress on the fillet weld

The sheer size for various electrodes is as follows:

Electrode	Shear Stress (psi)
E60 series	18,000
E70 series	21,000
E80 series	24,000
E90 series	27,000
E100 series	30,000

Example 1: What shear force can be handled by a 1/4-inch (0.25-inch) equal-legged fillet weld that is 1 foot long if it is welded with an E7018 electrode?

$$F = 0.707 \times 0.25 \times 21,000$$
$$F = 3,712 \text{ pounds per linear inch of fillet weld}$$

One foot (12 inches) of 1/4-inch fillet weld = 12 $\times$ 3,712 = 44,544 pounds for 12 linear inches of fillet weld.

Example 2: What shear force can be handled by a 3/8-inch (0.375-inch) equal-legged fillet weld that is 3 feet (36 inches) long if it is welded with an E6013 electrode?

$$F = 0.707 \times 0.375 \times 18,000 \times 36$$
$$F = 171,801 \text{ pounds over 3 feet}$$

Decimal Equivalents

Welding does not usually require sophisticated mathematics. However, welders should know the **decimal equivalents** for sixteenths and eighths or, at least, how to get them.

Example 1: The decimal equivalent for 5/16 can be found by dividing 1 by 16, then multiplying by 5.

$$1 \div 16 = 0.0625$$

$$0.0625 \times 5 = 0.3125$$

Example 2: The decimal equivalent for 7/8 can be found by dividing 1 by 8, then multiplying by 7.

$$1 \div 8 = 0.125$$

$$0.125 \times 7 = 0.875$$

The following list provides the decimal equivalents of common fractions:

1/16 = 0.0625	13/16 = 0.8125
2/16 = 0.125	14/16 = 0.875
3/16 = 0.1875	15/16 = 0.9375
4/16 = 0.250	16/16 = 01.000
5/16 = 0.3125	1/8 = 0.125
6/16 = 0.375	2/8 = 0.250
7/16 = 0.4375	3/8 = 0.375
8/16 = 0.500	4/8 = 0.500
9/16 = 0.5625	5/8 = 0.625
10/16 = 0.625	6/8 = 0.750
11/16 = 0.6875	7/8 = 0.875
12/16 = 0.750	8/8 = 01.000

REVIEW

1. Name the eight types of groove welds, and cite one important detail about each.
2. What are the eight parts that make up a fillet weld?
3. How is the size of a fillet weld determined?
4. What is used to measure the strength of a fillet weld?
5. When might a double-bevel-groove weld or a double-V-groove weld be used?
6. Why is the flat position preferred over the other welding positions?
7. What does the coefficient of linear expansion tell about metal?

8. What can be done to control distortion?
9. List and explain several of the defects that can occur as the result of welding.
10. Without looking back at the text, give the decimal equivalents of 1/16 and 1/8. Do the mathematics, if necessary.

Create Three Questions

1.
2.
3.

Related Math and English Questions

1. If the groove angles are 60°, 45°, and 30°, what are the bevel angles?
2. The strength, or allowable shear force (F), of equal-legged fillet welds 1 inch long can be determined by inserting the leg size (L) into this formula: $0.707 \times L \times 21,000 = F$. (For this question, E70 series electrodes are selected, and thus 21,000 is used as the sheer stress.) Find F (in psi) for 1-inch welds with the following leg sizes:
   a. 3/16"
   b. 0.25"
   c. 1/2"
3. Using your answers for the preceding question, determine the total strength for over 8 feet of weld.
4. In a paragraph, explain the decision in choosing between using a square-groove weld and a bevel-groove weld.

5. Why is the size of the weld important to the welder? Explain in a paragraph.

For Further Thought

1. The base metal composing a butt joint is 1 inch thick (with several feet of welding) and could be welded from both sides. What joint design would you choose, and why?
2. A lap joint is made of 1/2-inch material. What leg size for the fillet weld would given maximum strength?
3. Suppose you are fabricating a cylindrical tank out of 18-gauge steel and you are concerned about distortion and melt-thru welding up the seam. How can you handle these potential problems?
4. A failed weldment comes back to the shop. A fillet weld peeled right off the bottom member of a tee joint. What could have caused this weld to fail? Does this failure tell you something about the base metal?

SUGGESTED ACTIVITIES

1. Design (sketch out) a project, such as a trailer. Describe the joint designs, the welding processes, and so on.
2. Consider the welding equipment in your shop. Besides groove welds and fillet welds, what welds can be made?
3. Using a gauge, measure fillet welds located in the shop.
4. Look around the shop for welded structures, and figure out the strength of a given joint.

UNIT 8: EXERCISES

1. Sketch out a bevel-groove weld for a typical butt joint, using 3/4-inch plate without backing. The root opening is 1/16 inch with a 1/8-inch root face and a 45° bevel angle.
2. Sketch out a V-groove weld for a typical butt joint, using 1/4-inch plate without backing. The root opening is 1/16 inch with a 1/16-inch root face and a 30° bevel.
3. Sketch out a fillet weld for a typical joint, fastening a 2-inch pipe (4 inch long) centered on a 1/2-inch plate (4 inch × 4 inch). The leg size is 1/4 inch.
4. Calculate the minimum size of a fillet weld (intervals of 1/16 inch) required to support a weight of 7,000 pounds per linear inch using an E70-series electrode.
5. Sketch out a joint design that could be used to weld butt joints constructed of 2-inch thick plate using the gas metal arc welding process. The goal is to save both time and money.
6. Welders in the shipyard doing repair welding on the hull of a ship using the gas metal arc welding process complain that they are having trouble with porosity. Yesterday, the same work went smoothly, and there was no problem with porosity. What could be causing the problem, and what can be done as a possible solution?
7. Calculate the total weight of the A36 structural steel required for a hoist designed for a boom. The specifications are as follows:

4- 4″ × 4″ × 1/2″ angle × 15′ weight 12.8 lb. per foot

where: 4- = number of pieces
        4″ × 4″ = leg width
          1/2″ = thickness of angle
           15′ = length of angle

88- 1/2″ square bar × 15 1/2″ weight 0.85 lb. per foot

where: 88- = number of pieces
           1/2″ = width and thickness
         15 1/2″ = length of square bar

A36 steel has structural applications that require a minimum **yield** strength of 36,000 psi and a minimum tensile strength of 58,000 psi. (The information on weight in this exercise can be found in pamphlets furnished by many metal suppliers.)

Symbols for Welding

"He brings out photographs of things he has welded and these show beautiful birds and animals with flowing metal surface textures that are not like anything else."

—Robert M. Pirsig, *Zen and the Art of Motorcycle Maintenance*

GOAL

- Develop an understanding of the symbols for welding and how to use them

QUESTIONS

- What is a welding symbol?
- What is a weld symbol?
- What do the contour symbols mean?
- What do the finish symbols mean?

INSTRUCTIONS

Welders need instructions on how to execute the **welding** for any given fabrication project. If you give 25 welders a set of **blueprints** without welding instructions, they might complete the project 25 different ways. The **American Welding Society** has developed a set of symbols to standardize welding instructions. A complete description of these **welding symbols** can be found in publication AWS A2.4, *Symbols for Welding and Nondestructive Testing*. The use of welding symbols simplifies the instructions required to meet the specifications of a welded assembly.

THE WELDING SYMBOL

The foundation of all welding instructions is the welding symbol (Figure 9–1). The welding symbol is made up of three parts: the **arrow,** the **reference line,** and the **tail.**

The arrow points to the **joint** (Figure 9–2). The side to which the arrow points is the **arrow side** of the joint. The side opposite is the **other side**

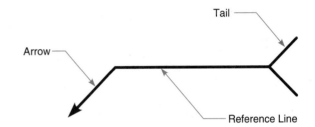

FIGURE 9–1   The welding symbol.

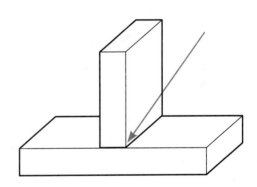

FIGURE 9–2   Arrow pointing at a joint.

of the joint. The reference line separates the arrow side from the other side.

The reference line provides the welding instructions (Figure 9–3). Instructions below the reference line are for the arrow side of the joint. Instructions above the reference line are for the other side of the joint.

The tail holds information for welding that cannot be provided another way (Figure 9–4). The welding **process** abbreviation SMAW, or "See Detail A." are examples of instructions found in the tail.

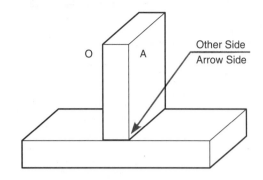

FIGURE 9–3   Reference line.

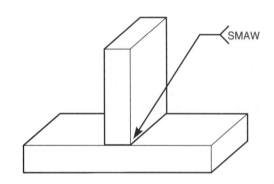

FIGURE 9–4   Tail with additional information.

WELD SYMBOLS

The previous unit described 19 different types of **welds.** There is a **weld symbol** for each of these welds (Figure 9–5).

Groove Welds

Groove welds drawn above or below the reference line indicate the side of the joint where the welding is to be done (Figure 9–6). Note the difference between the correct and incorrect displays of how a **bevel-groove weld** should be prepared (Figure 9–7). The **broken arrow** points to the piece of the joint that should receive the **bevel.**

If the weld symbol is drawn both above and below the reference line, welding should be completed on both sides of the joint (Figure 9–8). For example, a **plate** that is 2 inches thick may require preparation for **V-groove welds** on both sides of the plate to save welding time. In this case, welding is to be completed on both sides of the **butt joint.**

The **square-groove weld** might seem confus-

ing when the symbol is drawn both above and below the reference line (Figure 9–9). Just remember that welding on thick plate might require preparation and welding from both sides of the joint to achieve the necessary **root penetration.**

Size, effective throat, root opening, contour, and **finish** are five more instructions that may or may not be drawn with a groove weld (Figure 9–10). This information is not drawn if the shop has standards that are always applied.

For most groove welds, size is a measurement of the depth of the weld, extending from where the edges forming the bevel or groove are not parallel to each other. The size of a groove weld is the **groove angle,** which is given in degrees when required.

The **flare-V-groove weld** and the **flare-bevel-groove weld** are two exceptions. For the flare-groove welds, size is measured to where the pieces forming the joint can touch (point of tangency). For the bevel-groove weld, the bevel is the groove angle. For the V-groove weld, the groove angle is a measurement of twice the bevel angle (Figure 9–11).

The size is never drawn with the square-groove weld or the **scarf-groove weld;** the edges forming the joint are always parallel to each other (Figure 9–12). For other groove weld symbols in which size could be drawn but is not, a single detail drawing might provide instructions for all welds of a certain type (Figure 9–13).

The effective throat is the depth of penetration of a weld, measured to the **weld face** at the point of **flush contour.** The effective throat should always be drawn when the size is required. However, when the size equals the effective throat, only the effective throat should be drawn (Figure 9–14).

The root opening is the separation between the pieces forming the joint. The root opening may be drawn when **complete root penetration** is necessary. Note the position of the root opening in relation to the weld symbol in Figure 9–15.

The contour is the shape of the weld face. The contour can be flush, convex, or concave (Figure 9–16). Finally, the finish is the method used to give the completed weld its contour (Figure 9–17). Five finish methods are listed:

C	**Chipping**
G	**Grinding**
H	**Hammering**
M	**Machining**
R	**Rolling**

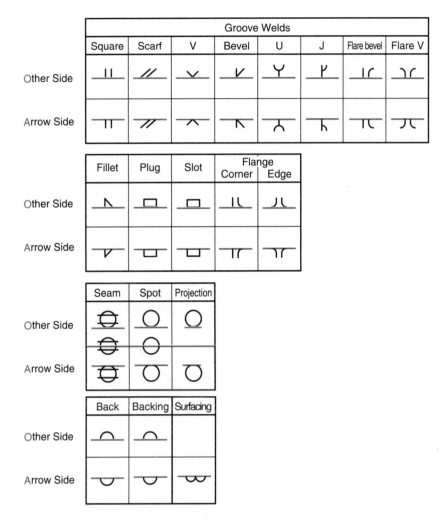

FIGURE 9–5   Symbols for the 19 types of welds.

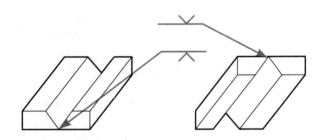

FIGURE 9–6   Direction of joint preparation for groove weld.

Fillet Welds

Fillet welds drawn above or below the reference line indicate the side of the joint where the welding is to be done. The fillet weld symbol can be drawn on both sides of the reference line (Figure 9–18). Size, **length, pitch,** contour, and finish are five more instructions that may or may not be drawn with a fillet weld (Figure 9–19).

The size of a fillet weld is a measurement of **leg** size. It is always drawn to the left of the fillet weld symbol. For unequal-legged fillet welds, the two leg sizes should be drawn within parentheses. Note that each leg must be dimensioned on the blueprint at the joint to avoid confusion (Figure 9–20). The length is always drawn to the right of the fillet weld symbol. The length of a fillet weld is only drawn if there is no continuous weld along the joint. If the length is required, dimension lines must be drawn on the print or marked by hatching (Figure 9–21).

The pitch is the center-to-center spacing for **intermittent** fillet welds. Note that there is a difference between **chain** and **staggered** intermittent welding (Figure 9–22). The contour symbol is used with the fillet weld symbol when welding a flush contour, a **convex contour,** or a **concave contour.** When a flush contour is required by finishing, a finish symbol is added (Figure 9–23).

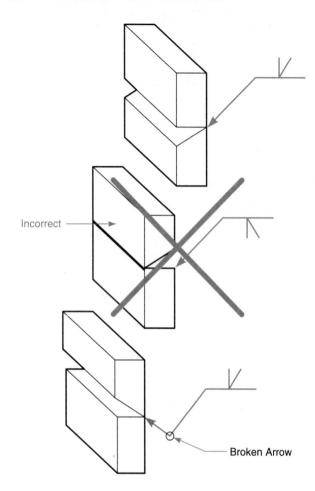

FIGURE 9–7   Correct and incorrect symbols for bevel-groove weld.

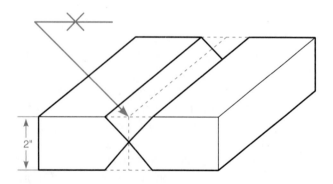

FIGURE 9–8   V-groove weld on reference line.

Plug and Slot Welds

Plug welds and **slot welds** share the same symbol. The size of the plug weld is always drawn to the left of the plug weld symbol (Figure 9–24). Size of a plug weld is always a measurement of the diameter of the hole before the angle of a

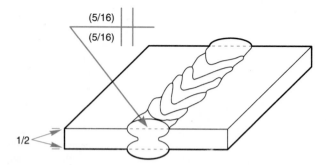

FIGURE 9–9   Square-groove weld on reference line at a joint of some thickness.

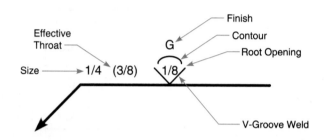

FIGURE 9–10   V-groove weld on welding symbol with additional instructions.

countersink. The degree of the angle is always the groove angle, and the bevel is half the groove angle (Figure 9–25).

The depth of fill is always complete unless drawn to indicate otherwise (Figure 9–26). The pitch of a plug weld is the center-to-center spacing. Pitch is drawn to the right of the plug weld symbol (Figure 9–27). The flush contour is drawn with plug welds filled by welding. The flush contour with finish symbol is drawn when finishing is required (Figure 9–28).

The same symbol is drawn for slot welds. The depth of fill is also drawn within the slot weld symbol, but all other information about the slot weld is drawn at the joint or in a detail (Figure 9–29).

Spot and Projection Welds

Spot welds and **projection welds** share the same symbol. The spot weld can be drawn either above or below the reference line. The spot weld can also be drawn on the reference line (Figure 9–30). The welding process to be used shall be noted in the tail.

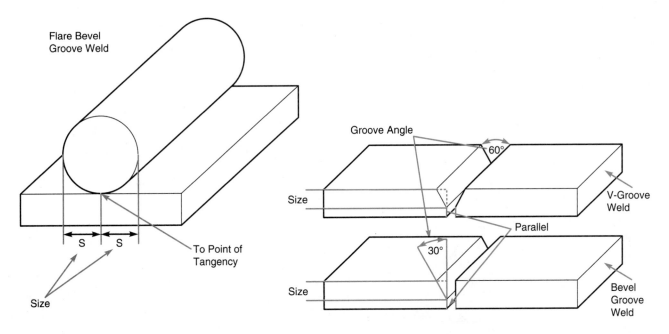

FIGURE 9–11   V-groove weld, bevel-groove weld, and flare-bevel-groove weld.

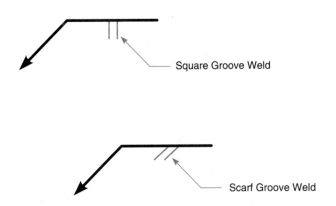

FIGURE 9–12   Square-groove weld and scarf-groove weld.

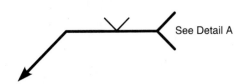

FIGURE 9–13   V-groove weld without size.

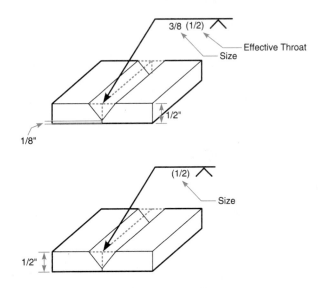

FIGURE 9–14   Size and size with effective throat.

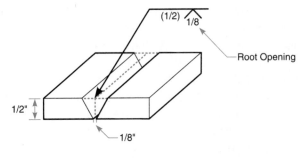

FIGURE 9–15   Symbol with root opening indicated.

The size of the spot weld is the diameter of each weld. Size is drawn to the left of the spot weld symbol (Figure 9–31). Instead of size, the **strength** can be drawn to the left of the spot weld symbol (Figure 9–32). The strength of a spot weld is determined by a two-step procedure. First, multi-

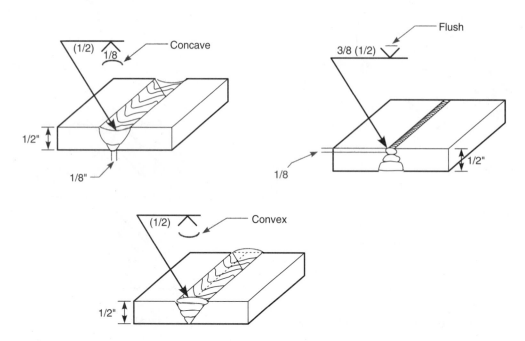

FIGURE 9–16   Three contour symbols.

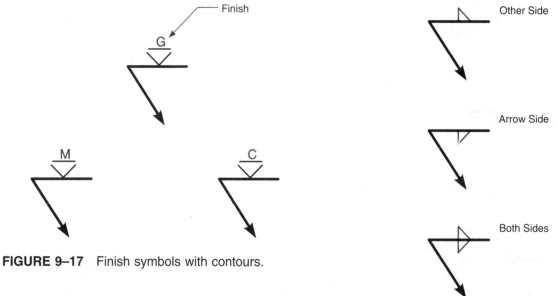

FIGURE 9–17   Finish symbols with contours.

FIGURE 9–18   Symbols for fillet weld.

ply 3.14 by R squared (R × R) to find the area where R is the radius of the spot weld. Second, multiply the area by the shear strength.

The pitch of a spot weld is the center-to-center spacing. Pitch is drawn to the right of the spot weld symbol. The exact placement for each spot weld must be dimensioned on the blueprint (Figure 9–33). The number of spot welds is given in parentheses (Figure 9–34).

The same symbol is used for the projection weld as for the spot weld (Figure 9–35). Unlike the

spot weld, however, the projection weld is never drawn on the reference line. The projection weld symbol is drawn above or below the reference line to indicate the side of the joint where the embossment (raised surface) is placed. Size, pitch, and number are determined by the embossment, and the projection weld process is given in the tail.

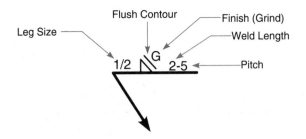

FIGURE 9–19 Fillet weld on the welding symbol with additional instructions.

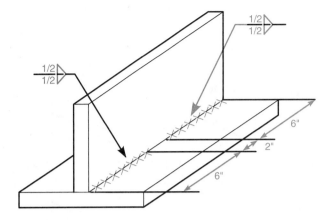

FIGURE 9–21 Length given with the fillet weld symbol.

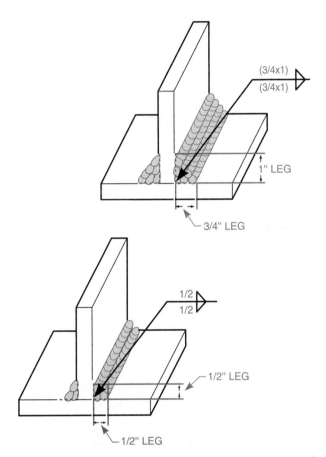

FIGURE 9–20 Fillet weld symbol at a joint.

Seam Welds

A **seam weld** can be a series of overlapping spot welds. The seam weld is drawn above, below, or on the reference line. The seam weld is indicated by either size or strength (Figure 9–36). The welding process to be used shall be noted in the tail.

The length is drawn to the right of the seam weld symbol. When the length is required, the seam weld is dimensioned on the blueprint (Figure 9–37). The pitch is the center-to-center spacing for intermittent welds. The pitch is drawn to the right of the seam weld symbol (Figure 9–38).

Backing and Back Welds

As noted in Unit 8, the difference between these two welds is the order in which they are made. The **backing weld** is made first on the other side of a joint prepared for a groove weld. The **back weld** is completed after the groove weld has been completed (Figure 9–39). The contour can be flush, convex, or concave. The finish is drawn when required to complete the contour (Figure 9–40).

Surfacing Welds

The symbol for a **surfacing weld** is drawn to give the height of **buildup** by welding. The surfacing weld symbol is always drawn on the arrow side of the reference line. The amount of buildup is drawn to the left of the surfacing weld symbol. If the entire surface does not require buildup, the appropriate width and length must be dimensioned on the blueprint (Figure 9–41).

Flange Welds

The **corner flange** and the **edge flange** are the final welds. Note how these two **flange welds** differ (Figure 9–42). Weld thickness (x), flange height (y), and flange radius (z) are the three dimensions given if the joint is detailed. The flange height is measured from the point of tangency to the edge of the joint before welding (Figure 9–43).

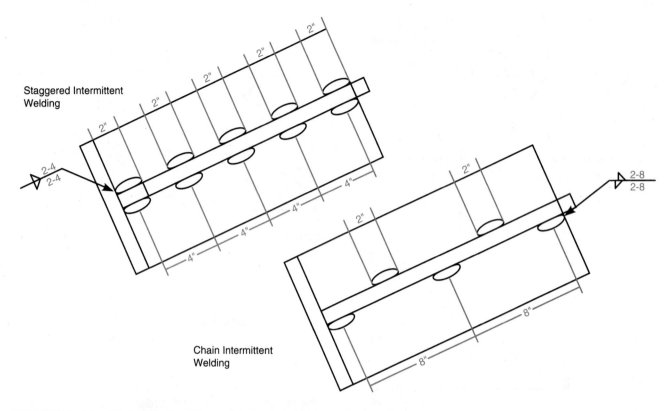

FIGURE 9–22 Pitch with the fillet weld symbol; chain and staggered intermittent welding.

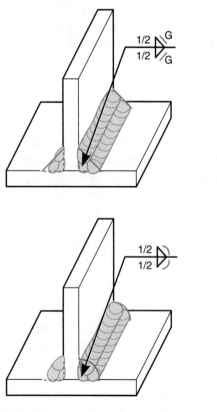

FIGURE 9–23 Fillet weld symbol and contour with a finish symbol.

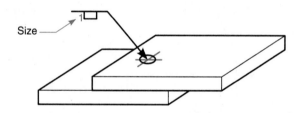

FIGURE 9–24 Plug weld symbol with size.

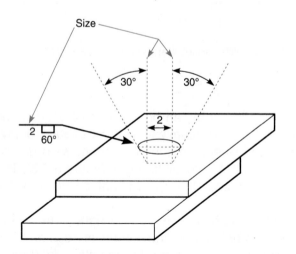

FIGURE 9–25 Plug weld with beveled angle.

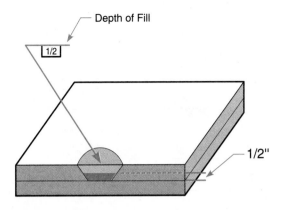

FIGURE 9–26 Plug weld with depth of fill.

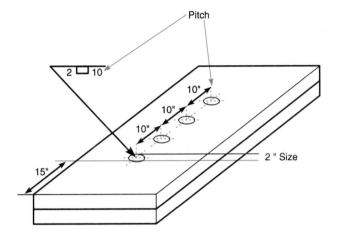

FIGURE 9–27 Plug weld with pitch.

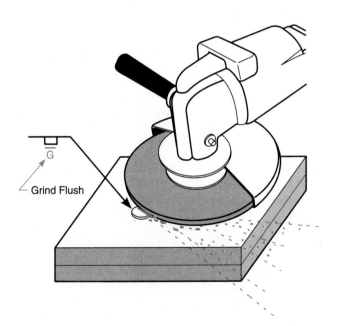

FIGURE 9–28 Plug weld with flush contour and finish symbol.

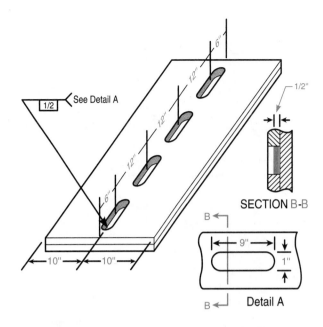

FIGURE 9–29 Slot weld with depth of fill, detail, and dimensions.

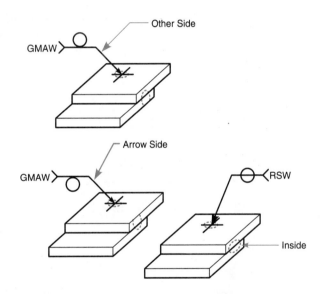

FIGURE 9–30 Spot weld drawn above, below, and on the reference line.

SUPPLEMENTARY SYMBOLS

The symbol indicating **weld all around** is a circle at the point where the arrow meets the reference line of the welding symbol. *Weld all around* means that the joint is to be welded completely (Figure 9–44). **Field weld** is an indication that welding will be completed on site (away from the shop) (Figure 9–45). The **melt-thru symbol** indicates 100% joint penetration from one side. If re-

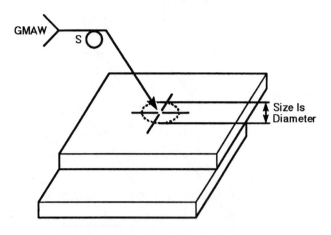

FIGURE 9–31　Spot weld with size.

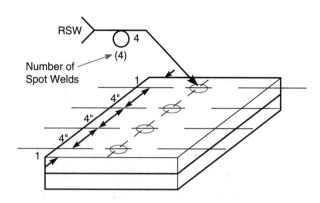

FIGURE 9–34　Spot weld with number given in parentheses.

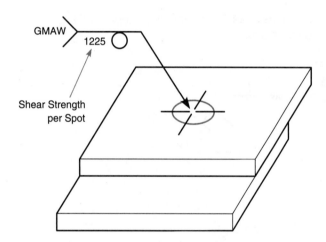

FIGURE 9–32　Spot weld with strength.

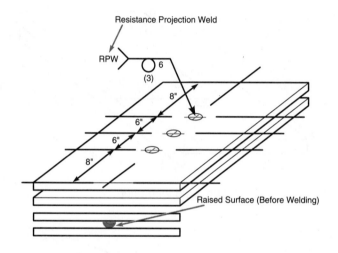

FIGURE 9–35　Projection weld symbol with process noted in the tail.

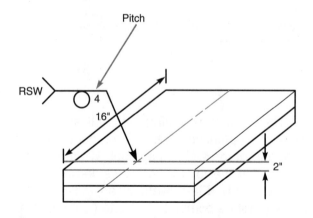

FIGURE 9–33　Spot weld with pitch and dimensions.

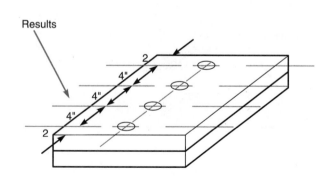

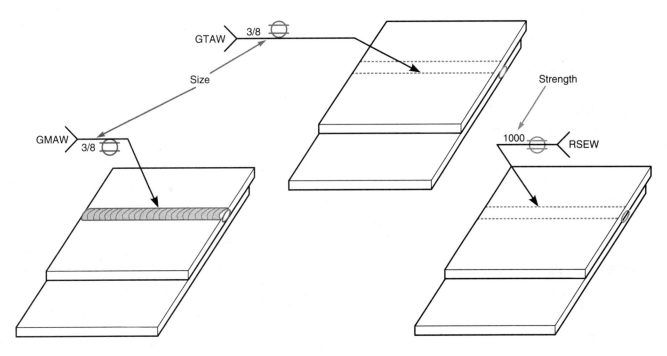

FIGURE 9–36 Seam weld symbol drawn above, below, and on the reference line with size and strength indicated.

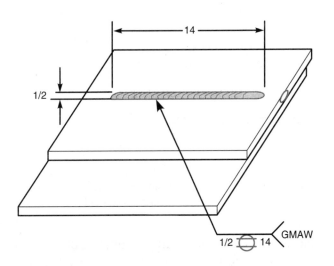

FIGURE 9–37 Seam weld with length and joint dimensions.

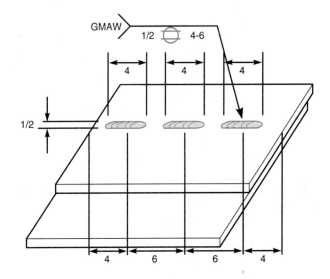

FIGURE 9–38 Seam weld with pitch and joint dimensions.

quired, the height is drawn to the left of the melt-thru symbol (Figure 9–46).

BACKING MATERIAL AND SPACER MATERIAL SYMBOL

The same symbol is used for **backing material** and **spacer material.** Backing material is used with the groove weld on the back side of the joint. *M* means "material as specified." *MR* means "remove material after welding" (Figure 9–47).

Spacer material is used to set a space or separation in the joint of a groove weld (Figure 9–48).

SUMMARY

This unit has covered the information that you will need to know to become competent in reading welding symbols. Unless used on a regular basis in conjunction with blueprints or drawings, you will

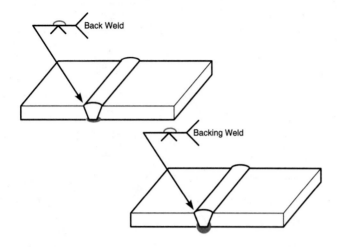

FIGURE 9–39   Back weld and backing weld symbols.

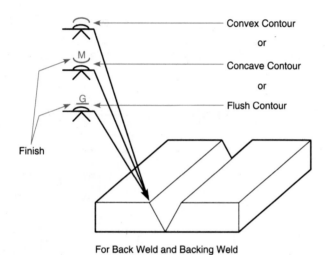

FIGURE 9–40   Back weld and backing weld with contour and finish symbols.

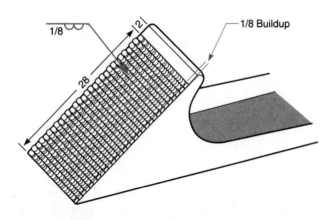

FIGURE 9–41   Surfacing weld symbol.

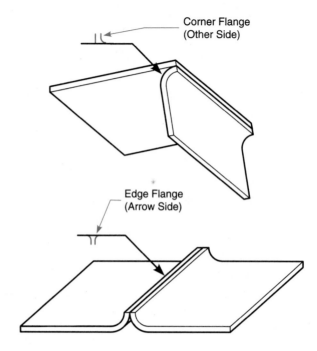

FIGURE 9–42   Corner flange and edge flange symbols.

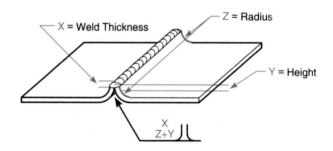

FIGURE 9–43   Flange weld with weld thickness, flange height, and flange radius indicated.

quickly forget this information. Remember that you can always reread this unit to refresh your memory.

REVIEW

1. Who is responsible for standardizing the welding symbol?
2. When is the tail added to the welding symbol?
3. If the bevel angle is 30°, what is the groove angle?
4. Why would a groove weld be indicated on both sides of a butt joint on thick plate?
5. How is the effective throat measured?
6. When is a finish symbol drawn?

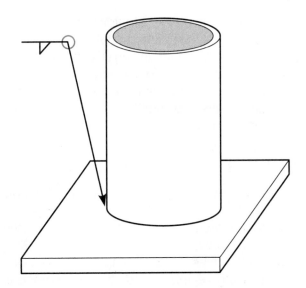

FIGURE 9–44 Weld all around symbol.

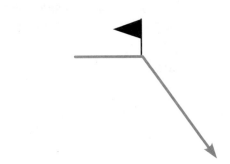

FIGURE 9–45 Field weld symbol.

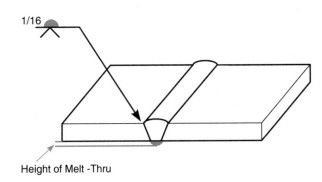

Height of Melt -Thru

FIGURE 9–46 Melt-thru symbol with height of melt-thru.

7. What is pitch, and when is it used?
8. What determines the size of a plug weld?
9. Explain the difference between a backing weld and a back weld.
10. What does a field weld indicate?

Note: Material Backing
MR= Material Removed
When Used

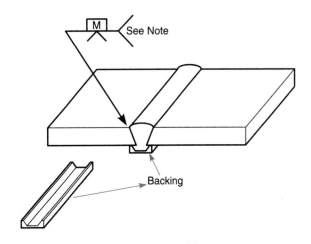

Backing

FIGURE 9–47 Backing material symbol with symbols for material and material removed.

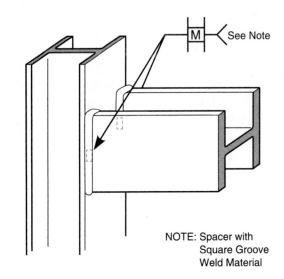

NOTE: Spacer with
Square Groove
Weld Material

FIGURE 9–48 Spacer material symbol.

Create Three Questions

1.
2.
3.

Related Math and English Questions

1. Using the formula $3.14 \times R^2 = A$ (R = the radius of the spot weld, A = the area of the spot weld), find the area of the following spot welds:

a. size of weld = 1/4 inch
b. size of weld = 1/2 inch

2. The shear stress for E70-series electrodes is 21,000 psi. Using the areas determined for the spot welds in the preceding problem, find the strength of each spot weld.
   a. A × 21,000 =
   b. A × 21,000 =

3. Write a paragraph using the following terms: *pipe, weld all around, groove weld, field weld, melt-thru, grinding, convex contour.*

4. Write an explanation giving reasons for using the welding symbol.

For Further Thought

1. After looking over a blueprint, you realize that all of the fillet welds were drawn without a given size. Why might this be?

2. Can a groove weld symbol be drawn at a tee joint? Explain.

3. Can a fillet weld symbol be drawn at a butt joint? Explain.

4. A butt joint calls for welding together 1-inch plate with a tensile strength that exceeds that of the base metal. Design a joint that gives complete penetration but requires the least amount of filler metal.

5. What is the minimum leg size of a fillet weld on a 1/2-inch lap joint using 70,000-tensile filler metal that will achieve 3,700 pounds shear strength per inch? See Unit 8 for assistance.

SUGGESTED ACTIVITIES

1. Examine blueprints with welding symbols on them.

2. Make flash cards and practice identifying welding symbols.

3. Sketch a working drawing of a project with welding symbols as though someone would do the welding.

4. Do a report on welding from an article in *Welding Journal* or *Welding Design & Fabrication.* For subscription information, write

Welding Journal
550 N.W. LeJeune Rd.
Miami, FL 33126

Welding Design & Fabrication
P.O. Box 91368
Cleveland, OH 44101

UNIT 9: EXERCISES ON WELDING SYMBOLS

For Exercises 1 to 10, sketch in the weld and include the necessary dimensions. An example is shown in Figure 9–49.

1. Figure 9–50.
2. Figure 9–51.
3. Figure 9–52.
4. Figure 9–53.
5. Figure 9–54.
6. Figure 9–55.
7. Figure 9–56.
8. Figure 9–57.
9. Figure 9–58.
10. Figure 9–59.

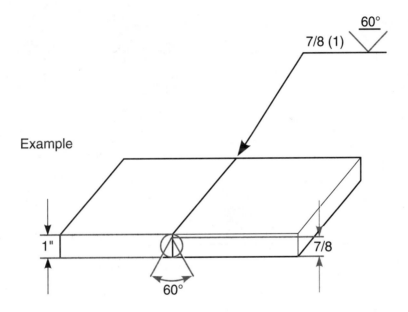

FIGURE 9–49   Example of completed exercise.

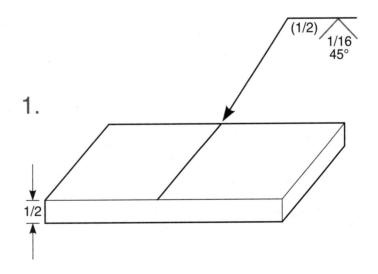

FIGURE 9–50   Exercise 1.

2.

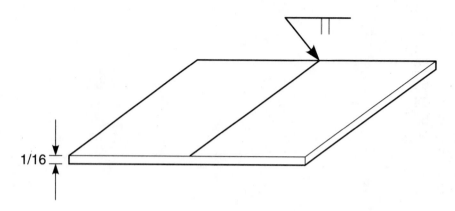

FIGURE 9–51   Exercise 2.

3.

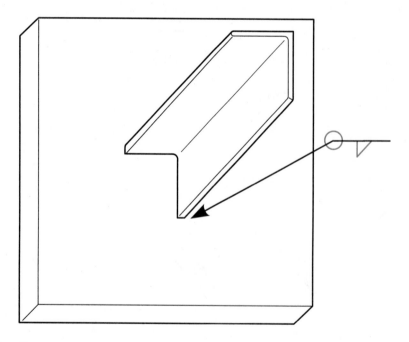

FIGURE 9–52   Exercise 3.

4.

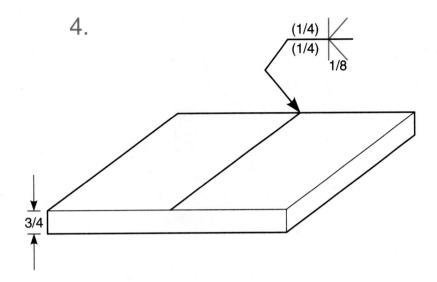

FIGURE 9–53   Exercise 4.

5.

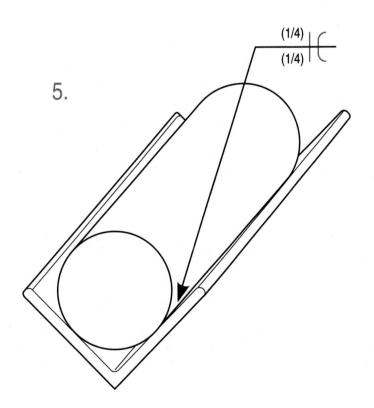

FIGURE 9–54   Exercise 5.

6.

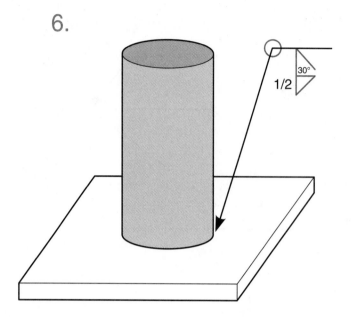

FIGURE 9–55   Exercise 6.

7.

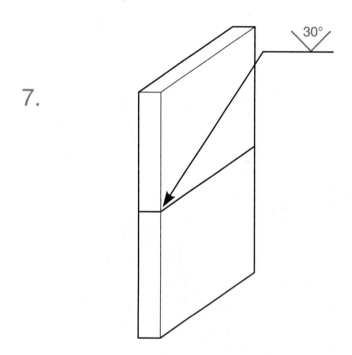

FIGURE 9–56   Exercise 7.

8.

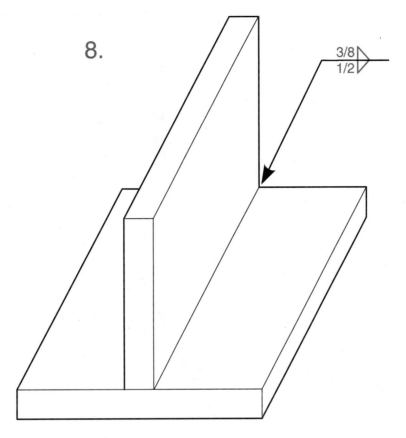

3/8
1/2

FIGURE 9–57 Exercise 8.

For Exercises 11 to 14, sketch in the welding symbol, and provide the necessary information.

    11. Figure 9–60.
    12. Figure 9–61.
    13. Figure 9–62.
    14. Figure 9–63.

For Exercises 15 and 16, complete the instructions provided by the drawings.

    15. The lattice support bracket is a drawing for carrying out welding instructions in a project using welding symbols (Figure 9–64). The welding and cutting processes are to be decided in the shop.

    16. The assembly support is another drawing for carrying out welding instructions in a project using welding symbols (Figure 9–65). The welding and cutting processes are to be decided in the shop.

9.

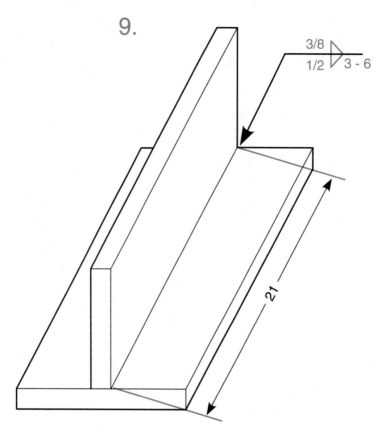

FIGURE 9–58 Exercise 9.

10.

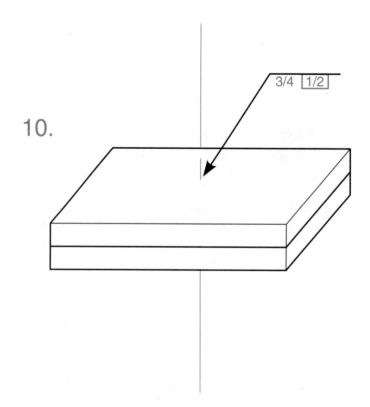

FIGURE 9–59 Exercise 10.

11.

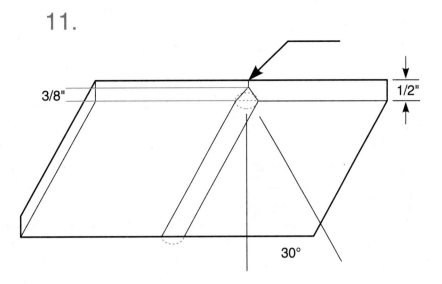

FIGURE 9–60   Exercise 11.

12.

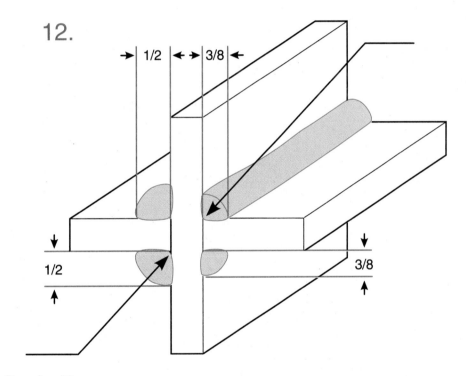

FIGURE 9–61   Exercise 12.

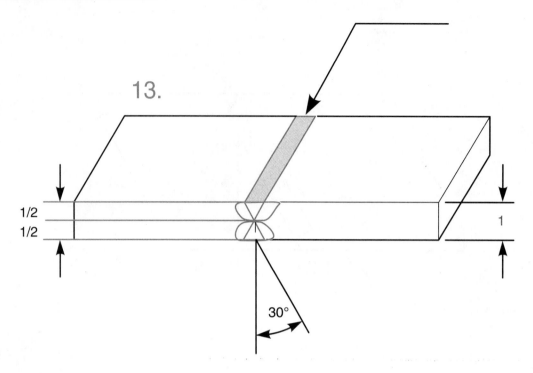

FIGURE 9–62 Exercise 13.

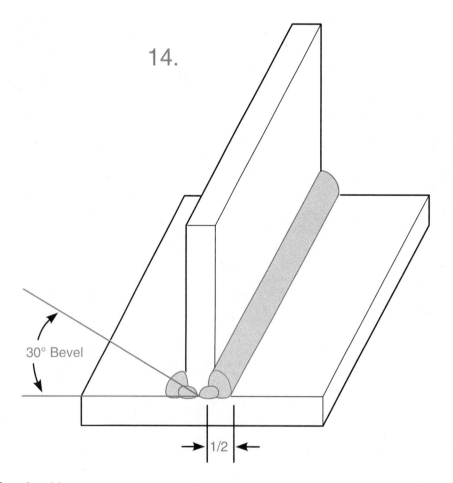

FIGURE 9–63 Exercise 14.

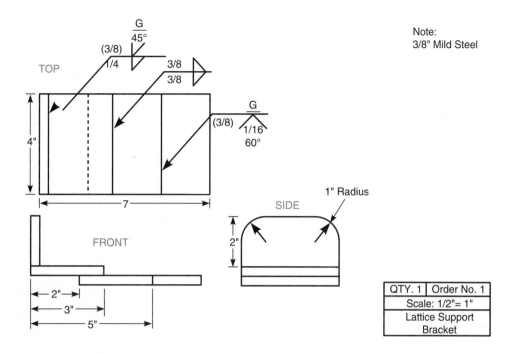

FIGURE 9–64 Lattice support bracket.

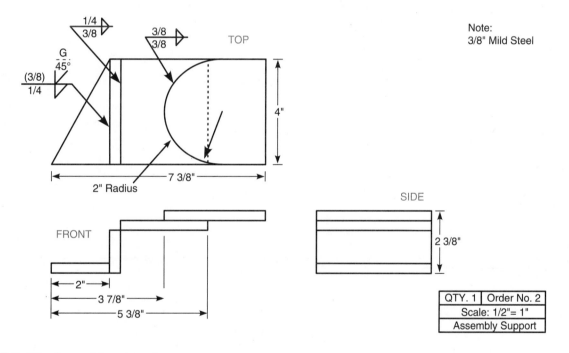

FIGURE 9–65 Assembly support.

Inspection, Testing, and Qualification

"He acts like I'm a detective trying to get something on him, and it isn't until he discovers I do a lot of welding that I become okay."

—Robert M. Pirsig, *Zen and the Art of Motorcycle Maintenance*

GOAL

- Develop an understanding that both inspection and testing are necessary for quality welding

QUESTIONS

- What is a nondestructive test?
- What is a destructive test?
- What are some methods for nondestructive testing?
- What are some methods for destructive testing?

QUALITY CONTROL

Inspection, testing, and qualification are a growing part of **quality control** in manufacturing and maintenance. Quality control has become very important because of competition and liability. To compete in what has become a gobal marketplace, companies must produce the best products possible. The cost of inspection and testing products can save greater expenses later on when the consumer becomes involved. Money that does not have to be spent defending the company can be used for research and development (R & D) of new products. The **welder** is the person who begins quality control by being satisfied only with **quality welding.**

VISUAL INSPECTION

Inspection begins with a visual examination of the **weld.** The welder starts the inspection in the process of welding, making immediate adjustments in response to what is being observed in the **weld pool.** This may require changing the **amperage,** the travel speed, the **arc length,** or any of the other factors that can affect the quality of the **weld bead.** Finally, the welder gives the completed weld a final **visual inspection** to look for any **defects** that could not be seen while welding. **Porosity, incomplete fusion,** or **poor root penetration** with the **base metal** can be observed by examining the weld itself. A visual inspection done immediately to uncover defects can prevent an increase in cost later on. A visual inspection can provide an indication of trouble so more extensive testing might be completed.

TESTING

Testing becomes necessary for several reasons. If new **welding procedures** are employed, the manufacturer might be interested in finding out if the welding **process** is producing quality welds. For example, a product welded by **shielded metal arc welding** may be produced more cost-effectively by a **gas metal arc welding** process, but will the quality be the same? The manufacturer may be interested in finding out if both the gas metal arc welding process and the filler metal are compatible with the **steel** used as the base metal. If the steel does not match up with the filler metal, no welding will make any difference.

There might also be the question of whether the welders have enough skill at some other welding process to make welds as sound as those made with shielded metal arc welding. No manufacturer wants to risk the company's reputation with welds that fail to hold up. The least serious outcome might be a product that fails; the worst could be injury or death.

Even if a manufacturer has no interest in testing, an insurance company might require it. The insurance company may demand some assurance that the welding procedures used in the manufacturing process are sound. Testing is very important whenever a risk of failure could have dangerous consequences.

Testing companies provide an invaluable service when it is necessary to have welding examined outside the manufacturer's plant. Testing companies keep records and furnish the results of weld testing to insurance companies, government agencies, and other interested parties. Testing companies provide an important service because, by uncovering problems in welding, they help manufacturers produce safe and durable goods and structures that consumers can live with.

Destructive Versus Nondestructive Testing

By definition, **destructive testing** results in the destruction of the **weldment.** This is a costly undertaking because of the time and money involved, not to mention the destruction of a weldment that might have to be rewelded. On the other hand, **nondestructive testing** leaves the weldment intact, at least until the results uncover major defects in the welding. Consequently, nondestructive testing is often preferred when testing is necessary.

NONDESTRUCTIVE TESTING METHODS

There are five common methods of nondestructive testing (NDT): the **magnetic particle test,** the **dye penetrant test,** the **fluorescent penetrant test,** the **ultrasonic test,** and the **radiographic test.** Each of these tests has advantages and disadvantages that become important in the selection of a nondestructive testing method.

Magnetic Particle Test

The magnetic particle test (MT) can locate surface and near-surface **cracks,** porosity, **slag inclusions,** and incomplete fusion (Figure 10–1). In one application, the magnetic particle test begins when direct current (DC) electric charge is passed between two poles. The forces line up until there is a break in these magnetic lines of force. Magnetic particles are applied to the weld in the form of powder. Nothing unusual results when the welding is sound, but when the welding might be defective, a break is created in the lines of force. The magnetic particles become attracted to the defect, which develops north and south poles at the edge of the defect, outlining the crack (Figure 10–2).

Dye Penetrant Test and Fluorescent Penetrant Test

The dye penetrant test (DPT) and the fluorescent penetrant test (FPT) can locate only cracks and porosity that have formed on the surface of the weld (Figure 10–3). With the dye penetrant test, a highly penetrative liquid is applied to a weld that has been thoroughly cleaned. Upon drying, a developer is applied to the weld. Under normal lighting, any defects are outlined, usually in red, where the dye has been absorbed. With the fluorescent penetrant test, the weldment is submerged in a fluorescent dye. It is removed from the dye and placed under a black light, where any defects in the weld appear greenish yellow.

FIGURE 10–1 Magnetic particle test unit. *(Courtesy Robert M. Bruno.)*

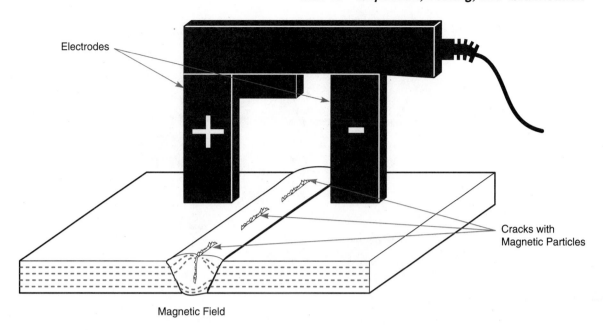

Electrodes

Cracks with
Magnetic Particles

Magnetic Field

FIGURE 10–2 Magnetic particles outlining a crack.

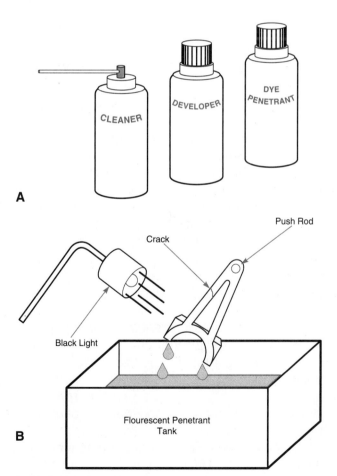

Crack

Push Rod

Black Light

Flourescent Penetrant
Tank

A

B

FIGURE 10–3 (A) Dye penetrant and fluorescent penetrant test liquids. (B) Setup for fluorescent penetrant
test.

Ultrasonic Test

The ultrasonic test (UT) can locate surface and
subsurface **discontinuities** too small to even be
considered defects in the weld (Figure 10–4). A
sound wave of high frequency is applied to the weldment. The sound wave travels through the metal,
reflecting back any breaks on the screen of an oscilloscope. The oscilloscope displays an electrical
signal graphically. Experienced personnel are required to interpret the readings on the screen.

Radiographic Test

The radiographic test (RT) can locate discontinuities within the metal that may or may not be
large enough to be considered defects (Figure
10–5). With this test, X rays penetrate the metal
and are absorbed by any defects. Exposure to a
strip of film records the results. Energy that is not
absorbed by defects will appear black on the film,
whereas cracks and other defects will be outlined
in white. The radiographic test is popular because
the film becomes a permanent record of the testing
which can be kept on file and referred to later, if
necessary. There are also ultrasonic test units that
have the capability to record graphically the images displayed on the oscilloscope.

The X rays given off by radiographic test equipment are only a danger when the equipment is
operating. Those who work in the area, an area

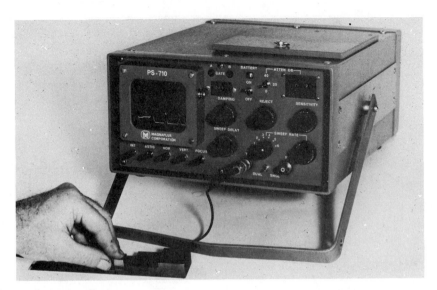

FIGURE 10–4   Ultrasonic test unit.

roped off or visually identified for safety purposes, should wear badges to record their exposure to the X rays.

DESTRUCTIVE TESTING METHODS

As mentioned earlier, a destructive test (DT) is a test that destroys the material being tested. Three popular types of destructive tests used in welding are the **nick-break test,** the **tensile test,** and the **guided bend test.**

Nick-Break Test

The nick-break test is a very simple **impact** test. It is a visual test made on the inside of a given break. With this test, a **coupon** (a strip of metal) is cut out of the **joint.** A **torch** or band saw is used to produce this coupon. Then a hacksaw (instead of a torch) is used to cut a 1/4-inch nick into each side of the strip (Figure 10–6). The strip is then placed in a vice, with the nicks just above the jaws. In this case, a hammer is used to break the strip in half (Figure 10–7). A press brake can also be used to accomplish the same results. A visual inspection is then made. A visual inspection of the break will uncover porosity, slag inclusions, or other defects within the break.

Tensile Test

The tensile test is more sophisticated (Figure 10–8). A tensile test measures the **strength** of a **material** to resist a force trying to pull it apart.

Tensile testing equipment is usually found in a laboratory setting with the means of applying controlled force in **psi** (pounds per square inch) for an accurate measurement. A calculated reading of what is happening to the metal under **load** is recorded on a graph. The graph provides such information as **elastic limit, yield point, ultimate tensile strength,** and **breaking point** (Figure 10–9).

The elastic limit is a measurement of the ability of a material to stretch under load and still return to its original shape. During this period as the load increases, the material stretches in proportion to the load. The yield point is a measurement of when the material first continues to stretch more than the load being applied. In other words, it shows when the relationship between the stretching of the material and the increase in load is out of proportion: less added load results in more stretching.

The ultimate tensile strength is a measurement of the maximum load that can act on a material without causing it to fail. The transition from yield point to ultimate tensile strength happens very quickly when observed on a graph in **real time.** When a material has reached the point of ultimate tensile strength, failure will follow.

The breaking point is a measurement of when the material fails. This happens shortly in real time after the ultimate tensile strength. The breaking point is accompanied by a loud pop as the coupon fractures (breaks apart).

FIGURE 10–5 Radiographic test unit. *(Courtesy Lorad Industrial Imaging, Danbury, Connecticut.)*

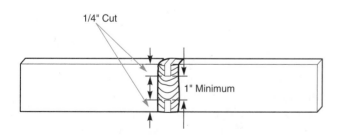

FIGURE 10–6 Nick-break coupon in vice.

In designing products, engineers try to keep the load on the weldment from ever reaching the proportional yield point. Beyond the proportional yield point, which is never going to occur under exactly the same load because the elastic limit has been exceeded. The material is now permanently deformed or stretched. Material permanently deformed becomes strain-hardened, producing an even higher elastic limit.

Weld coupons for use in a tensile test have to be prepared to meet the specifications of the testing machine (Figure 10–10). In other words, the testing machine can only produce a limited force.

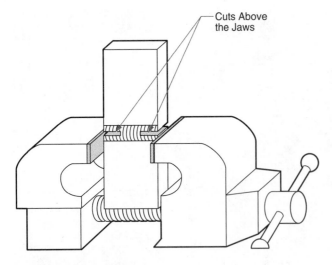

FIGURE 10–7   Setup for nick-break test.

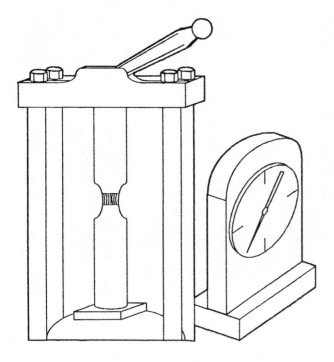

FIGURE 10–8   Tensile test machine.

If the strength of the coupon is greater than the limits of the tensile testing machine, there is no way to determine the strength of the weldment. The idea is to design a coupon that will fail the test so that measurements can be taken. A successful test means that the tensile strength was measured to be above the requirements of the joint.

For example, if a 3/8-inch-thick **plate** is welded

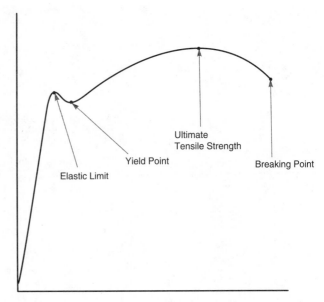

FIGURE 10–9   Graph illustrating tensile test results.

with a filler metal meeting the tensile strength of 70,000 psi, a tensile test can be used to determine the strength of the weld. Suppose a 1/2-inch-wide coupon from a weldment with these dimensions fractures at a measurement of 13,500 psi on the tensile testing machine. What is the tensile strength of the weld in psi? Does the weld pass the test? The answer is found by dividing 13,500 by the width of the coupon multiplied by the thickness of the coupon:

$$13,500 \div (1/2 \times 3/8) = 72,000 \text{ psi}$$

A quality weld is stronger than the base metal. The results of the tensile test should be a fracture in the coupon away from the weld (Figure 10–11).

Whereas the load, or force, in pounds per square inch is a measurement of a material under **stress, strain** is a measurement of the **elongation** (stretching) of the material that occurs before fracture. For example, in the design of the coupon used for tensile testing, suppose the distance between the two punch marks is measured at 2 inches before testing. After the coupon has failed, the distance between the two punch marks is measured again. If the measurement after failure is 2.4 inches, elongation is found by dividing 0.4 by 2 to get 0.2, or 20% elongation (the strain). A steel with a 20% elongation is within the range of **ductility** for **low-carbon steel.** Higher-carbon

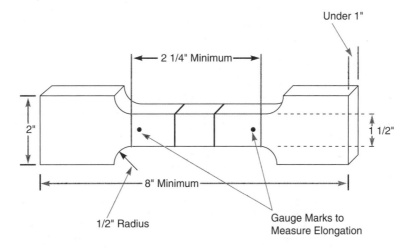

FIGURE 10–10   Tensile test coupon.

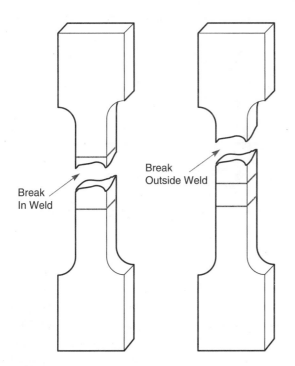

FIGURE 10–11   Results of two tensile tests.

steels and **cast irons** would not record as high a percentage of elongation. In fact, some cast irons would fracture without any measurable elongation whatsoever.

The tensile test can also be used to examine the properties of metals that have not been welded. Coupons designed with the dimensions described earlier can be used to test the base metal. Manufacturers of products made of steel might want to test the base metal to be sure that the mill from which they purchased the steel has met the specifications (strength, ductility, and so on) that a given product requires.

Guided Bend Test

The guided bend test is a popular test. It is used by many companies and schools to examine welders' skill, the welding process, the filler metal, or the base metal. The guided bend test stretches the weld, subjecting the weld to stress that will reveal defects in the welding.

The guided bend test involves a **jig,** which is made up of a plunger and a shoulder and has a seat custom-made to fit the plunger. The design for the jig illustrated in Figure 10–12 is specified by the AWS *Structural Welding Code.* Table 10–1 gives the dimensions for building three different jigs. A hydraulic jack can be used to push either the plunger or the shoulder onto the coupon, bending the coupon over (Figure 10–13).

The guided bend test is used for making **face bends, root bends,** or **side bends** to test welding (Figure 10–14). The face bend stretches the weld

TABLE 10–1: DIMENSIONS FOR THREE JIGS				
Yield Strength (psi)	**A**	**B**	**C**	**D**
50,000 and under	1 1/2″	3/4″	2 3/8″	1 3/16″
50,001–90,000	2″	1″	2 7/8″	1 7/16″
90,001 and up	2 1/2″	1 1/4″	3 3/8″	1 11/16″

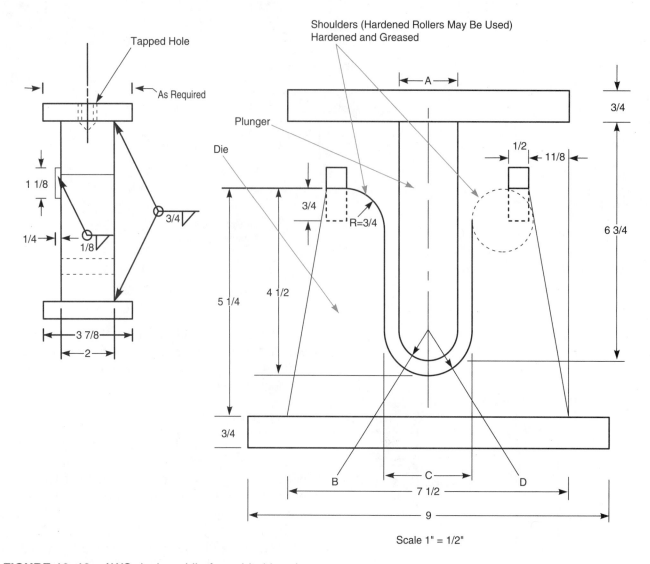

FIGURE 10–12   AWS-designed jig for guided bend test.

face of the coupon. The root bend stretches the root of the coupon. The side bend stretches the side of the coupon.

The coupons must be taken from designated places in the weldment for testing. The type of test will determine where the coupons are to be taken. The data (plate dimensions and coupon location) provided in Figures 10–15 and 10–16 are the requirements in the AWS *Structural Welding Code* for the **welder qualification** on **groove welds** of **limited thickness.**

Most of the exercises in this book are designed for 3/8-inch A36 steel. A36 steel has a yield strength of 36,000 psi. A jig with the dimensions required for a yield strength of 50,000 psi and

under will meet the specifications required for testing in all of the exercises in this book.

The AWS has developed testing that does not restrict the thickness of steel that the welder can become qualified to weld. Note that the dimensions are different for steel 1 inch in thickness and that side bends are required (Figure 10–17). A welder who passes a test given at a qualified testing facility under specifications that include 1-inch steel is qualified under that procedure to weld that specified steel of unlimited thickness.

Preparing Coupons for the Guided Bend Test

The coupons can be cut to size from the weld by a torch or a band saw. Reinforcement on the

FIGURE 10–13 Hydraulic jack. *(Courtesy Lincoln Electric, St. Louis, Missouri.)*

weld face, root face, or backing strip must be removed. A backing strip can be cut off with a torch. Grind flush (pedestal or hand grinder) the weld face and **root bead** with the grinding marks parallel to the length of the coupon. File the edges of the coupon to an 1/8-inch maximum radius (Figure 10–18). By keeping the grinding marks parallel and filing a radius on the edges of the coupon, you will prevent discontinuities forming in the bend that have nothing to do with the quality of welding.

Testing Coupons

In a root bend, the root is stretched. In a face bend, the face is stretched. The plunger should force the coupon into the seat until the coupon is bent into a U shape (Figure 10–19).

Measuring the Test Results

For the AWS's welder qualification test, the welder shall have passed the guided bend test. The weld on the **convex** side of the U-shaped coupon must meet the following conditions:

1. There must be no discontinuities of 1/8 inch or more in length, measured in any direction.
2. The sum of all discontinuities shall not equal or exceed 3/8 inch.
3. A crack along the edges that were filed shall measure no more than 1/4 inch.
4. A crack along the edges that were filed and that contains visible slag shall measure no more than 1/8 inch.
5. A coupon with corner cracks greater than 1/4 inch shall be replaced with another test coupon from the original weldment.

Passing the Test

If this test is taken and supervised by a licensed testing facility, there are certain benefits. The results will be recorded, and the welder will be issued proof of the accomplishment. If successfully completed, the test will allow the welder to make groove welds on plate and **pipe** under the procedure up to 3/4 inch in thickness (Tables 10–2 and 10–3).

This test is given for groove welds. **Fillet welds** can also be tested under the AWS *Structural Welding Code.* Welding a single side of a **tee joint** is one option. Unlike the testing of groove welds, successfully passing the fillet weld test limits the welder to fillet welds only (Figure 10–20).

TABLE 10–2: WELDING LIMITS FOR SUCCESSFUL TEST OF GROOVE WELDS ON 3/8-INCH STEEL[a]

Position Tested	Weld	Position Qualified	Thickness
Flat	Groove welds	Flat	Up to 3/4″
	Fillet welds	Flat and horizontal	Unlimited
Horizontal	Groove welds	Flat and horizontal	Up to 3/4″
	Fillet welds	Flat and horizontal	Unlimited
Vertical	Groove welds	Flat, horizontal, and vertical	Up to 3/4″
	Fillet welds	Flat, horizontal, and vertical	Unlimited
Overhead	Groove welds	Flat and overhead	Up to 3/4″
	Fillet welds	Flat, horizontal, and overhead	Unlimited
Vertical and overhead	Groove welds	All	Up to 3/4″
	Fillet welds	All	Unlimited

[a]Testing must be completed on steels such as A36 that are allowed under the AWS *Structural Welding Code.*

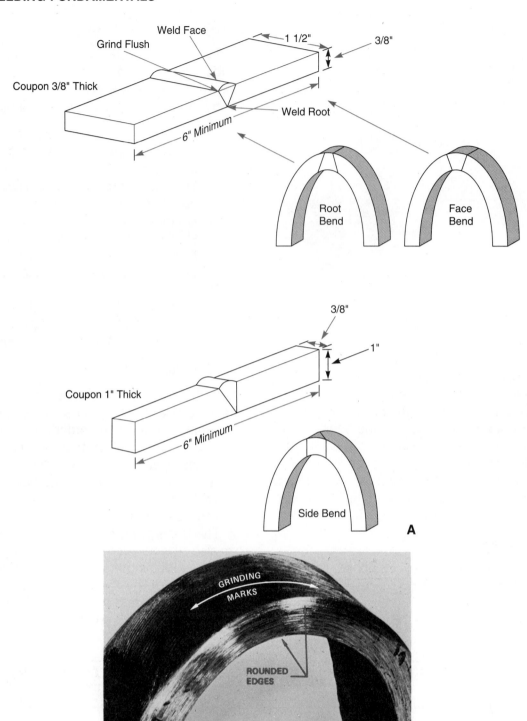

FIGURE 10–14 (A) Specifications for test coupons: face bend, root bend, and side bend. (B) Coupon after testing. *(Courtesy of Larry Jeffus.)*

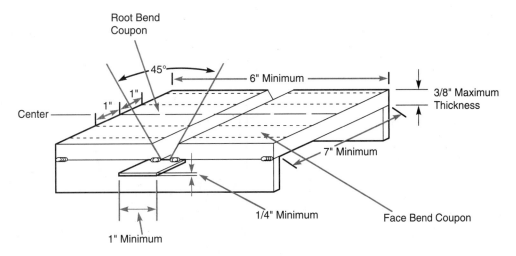

FIGURE 10–15 Test specifications for groove welds of limited thickness.

TABLE 10–3:	WELDING LIMITS FOR SUCCESSFUL TEST OF PIPE WELDS ON 3/8-INCH STEEL PIPE		
Position Tested	**Weld**	**Position Qualified**	**Thickness**
2G on 4" pipe	Groove welds	Flat and horizontal	Up to 4" diameter pipe, 1/8" minimum thickness, and 0.674 maximum thickness
	Fillet welds	Flat and horizontal	Unlimited
5G on 4" pipe	Groove welds	Flat, vertical, and overhead	Up to 4" diameter pipe, 1/8" minimum thickness, and 0.674 maximum thickness
	Fillet welds	Flat, vertical, and overhead	Unlimited

Successfully completing the fillet weld test allows welding fillet welds of unlimited thickness. Completing the test in the **flat position** qualifies the welder for fillet welds of unlimited thickness in the flat position. Completing the test in the **horizontal position** qualifies the welder for fillet welds of unlimited thickness in the flat and horizontal positions. Completing the test in the **vertical position** qualifies the welder for fillet welds of unlimited thickness in the flat, horizontal, and vertical positions. Completing the test in the **overhead position** qualifies the welder for fillet welds in the flat, horizontal, and overhead positions.

A visual test is given first. It should show a uniform appearance without excessive **undercut,** weld bead **overlap,** and with no porosity. The weldment is then fractured, using whatever means are reasonable. Hammering the joint over in a vise will work (Figure 10–21).

According to the AWS *Structural Welding Code,* a visual test of the fractured fillet weld should show no discontinuities greater than 3/32 inch.

The sum of all discontinuities shall be no greater than 3/8 inch in a weld that is 6 inches long.

AWS QUALIFICATION TESTS

The AWS *Structural Welding Code* has four different tests: **procedure qualification, tack welder qualification, welding operator qualification,** and welder qualification.

Procedure Qualification

The procedure qualification is a test of whether pieces of a joint for a specified base metal can be successfully welded together. In the test, all of the specifications are laid out: the type of base metal, type of filler metal, welding process, welding position, **joint design,** welding temperature, and any other factors that can possibly affect the outcome of the welding.

The AWS *Structural Welding Code* provides information on many procedures that need not

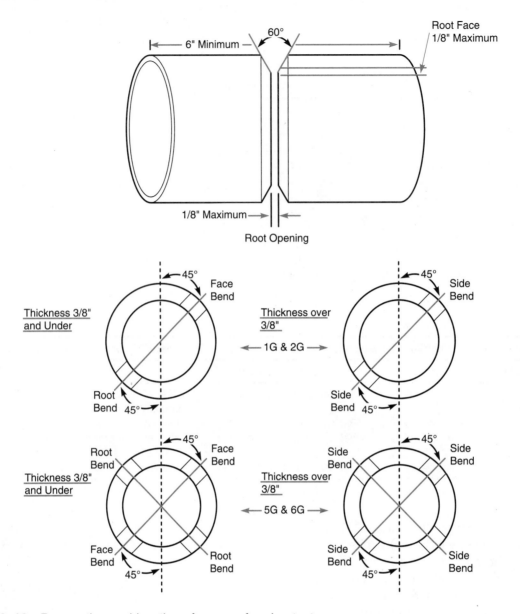

FIGURE 10–16 Preparation and location of coupon for pipe test.

be tested because they have been tested and used successfully over years of application. Any procedure not listed by the AWS or by another organization providing welding procedures must be tested.

Tack Welder Qualification

The tack welder qualification is a **performance test** of a tack welder's ability to make **tack welds** under a given procedure. A tack welder prepares joints to be welded by a qualified welder. Successful completion of the test qualifies the tack welder

to make welds within the limits established by the procedure.

Welding Operator Qualification

The welding operator qualification is a performance test of a welding operator's ability to weld under a given procedure. Successful completion of the test qualifies the welding operator to make welds within the limits established by the procedure. The welding operator is tested operating controls of **automatic** or robotic welding equipment. A radiographic examination can be used

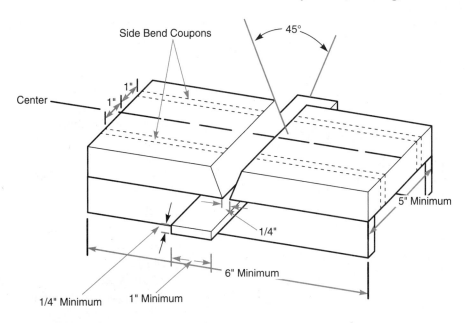

FIGURE 10–17 Test specifications for groove welds of unlimited thickness.

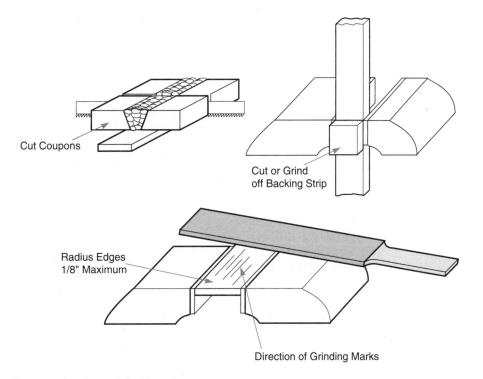

FIGURE 10–18 Preparation for guided bend test.

as a substitute for the guided bend test in this qualification.

Welder Qualification

The welder qualification is a performance test of a welder's ability to weld under a given proce- dure. Successful completion of the test qualifies the welder to make welds within the limits estab- lished by the procedure. A **procedure specifica- tion** and a welding procedure list the requirements that must be followed in the test (Figure 10–22).

Passing an examination qualifying a welder for a given procedure requires a testing agency.

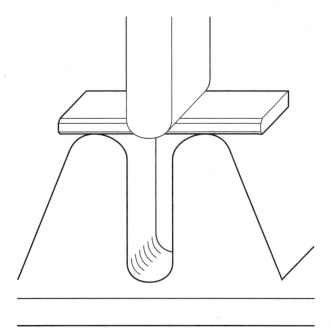

FIGURE 10–19 Setup for testing groove weld coupon.

Schools, government agencies, and independent testing companies give welding tests. It is important to remember that passing a given procedure results in welder qualification for *that* procedure. There are hundreds of procedures, so there are hundreds of welder qualifications. A welder qualified under one procedure must be qualified again if another procedure is required.

A welder qualified under a procedure has demonstrated an ability to make quality welds by passing a welding test. The welder should feel confident in applying for welding jobs because the skills needed to pass one test can be used to pass other tests. This book provides exercises to test and develop the welding skills you will need to become a skilled welder. The material in this book is a supplement to conscientious instruction. The sky is no limit for those who are motivated to succeed.

Welding Inspectors

The American Welding Society has developed a qualification program for **welding inspectors.** AWS qualification is based on experience, good vision, and the successful completion of an examination. The AWS sponsors the examination at locations throughout the United States at different times of the year. Qualified inspectors help to maintain standards of excellence in welding.

REVIEW

1. Which inspection is the most cost-effective?
2. Give four reasons for weld testing.
3. List the five nondestructive testing methods mentioned, describing each one.
4. List the three destructive testing methods mentioned, describing each one.
5. What happens to the coupon during a tensile test?
6. What is a metal's elastic limit?
7. What is a metal's yield point?
8. What is a metal's ultimate strength?
9. What happens to a metal beyond its elastic limit?
10. What is a coupon?
11. If, during a tensile test, a coupon breaks outside of the weld, what does that indicate?
12. Define the difference between stress and strain.
13. Describe what happens during a guided bend test.
14. What does the nick-break test reveal?
15. Why is it important to know that A36 steel has a yield strength of 36,000 psi? (Hint: It is something to do with the jig.)
16. Distinguish between a face bend, a root bend, and a side bend.
17. What kind of bend would be used to test steel that is more than 3/8 inch thick?
18. Give two of the defects that would cause a weld to fail the guided bend test.
19. If a welder passes a guided bend test for 3/8-inch-thick steel, what are the thickness limits for groove welds and fillet welds?
20. Name two of the four qualification tests under the AWS *Structural Welding Code.*

Create Three Questions

1.
2.
3.

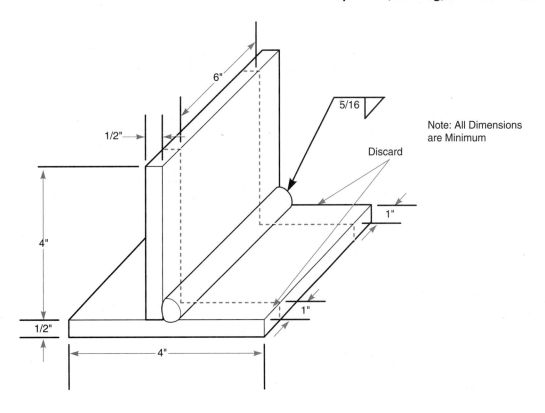

Note: All Dimensions
are Minimum

FIGURE 10–20   Specifications for coupon for fillet weld break test.

Related Math and English Questions

1. Given the following information, do coupons 1 and 2 pass or fail the tensile test? E60 (60,000 psi) filler metal is used to weld the plate from which the coupons are taken.

$$M \div (W \times T) = TS$$

where: W = width of coupon
T = thickness of coupon
M = measurement of coupon during testing
TS = tensile strength of the weld from which the coupons were taken

a. Coupon 1
M = 11,000 psi (when coupon broke)
W = 1/2 inch
T = 3/8 inch

b. Coupon 2
M = 11,800 psi (when coupon broke)
W = 1/2 inch
T = 3/8 inch

2. Explain the factors to consider in a visual inspection.
3. Write a paragraph on the difference between a tack welder, a welding operator, and a welder.

For Further Thought

1. A welder has passed a welder qualification test for the Acme Company. Should Jones Industries accept this welder's test results as proof of qualification for welding on their critical welding projects? Explain.
2. A welder produces a piece of paper that states a qualification to weld in all positions

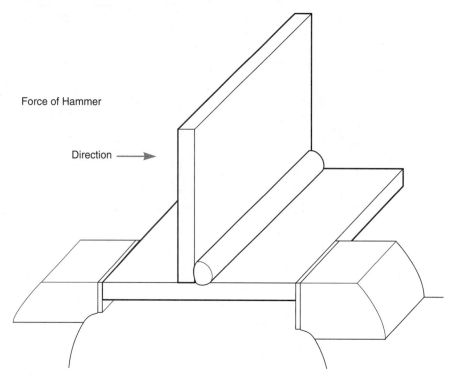

Force of Hammer

Direction ⟶

FIGURE 10–21   Setup for testing fillet weld.

and on steel of unlimited thickness. Can this welder do all kinds of welding?

3. A set of two coupons taken from A36 steel is given a tensile test after being welded with E7018. One coupon breaks in the weld and the other coupon breaks outside the weld. The dimensions of the coupons were 1/2 inch wide and 3/8 inch thick. One coupon broke at 14,000 psi, and one coupon at 13,200 psi. Did the welds fail the test?

4. In the preceding question, A36 steel was used for the base metal. What if the steel used for the base metal had a tensile strength of 80,000 psi? What do the results of the test tell us now?

SUGGESTED ACTIVITIES

1. Visit a testing facility.
2. Watch a demonstration of any of the five methods of nondestructive testing.
3. If a tensile-testing machine is available, run tests using different welding processes.
4. Use any of the three destructive testing methods to examine welds. Invite a welding inspector in to talk about the profession.

WELDER, WELDING OPERATOR OR TACK WELDER QUALIFICATION TEST RECORD

Type of Welder ___*Welder*___

Name ___*B. Smith*___ Identification No. ___*BS*___

Welding Procedure Specification No. ___*1A (File No.)*___ Rev _____ Date ___*11-15*___

Variables	Record Actual Values Used in Qualification	Qualification Range
Process/Type (5.16.2)	*Shielded Metal Arc Welding*	
Electrode (single or multiple)	*Multiple*	*SMAW*
Current/Polarity	*DCEP*	
Position (5.16.5)	*Vertical and Overhead*	
Weld Progression (5.16.7)	*Uphill*	*All Positions*
Backing (YES or NO) (5.16.18)	*Yes*	
Material/Spec. (5.16.1)	*P1* to	
Base Metal	*A 36*	*Carbon Steel*
Thickness: (Plate)	*3/8" (Thick)*	*3/4" (Thick)*
Groove	*V-Groove Weld*	
Fillet		*Fillet Welds (Size unlimited)*
Thickness: (Pipe/tube)		
Groove		
Fillet		
Diameter: (Pipe)		
Groove		
Fillet		
Filler Metal (5.16.3)		
Spec. No.	*A 5.1 (Mild Steel Electrodes)*	
Class	*E7018*	
F-No.	*F4 (Carbon Steel: EXX15; EXX16; EXX18)*	
Gas/Flux Type (5.16.4)		
Other		

VISUAL INSPECTION (5.12.6 or 5.12.7)
Acceptable YES or NO *Yes*

Guided Bend Test Results (5.28.1/5.29.1)

Type	Result	Type	Result
Root Bend	*Vertical Passed*	*Root Bend*	*Overhead Passed*
Face Bend	*Vertical Passed*	*Face Bend*	*Overhead Passed*

Fillet Test Results (5.28.2/5.28.3; 5.39.3/5.39.4)

Appearance _____ Fillet Size _____
Fracture Test Root Penetration _____ Macroetch _____
(Describe the location, nature, and size of any crack or tearing of the specimen.)

Inspected by ___*H. Jones*___ Test Number ___*20*___
Organization _____ Date ___*December 31st*___

RADIOGRAPHIC TEST RESULTS (5.28.4/5.39.2)

Film Identification Number	Results	Remarks	Film Identification Number	Results	Remarks

---

Interpreted by _____ Test Number _____
Organization _____ Date _____

We, the undersigned, certify that the statements in this record are correct and that the test welds were prepared, welded, and tested in accordance with the requirements of Section 5, Part C or D of ANSI/AWS D1.1, (_____) Structural Welding Code—Steel
year

Manufacturer or Contractor ___*Jiffy Corporation*___
Authorized By _____
Date _____

FIGURE 10–22 Sample qualification record.

Robotic Welding

"Technology presumes there's just one right way to do things and there never is."
—Robert M. Pirsig, *Zen and the Art of Motorcycle Maintenance*

GOAL

- Introduce robotic welding as a complement to the skilled welder

QUESTIONS

- What are some of the safety concerns of robotic welding?
- When is robotic welding justified?
- What are the parts of a robotic system?
- How do the parts of the system work?

SAFETY FIRST

Safety begins with a conscious effort to work safely. Most of the accidents involving robotic systems are the result of human error. Understanding that the technology itself is not the cause of accidents is important. Each worker must take responsibility for his or her own safety and for the safety of others.

Safety in robotics begins and ends with the **work envelope** (Figure 11–1). The work envelope is the area in which the **robot** operates. There are two kinds of safety devices for alerting people entering the work envelope: boundary devices and electronic warning devices. **Boundary devices** are fences or barriers that are brightly painted with warning signs. These devices should be used to mark off the work area used by the robot. The only equipment and materials allowed inside the work envelope should be used in the **welding** operation.

Electronic warning devices are activated when someone enters the work envelope. An alarm may sound to alert the workers, shutting down the robotic system. Some pressure-sensitive mats shut down the robot when the person loading or unloading a double-ended positioner steps off the mat (Figure 11–2).

The person responsible for operating the robot must use care whenever working inside the work envelope. The robot should be disabled so that it is impossible for any accidental startup. Care should be taken whenever stepping between the robot and the **positioner** (Figure 11–3). Everyone who works in the vicinity should receive safety training about the operation of the system.

Finally, in addition to the special safety measures that must be followed when operating a robotic system, there are the general safety precautions for welding. The welding fumes, **ultraviolet rays,** sparks, and heat given off by robotic welding deserve the same safety measures that should be taken with any welding **process.** Curtains should be in place to protect against ultraviolet rays and sparks. Exhaust systems should be designed to remove welding fumes.

ROBOTICS AND THE FUTURE

Robotics is one more tool industry has at its disposal for competing in the worldwide marketplace. The need for robotics will grow along with the need for skilled welders. In fact, robotics makes skilled welders even more valuable. The fact that robots will eliminate semiskilled welding jobs on production lines only raises the skill level of all welders. Many more welding jobs will require the services of welders who are competent at more than one or two welding processes. Skilled welders will be employed in more varied welding jobs in which robots are not practical. In addition, mo-

295

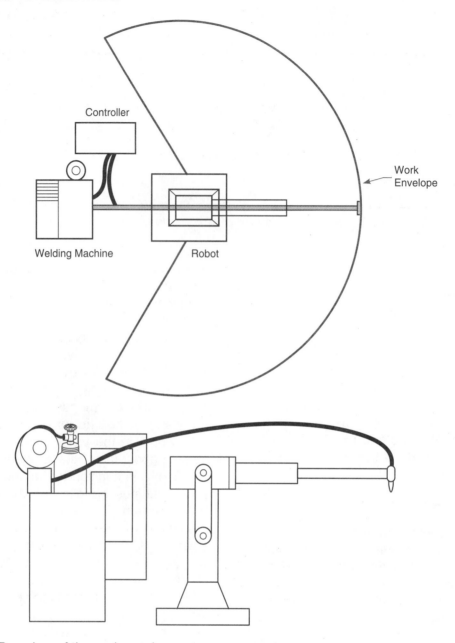

FIGURE 11–1 Boundary of the work envelope.

notonous or dangerous welding jobs will not be missed by anyone.

The world of tomorrow is here. The future will continue to show growth in the high-tech area of industrial robotics, and the future for skilled **welders** in the field of robotics is bright. The cost of setting up one or several stations employing **robotic welding** means job security for the welders. The cost of the investment in robotics guarantees that these jobs will go to skilled welders, who will continue to be trained as **welding operators** (Figure 11–4).

Not only can skilled welders recognize **quality**

welds that meet the standards of industry, but they know how to make quality welds. Even with some of the best robotic systems, **manual** welding is still necessary to complete many jobs. Skilled welders are the only people with the ability to match the quality welding of robotic systems. In fact, it is less expensive to train skilled welders how to operate robotic systems than to train computer specialists how to weld. The future of robotics in welding offers an opportunity for skilled welders.

Consistency in welding and hazards in the welding environment are two reasons for choosing robotics. In fact, there is a connection between

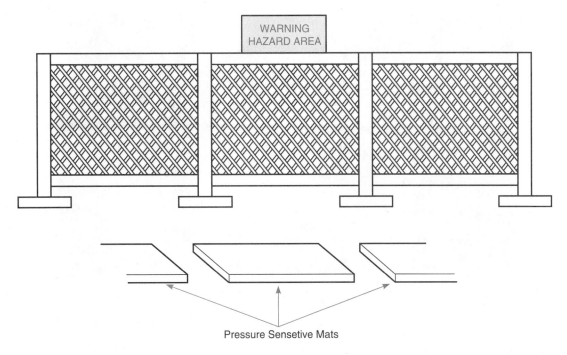

FIGURE 11–2   Warning device.

FIGURE 11–3   Danger between the robot and the positioner. *(Courtesy of Cloos International, Inc., Elgin, Illinois.)*

FIGURE 11–4   Welder and robot working together. *(Courtesy of Larry Jeffus.)*

these two factors. Welders, no matter how skilled, will have trouble making quality welds with consistency if exposed to hazards in the welding environment. Heat, smoke, and intensity of the **arc** are three factors that can cause **physical stress** on the body. Physical stress can cause overwelding or inconsistent work, which might require extra grinding or cutting out the substandard **welds.** Robotics can overcome physical stress, overwelding, inconsistent work, and other problems that can cause the quality of the welding to drop. A robotic

system can produce quality welds consistently in hazardous welding environments (Figure 11–5).

THE ROBOTIC SYSTEM

The **robot controller,** the robot, the positioner, and the welding equipment are the four basic components of a robotic system. The robot controller is the communication center of the entire system (Figure 11–6). The robot controller runs the programming, relaying instructions to and from the robot, to the positioner, and to the welding equip-

FIGURE 11–5   Robot offers protection from dangerous welding. *(Courtesy of Miller Electric Mfg. Co., Appleton, Wisconsin.)*

FIGURE 11–6   Robot controller. *(Courtesy of Motoman, Inc., West Carrollton, Ohio.)*

ment. The welding operator oversees the program in operation by the robot controller. The welding operator programs the robot controller with the instructions required for a given **welding procedure.** The welding operator can override the robot if trouble develops.

The robot, the second component of the system, carries out the commands set by the program (Figure 11–7). The robot might be powered by

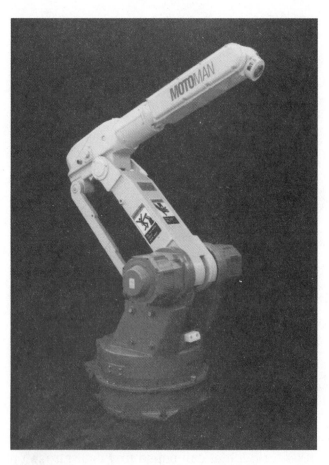

FIGURE 11–7   Robot. *(Courtesy of Motoman, Inc., West Carrollton, Ohio.)*

electric motors in conjunction with hydraulics or pneumatics to carry out the operations of the welding process. Electric drive systems are preferred over pneumatics because of their smoother action in the motion of the robot. The robot can be programmed for a series of simple or complex movements to be repeated continuously.

The positioner holds the weldment in place and moves when commanded by the program in coordination with the movements of the robot (Figure 11–8). A positioner increases the versatility of a robot when it can be programmed to rotate in several positions. A rotating positioner makes the **joints** of the assembly more easily accessible for welding. Rotation by the use of the positioner offers the advantage of moving the weldment for welding in the **flat position.** The

flat welding position permits higher **amperage** settings and faster **wire feed.** This means that more **filler metal** is deposited in less time.

The welding equipment is the final component of the system. It consists of the **power source,** the **torch,** the wire feed system, and the **shielding gas** (Figure 11–9). The welding can be carried out using any one of several welding processes. Welding equipment that is designed to respond to sensors works in coordination with the robot and the positioner for starts, continuous welding, and stops, as required by the welding operation.

A

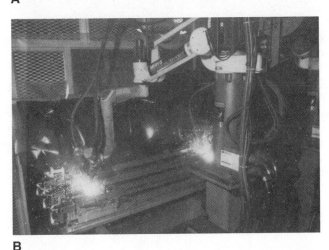

B

FIGURE 11–8   (A) Positioner. *(Courtesy of Motoman, Inc., West Carrollton, Ohio.)* (B) Robot and positioner in a fabrication shop. *(Courtesy of Miller Electric Mfg. Co., Appleton, Wisconsin.)*

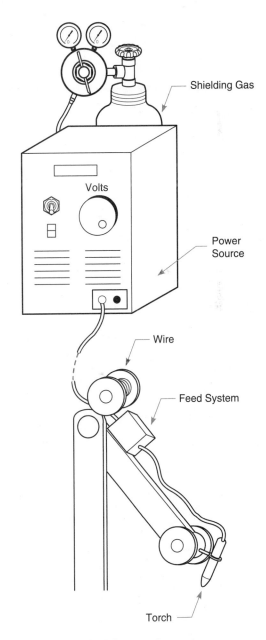

FIGURE 11–9   Welding equipment.

ROBOTIC MOVEMENT

The robot is capable of moving in the three planes (dimensions). The **x-plane** allows movement from side to side, the **y-plane** allows movement back and forth, and the **z-plane** allows movement up and down. The design of the robot determines its ability to carry out a given program within these three planes (Figure 11–10). Within these three planes more complex movements can be designed, allowing the robot to pivot on axes within the three planes. A five-axes robot is designed to pivot in five different directions (Figure 11–11). A six-axes robot is designed to pivot in six different directions (Figure 11–12).

The programming of a robot can be carried out by several different methods. One method is to physically guide the robot along the path required to complete the welding. A **teach pendant** is a second method (Figure 11–13). A teach pendant is a remote-control panel used by the welding operator to walk the arm of the robot through a series of points in space required to complete the welding. A third method calls for a specialized camera to record the width, depth, and length of the joint before welding. These measurements are analyzed by the robot controller. If the camera is off during the welding operations, the robot responds to instructions stored as memory in the program.

ROBOTIC SENSORS

One major advantage that the skilled welder has over the robot is the ability to use eyesight, touch, and hearing to make immediate changes to accommodate welding conditions. For example, the welder can speed up or slow down if the size of the **root opening** is changing. The ability of the welder to adjust instantly happens in **real time.** This is an aspect of robotics in which some of the most important developments have been made.

Robots have two basic kinds of sensors: **contacting sensors** and **noncontacting sensors.** A contacting sensor is a probe that touches the joint

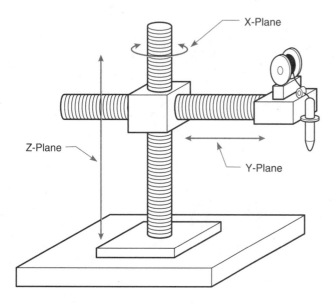

FIGURE 11–10   The x-, y-, and z-planes.

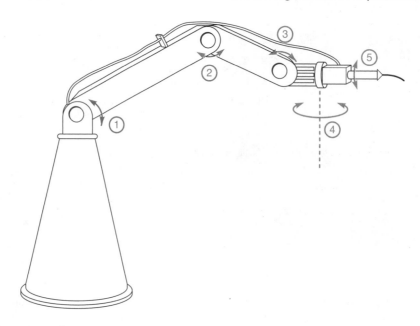

FIGURE 11–11   Five-axes robot.

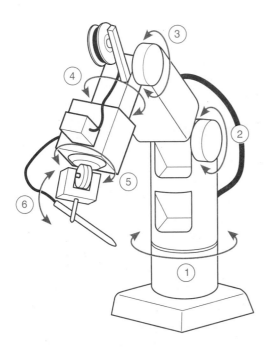

FIGURE 11–12   Six-axes robot.

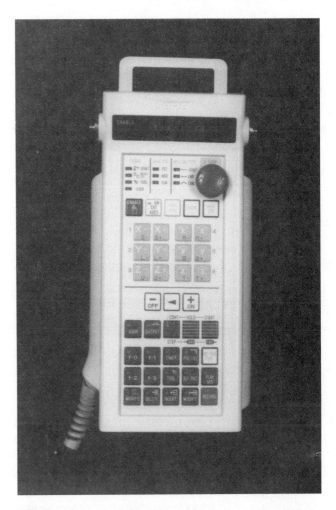

FIGURE 11–13   Teach pendant. *(Courtesy of Moto-man, Inc., West Carrollton, Ohio.)*

ahead of the weld pool. It is mounted off the end of the torch or gun and is usually limited to straight-line welding. The robot's job is to put the sensor in position within the **joint root.** A major problem is keeping the sensor within the joint root. Because the sensor is mounted ahead of the **gun** or torch, there is a time delay in correcting for changes. This mechanical sensor is primitive in comparison with other types of sensors that operate in real time and do not need to stay within the joint root.

A noncontacting sensor does not require physical contact with the **weldment. Through-the-arc sensors** are one type of noncontacting sensor. They use an oscillating (weaving right and left) welding torch or gun (Figure 11–14). These sensors measure welding variables, such as the voltage and the amperage, as the torch or gun oscillates. At the robot controller, measurements can be taken from the frequency oscillations recorded in the weld pool. A frequency approaching 100 hertz (Hz) can tell the welding operator that there is no penetration, whereas a frequency as low as 20 Hz might be evidence that **complete root penetration** is indeed taking place. Through-the-arc sensors are able to adjust for changes within the weld pool in real time, correcting for the **distortion** created by the welding process.

Vision sensors are another kind of noncontacting sensor for real-time welding applications.

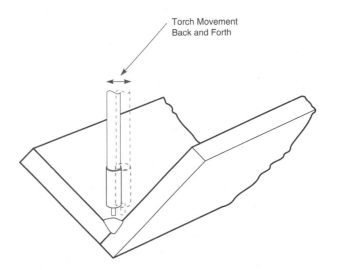

FIGURE 11–14   Oscillating through-the-arc sensor with welding gun.

Vision sensors to be effective must be designed for applications that completely overcome the hazards of the welding environment, such as smoke and heat. They are capable of accumulating large amounts of information about the weld pool.

A noncontacting viewing system mounted into the torch of the robot can project an image on a television monitor for real-time tracking of the welding process (Figure 11–15). Tracking the weld pool in relation to the joint is more important than following the torch or gun. A constant level of luminance with a high contrast can outline the edge of the joint. The visual information communicated to the robot controller enables it to correct for the location of the torch or gun in relation to the edge of the joint.

Sensors can judge width, depth, and length, measured from an imaginary centerline of the root opening. The movement of the robot arm, the voltage, the amperage, the wire feed, and the gas flow can be controlled to accommodate changes that develop ahead of the weld pool (Figure 11–16). Sensors can pick up **tack welds** and adjust the welding **parameters** to remelt the tack welds, if necessary. They can also sense the need for additional **passes,** if required.

For quality results, the robot must position the welding gun correctly at the start of the weld. Lasers are used in robotic welding to overcome the problem of locating the spot to begin welding. A low-power **laser sensor** mounted on the front of the welding gun oscillates to search out the beginning of the weld. The robot controller analyzes the

signal created by the laser, correcting the position of the gun for any deviations when beginning to weld the joint.

REVIEW

1. What is the work envelope?
2. Why does robotics require skilled welders?
3. Give three reasons for using robotic welding.
4. Define the four basic components of a robotic system.
5. What makes a positioner valuable in robotic welding?
6. What methods can be used to program a robot?
7. What is the one major advantage that the human welder has over the robot?
8. Name the two kinds of sensors.
9. What are two types of noncontacting sensors? Describe the differences between them.
10. Describe the use of lasers in robotics.

Create Three Questions

1.
2.
3.

Related English Questions

1. Write a paragraph using the following terms: *work envelope, boundary devices, robot, positioner, welding equipment.*
2. Write a paragraph on the safety concerns of robotic welding.

For Further Thought

1. In what types of product manufacturing might the use of robots be beneficial?
2. What types of business would not find robots beneficial?
3. The word *robot* comes from Czech, meaning "worker." Science fiction writer Isaac Asimov wrote a collection of short stories in 1950, *I Robot,* in which he laid down rules for robots to exist as workers serving human beings. His rules state that a robot will protect people at the risk of its own life,

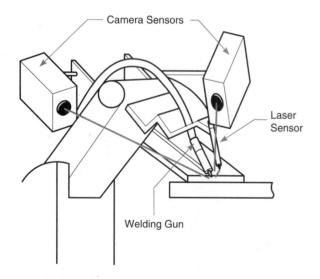

FIGURE 11–15   Direct arc sensor on robot.

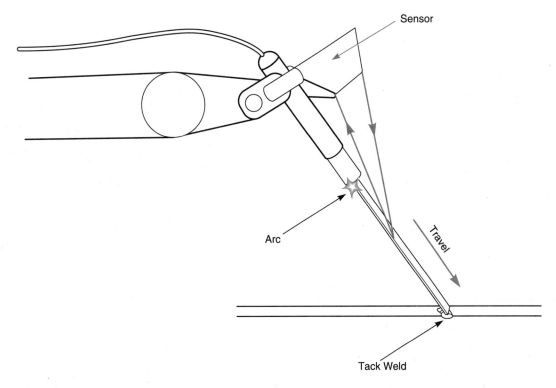

Sensor

Arc

Travel

Tack Weld

FIGURE 11–16   Sensor measuring root opening in front of the torch.

listen to authority, and obey all laws, even at the cost of its own safety. Do these rules apply to the robotic systems we are developing in the real world? Explain.

SUGGESTED ACTIVITIES

1. Invite someone working in the area of robotics to speak to the class (the discussion need not focus on welding).

2. Tour an industry that uses robotics for welding.

3. As a group or individually, discuss and design a product that could be welded by a robot.

4. Design a robot for a given welding task.

UNIT 12

Pipe Welding

"I have heard that there are two kinds of welders: production welders, who don't like tricky setups and enjoy doing the same thing over and over again; and maintenance welders, who hate it when they have to do the same job twice."

—Robert M. Pirsig, *Zen and the Art of Motorcycle Maintenance*

GOAL

- Develop an ability to weld pipe safely

QUESTIONS

- What is pipe welding?
- What does its specialized terminology include?
- How do pipe welding techniques differ?
- What are some of the ways to prepare pipe for welding?

SAFETY FIRST

The safety procedures listed in this book for **oxyacetylene welding** and the electric **arc welding processes** apply to the **welding** of **pipe.** Additional safety procedures should be followed with maintenance pipe welding for industry and transportation. High-pressure piping is used in the nuclear power and chemical industries. Pipelines transport **volatile** and hazardous substances across the country.

The **American Petroleum Institute** (API), the **American Society of Mechanical Engineers** (ASME), and the **American Welding Society** issue **procedure qualifications** and **procedure specifications** for pipe welding. More information can be found in *Qualification of Welding Procedures and Welders for Pipe and Tubing* (D10.9-80) by the AWS, *Welding and Brazing Qualifications* (ISBN 0-7918-2019-X) by the ASME, and *Welding of Pipelines and Related Facilities* (831-11040) and *Pipeline Maintenance Welding Practices* (831-11070) by the API.

TRAINING FOR PIPE WELDING

Pipe welding requires an advanced welding skill. Although many of the techniques developed in welding **plate** apply, pipe welding goes beyond them. Try to visualize the center of the pipe to understand the challenge that pipe welding presents in maintaining the **travel angle.** The **electrode, torch,** or **gun,** must move quickly to remain pointing at the center of the pipe (Figure 12–1). The smaller the diameter of the pipe, the faster the movement required to keep the focus on the center.

Critical pipe welding involves high-pressure systems that are subject to specifications that gov-

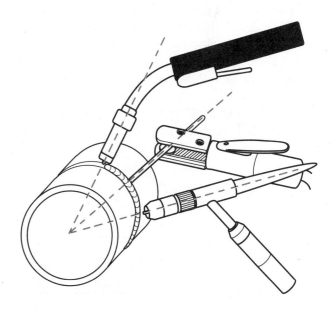

FIGURE 12–1 Focusing travel angle on the center of the pipe.

305

ern welding procedures. These specifications are issued by various organizations, such as the American Society of Mechanical Engineers and the American Petroleum Institute. Welder performance qualification is required for critical pipe welding. In contrast, **noncritical pipe welding** involves low-pressure pipes used in water, sanitary, and heating systems, to name three. Welder performance qualification may or may not be required for noncritical pipe welding.

Pipe welding is at the top of the welding profession because of three factors: the cost of pipe for training exercises, the time required to learn pipe welding, and the coordination skills that pipe welding demands. If you have successfully completed the exercises to this point, congratulations are in order. You have developed the skills you will need to seek out rewarding employment in welding. The ability to do quality pipe welding will add to the arsenal, making you even more valuable as an employee.

POSITIONS OF THE PIPE

There are five basic positions of the pipe for welding. These positions are described here in order of difficulty, with the 1G position being the easiest and the 6GR position the most difficult.

In the **1G position,** the pipe is laid out horizontally, with 0° to 15° of slope (Figure 12–2). The pipe is not fixed in place and can be rolled. In the **2G position,** the pipe is laid out vertically, with 0° to 15° of slope (Figure 12–3). The **5G position** means that the pipe is laid out horizontally, with 0° to 15° of slope (Figure 12–4). The pipe is fixed in place and cannot be rolled. In the **6G position,** the pipe is laid out on an incline of 45° (plus or minus 30°) of slope (Figure 12–5). The final position, the **6GR position,** is similar. The pipe is laid

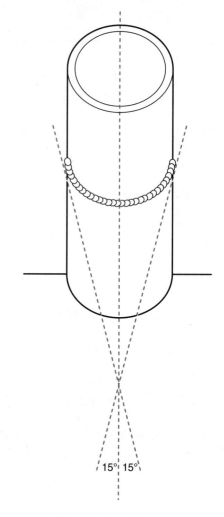

FIGURE 12–3 Pipe in the 2G position.

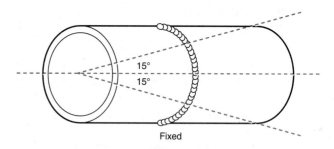

FIGURE 12–4 Pipe in the 5G position.

on an incline of 45° (plus or minus 30°) of slope, and a ring is positioned around it, causing a restriction making the welding more difficult (Figure 12–6).

DOWNHILL AND UPHILL WELDING

Downhill pipe welding is used on pipelines that transport oil and gas across the country.

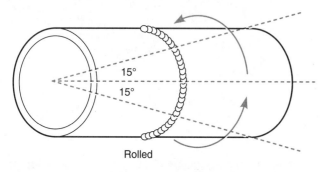

FIGURE 12–2 Pipe in the 1G position.

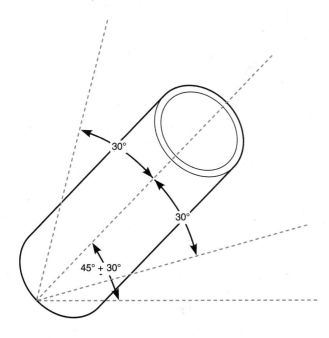

FIGURE 12–5   Pipe in the 6G position.

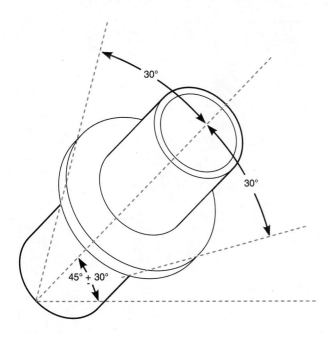

FIGURE 12–6   Pipe in the 6GR position.

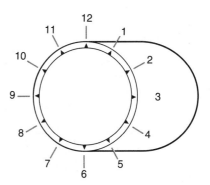

FIGURE 12–7   Pipe as a clock face.

TABLE 12–1:	PASSES REQUIRED FOR DOWNHILL PIPE WELDING
Wall Thickness of Pipe	**No. of Passes**
1/4″	3
5/16″	4
3/8″	5
1/2″	7

Downhill welding is also used on thin-walled pipe. This is considered to be the fastest method when using **shielded metal arc welding** for pipe with a wall thickness of up to 1/2 inch.

Looking at the end of a pipe, imagine the face of a clock. Downhill pipe welding begins at 12 o'clock and finishes at 6 o'clock (Figure 12–7).

Imagine two welders working together as a team, one on each side of a 4-foot-diameter pipe. The **root bead** is followed by the **hot pass** (additional amperage) to remove **defects.** The hot pass should eliminate incomplete fill along the **weld interface,** where the **base metal** meets the **filler metal.** The hot pass should also remove **slag** left by a less than perfect cleaning job (Figure 12–8).

Additional **filler passes** or **stringers** are made without removing the top edges of the **joint.** The edges of the joint act as boundaries to guide each pass. The **cover pass** takes out the edges, leaving a **convex** contour to the **weld,** which overlaps the edge on each side by 1/16 inch (Figure 12–9).

Downhill welding requires more passes than **uphill** welding because gravity requires a faster travel speed to keep the **weld pool** from flowing below the **arc.** Table 12–1 indicates the number of passes required in downhill welding for pipes with various wall thickness up to 1/2 inch.

Uphill pipe welding is used in industries such as power, paper, and chemical. It requires more demanding **welding procedures** than downhill welding. Instead of beginning at the top (the 12 o'clock position) like downhill welding, uphill welding begins at the bottom (the 6 o'clock position). Uphill welding proceeds more slowly but requires fewer passes than downhill welding (Table 12–2).

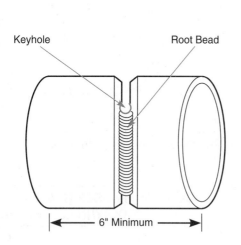

FIGURE 12–8 Root bead and hot pass.

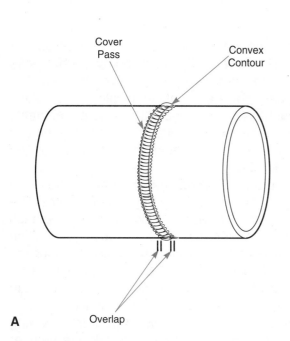

FIGURE 12–9 (A) Cover pass with overlap. (B) Stringer beads. *(Courtesy of Larry Jeffus.)*

Backing rings are strips of metal that are positioned inside the pipe mainly to keep the root pass from sagging. However, backing rings also restrict the flow of fluids. There are two basic types of backing rings. The **split ring** is a flexible ring that allows some adjustment and a closer fit. The more expensive **solid ring** allows no adjustment. The diameter must be exact for very close fitting (Figure 12–10).

PIPE AND TUBING

The differences between pipe and **tubing** can be a little confusing. A major use of pipe is the

TABLE 12–2: PASSES REQUIRED FOR UPHILL PIPE WELDING

Wall Thickness of Pipe	NO. OF PASSES	
	With Backing Ring	Without Backing Ring
1/4″	2	2
3/8″	3	3
1/2″	3	4
5/8″	4	5
3/4″	6	7
1	7	10

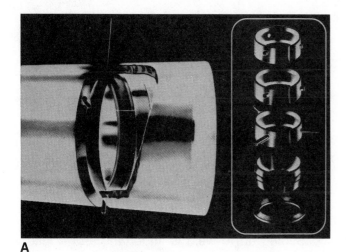
A

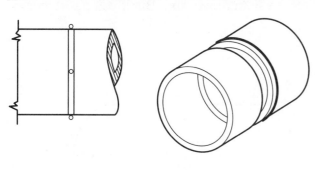

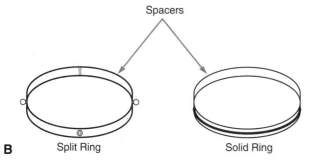
Spacers

B   Split Ring   Solid Ring

FIGURE 12–10 (A) Backing rings. *(Courtesy of Robvon Backing Ring Co.)* (B) Split and solid backing rings.

transportation of fluids. This is generally not the case with tubing. Round mechanical tubing is used in axles and truck and trailer parts. Square structural tubing can be easily welded, formed, and drilled for building and fabrication. Hydraulic tubing is used in machine tool hydraulic lines, air lines, and spray equipment (Figure 12–11).

Another difference between pipe and tubing is how they are measured. Pipe is measured by its inside diameter, up to and including 12-inch-diameter pipe. Tubing, which can be square or round, is measured by its outside diameter. A pipe might be specified as 2-inch Schedule 40. In this case, the inside diameter of pipe is 2 inches, and Schedule 40 indicates a 0.068-inch wall thickness. Schedule 5 to Schedule XX (double extra strong) provides over 14 different **schedules** for various wall thicknesses. Table 12–3 lists some common Schedule 40 and Schedule 80 pipe sizes. Compare Schedule 40 to Schedule 80. Which has a thicker wall?

PIPE WELDING PROCESSES

Several processes exist for welding pipe together. All of the welding processes described with some detail in this book can be used to weld pipe. For example, the oxyacetylene welding process works with pipe, but it should be limited to 4-inch pipe. This does not mean that pipe of more than 4 inches in diameter cannot be welded by oxyacetylene welding. Wall thickness must also be considered; acceptable oxyacetylene welding is limited to pipe with a wall thickness of 3/16 inch. Other welding processes become more cost-effective as the size of the pipe increases.

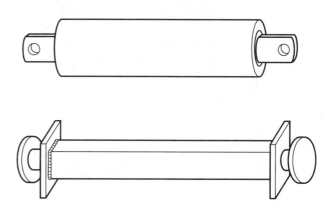

FIGURE 12–11 Axle and hydraulic cylinder.

TABLE 12–3:   COMMON SCHEDULE 40 AND SCHEDULE 80 PIPE SIZES

Pipe Size	SCHEDULE 40		SCHEDULE 80	
	Wall Thickness	Weight per Foot	Wall Thickness	Weight per Foot
Inside diameter				
1″	0.133″	1.68 lbs.	0.179″	2.17 lbs.
2″	0.154″	3.65 lbs.	0.218″	5.02 lbs.
3″	0.216″	7.58 lbs.	0.300″	10.25 lbs.
4″	0.237″	10.79 lbs.	0.337″	14.98 lbs.
6″	0.280″	18.97 lbs.	0.432″	28.57 lbs.
8″	0.322″	28.55 lbs.	0.500″	43.39 lbs.
10″	0.365″	40.48 lbs.	0.594″	64.40 lbs.
Outside diameter				
14″	0.438″	63.37 lbs.	0.750″	106.13 lbs.
18″	0.562″	104.76 lbs.	0.938″	170.84 lbs.
24″	0.688″	171.17 lbs.	1.219″	296.53 lbs.

Gas tungsten arc welding is used for critical applications and for joining specialty metals that are not readily joined by other welding processes. Shielded metal arc welding is commonly used for joining pipe. **Gas metal arc welding** and **flux cored arc welding** also have pipe welding applications.

When gas tungsten arc welding or gas metal arc welding is chosen, it may be necessary to provide **shielding gas** to the inside of the pipe. A **Y-valve** can be used to supply shielding gas from a separate line to the inside of the pipe (Figure 12–12).

The ends of the pipe must be closed off to keep the shielding gas at the joint. **Plugs** can be purchased for this purpose, or handmade plugs can be constructed (Figure 12–13). Cardboard and adhesive tape will do the job in some applications.

PREPARATION FOR PIPE WELDING

Setting up to weld is important, and it takes time away from welding. But if you patiently pre-

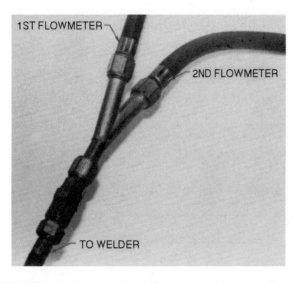

FIGURE 12–12   Y-valve. *(Courtesy of Larry Jeffus.)*

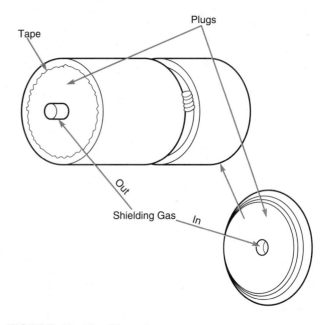

FIGURE 12–13   Pipe plugs.

pare yourself by doing all of the little things to get ready, more than likely you'll do **quality welding.**

Pipe Cutting

Pipe cutting can be accomplished in many different ways in preparation for welding. Hacksaw, oxyacetylene torch, power band saw (Figure 12–14), abrasive cutoff saw, and lathe are five ways to make the initial cut.

A **butt joint** generally requires a **bevel** for **complete root penetration.** A grinder or **freehand** cutting torch can be used to make the bevel. A torch with special pipe-cutting equipment can be adjusted for accurate bevels in wall thicknesses up to 3/4 inch (Figure 12–15). Bevels of 30° and 37.5° are common (Figure 12–16).

Pipe Marker

Freehand cutting might require some assistance. A mechanical pipe-marking tool can set the degrees of the cut angle (Figure 12–17). Pipe cuts can be also be laid out on the pipe with the help of mathematics. The key for using a pipe-marking tool or mathematics is that all measurements be taken from the diameter of the pipe.

To locate the pipe diameter, wrap a piece of paper around the pipe equal to the circumference. Fold the paper in half, and put it back on the pipe, marking it with soapstone in two places at the crease and where the edges meet (the diameter of the pipe) (Figure 12–18).

Measure the outside diameter of the pipe. Find the cut angle, which is half of the pipe angle (the included angle if there is only one joint). Multiply the outside diameter (OD) by the tangent of the cut angle and then divide by two to find the measurement of the cut angle. Mark it on the pipe, and connect the marks (Figure 12–19). (A more thorough discussion of the process can be found in pipe-fitting manuals.)

For example, outside diameter of 6-inch pipe is 6.625 inch $\times$ 0.5773 (tangent for 30° cut angle) = 3.8 inch ÷ 2 or approximately 1 7/8 inch (Figure 12–20). Table 12–4 provides a list of tangents for various cut angles. Remember that the cut angle is the cut through the pipe; this is not the bevel of the pipe wall.

Freehand pipe cutting requires keeping the **cutting tip** of the torch on the imaginary center of the pipe (Figure 12–21).

Alignment for Welding

The larger the diameter of the pipe, the more difficult the job of aligning the pieces for welding.

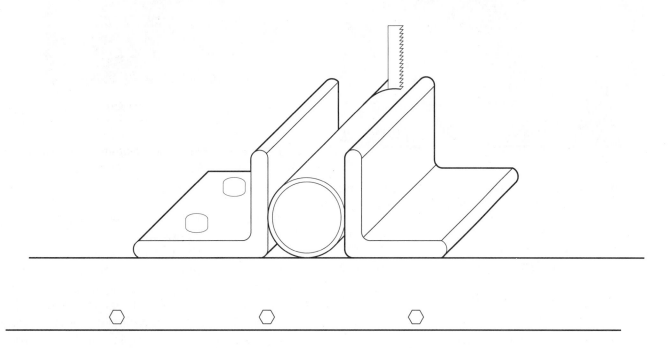

FIGURE 12–14   A band saw cutting pipe.

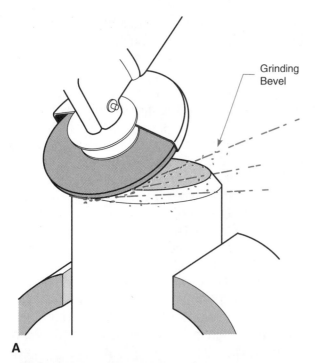

Grinding
Bevel

A

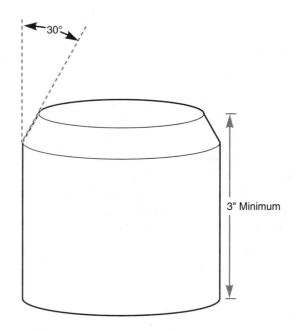

FIGURE 12–16   Bevel angle of 30°.

B

FIGURE 12–15   (A) Beveling pipe. (B) Cutting and beveling pipe in one action. *(Courtesy of Larry Jeffus.)*

Pipe clamps help in some applications (Figure 12–22). Setting the pipe on a length of angle iron works for short pieces of small diameters (Figure 12–23).

The welding specifications determine the welding electrodes. 6010 and 6011 are popular choices

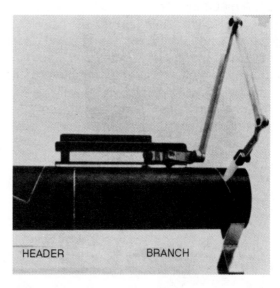

HEADER   BRANCH

FIGURE 12–17   Mechanical pipe-marking tool. *(Courtesy of Larry Jeffus.)*

for applications requiring pipe welding (Figure 12–24). When *10* appears as the last two digits, the electrode can be used for quality work in pipe welding. A pipe welding electrode leaves a small residue of slag, which makes the job of controlling the arc easier for downhill welding. Electrodes 6010 and 6011 demand a **whipping** motion during welding. Whipping is the motion of the electrode in and out of the weld pool.

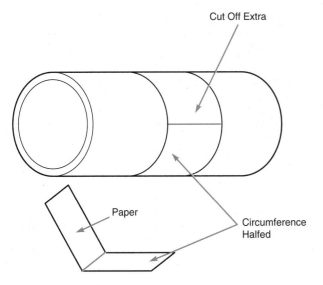

FIGURE 12–18   One method of locating pipe diameter.

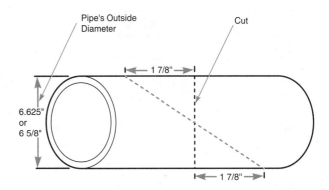

FIGURE 12–20   Cut angle for 6-inch pipe.

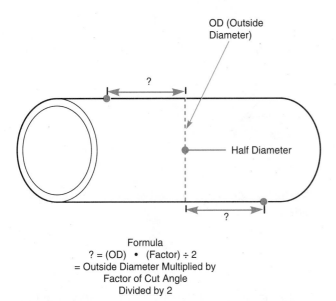

Formula
? = (OD) • (Factor) ÷ 2
= Outside Diameter Multiplied by
Factor of Cut Angle
Divided by 2

FIGURE 12–19   Measuring cut angle (connect the dots).

Setup for Pipe Welding

The pieces of the joint must be prepared. This may require cutting and beveling. (The degree of bevel is set by the welding procedure.) A grinder and **emery cloth** should be used to remove **mill scale,** varnish, and so on, from the joint, at least 1/4 inch outside and inside.

A grinder can be used for making a **root face,**

if necessary. The measurement of the root face will vary. Its purpose is to achieve complete root penetration without the formation of oxidized metal on the inside of the pipe.

A **welding rod** of the required diameter can be used to set the **root opening,** which can vary (Figure 12–25). Travel speed and amperage help to determine the root opening.

A pipe joint should be **tack welded** (outside the testing area) to maintain the root opening through the effects of welding. It may be necessary to **feather** the tack welds with a grinder for complete penetration and slag removal. To feather is to thin the tack welds (Figure 12–26).

The root bead is the most important pass. The root bead should be made only by stopping outside of the testing area. The **keyhole** melts the root opening on both sides of the joint just ahead of the weld pool. If the keyhole forms and is then backfilled, the welder has achieved complete root penetration (Figure 12–27). If the **visual inspection** after welding shows that the root pass does not have complete penetration, the joint will fail the **guided bend test.**

The hot pass by itself may not be enough to ensure a slag-free weld. If possible, the root pass should be touched with the grinder. Using the edges of the joint as a guide, lay as many filler passes as necessary to prepare for the cover pass (Figure 12–28).

The cover pass can be a single pass, using a **weave,** or a series of stringers. The size of the weld pool can be used to determine whether to weave the cover pass. The cover pass takes out the edges, leaving a **convex** weld (Figure 12–29).

TABLE 12–4: TANGENTS FOR VARIOUS CUT ANGLES

Cut Angle	Tangent	Cut Angle	Tangent	Cut Angle	Tangent
10°	0.1763	22°	0.4040	34°	0.6745
11°	0.1944	23°	0.4245	35°	0.7002
12°	0.2126	24°	0.4452	36°	0.7265
13°	0.2309	25°	0.4663	37°	0.7536
14°	0.2493	26°	0.4877	38°	0.7813
15°	0.2679	27°	0.5095	39°	0.8098
16°	0.2867	28°	0.5317	40°	0.8391
17°	0.3057	29°	0.5543	41°	0.8693
18°	0.3249	30°	0.5774	42°	0.9004
19°	0.3443	31°	0.6009	43°	0.9325
20°	0.3640	32°	0.6249	44°	0.9657
21°	0.3839	33°	0.6494	45°	1.0000

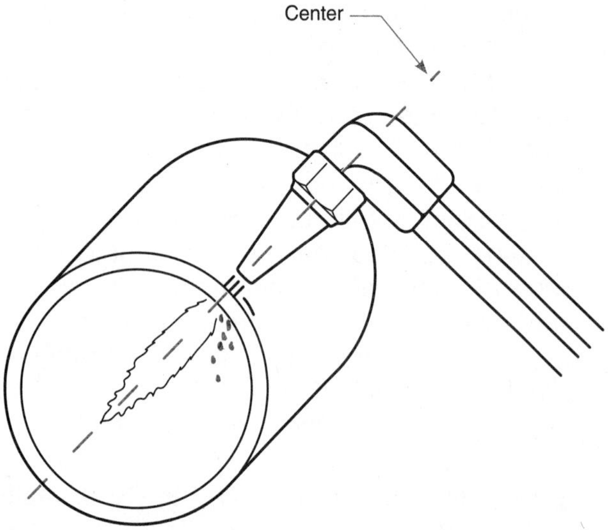

FIGURE 12–21 Torch cutting on center of pipe.

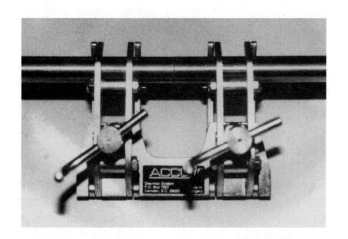

FIGURE 12–22 Pipe clamps. *(Courtesy of Larry Jeffus.)*

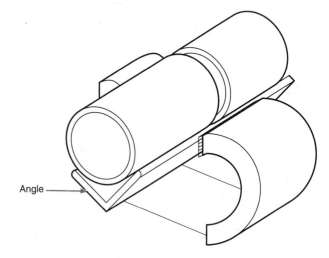

FIGURE 12–23 One method of positioning pipe to be welded.

Practice

Be patient. Learning to pipe weld will take some time. The key to success is practice. If you practice and understand the basic welding principles, you will have success.

REVIEW

1. Give the abbreviations of three organizations that are concerned with pipe welding.
2. Describe the difference between the 1G and 5G pipe positions.
3. In what industry is downhill welding commonly used?
4. What is the purpose of the hot pass?

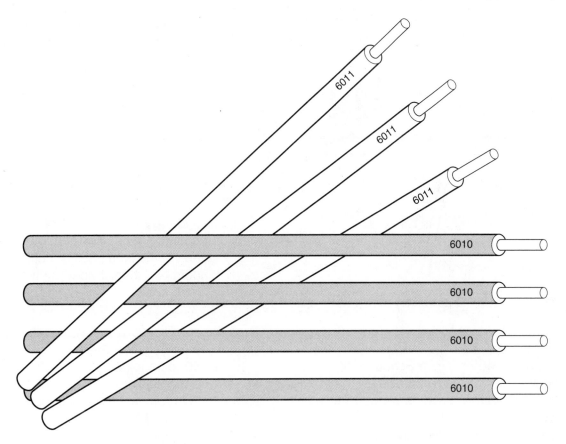

FIGURE 12–24   Commonly used electrodes for pipe.

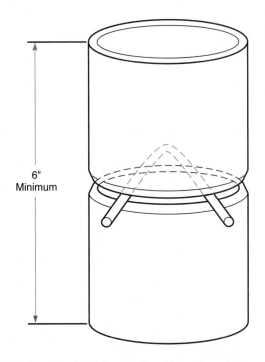

FIGURE 12–25   Welding rod setting pipe root opening.

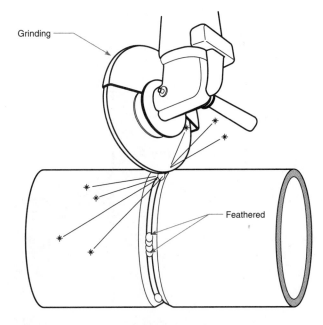

FIGURE 12–26   Feathering tack welds.

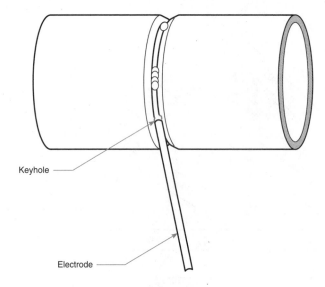

Keyhole

Electrode

FIGURE 12–27   Keyhole necessary during welding.

5. Which pipe has the thicker wall: 8-inch Schedule 40 or 8-inch Schedule 80?
6. How might a Y-valve be used?
7. What degree of bevel is common to pipe welding?
8. Three pipes (two joints) have to be cut for butt welding in an 80° pipe joint. Four-inch Schedule 40 pipe is being used. What is the cut angle? What is the measurement of the cut for use in laying out the cut angle on the pipe with soapstone?

Create Three Questions

1.
2.
3.

Related Math Questions

1. To find the measurement of the cut angle for cutting a pipe freehand, the outside diameter of the pipe (OD) is multiplied by the tangent (T), then divided by two:

OD × T ÷ 2 = measurement of the cut

Given the specifications below, find the measurement of the cut angle in each case.

    a. 50° pipe joint
       2-piece joint
       4-inch pipe, OD = 4.5 inch
       T = 0.4663
    b. 60° pipe joint
       3-piece joint
       6-inch pipe, OD = 6.625 inch
       T = 0.2679

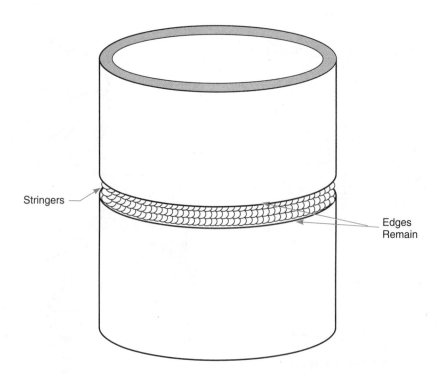

Stringers

Edges Remain

FIGURE 12–28   Filler passes.

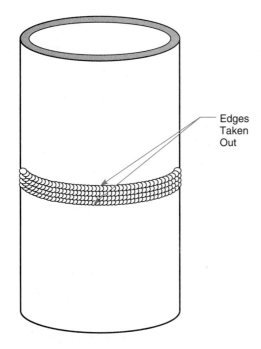

Edges
Taken
Out

FIGURE 12–29 Cover pass.

   c. 80° pipe joint
      4-piece joint
      8-inch pipe, OD = 8.625 inch
      T = 0.2370

For Further Thought

1. What could cause filler metal to pile up in a joint with poor penetration?
2. Name one problem that might arise when using shielded metal arc welding on 2-inch pipe that would not be a problem on 4-inch pipe?

3. The angle of a three-piece pipe joint is 130°. What are the cut angles for the pipe?
4. A job calls for bending tubing into a 90° angle. On the first attempt, the wall of the tubing collapses. What can be done to prevent this problem on the second attempt?

SUGGESTED ACTIVITIES

1. Practice laying out different cut angles on pipe.
2. Visit a facility where pipe welding is done.
3. Become acquainted with information put out by the API or the ASME.
4. Invite a pipe welder to speak to the class.

UNIT 12: WELDING EXERCISES

Review the safety procedures for the welding process to be used before beginning each of the following exercises. If a visual examination reveals **incomplete fusion** on the root bead or the cover pass, cut out the weld and start over. Pipe is expensive, so take every opportunity to reuse the materials.

It is not necessary to complete all of these exercises for pipe welding. Concentrate on the exercises that will be helpful in developing the skills you will need to get that job. Adapt these exercises to welding procedures that are being used in your area. Call local industries to find out what these procedures are, if need be. The AWS *Structural Welding Code* D1.1 is indicated where it was used to design the exercise.

UNIT 12: EXERCISE 1

Oxyacetylene Welding (OAW): Butt Joint in 2G Position, Square-Groove Weld

This exercise is designed to provide practice in welding small-diameter pipe. Small-diameter pipe can be easily welded by using the oxyacetylene welding process. Shielded metal arc welding or gas metal arc welding requires quicker movements and more adjustments at the power source. With oxyacetylene welding, the welder can control the heat input simply by pulling the torch away from the pipe. Keep the torch on the center of the pipe. Push the **rod** into the weld pool, using the same techniques you used for welding sheet steel in the **horizontal position.** Figure 12–30 shows the setup for this exercise.

Necessary Material and Equipment

Safety glasses	Goggles	2 pieces 2″ Schedule 40	Welding rod, 1/8″
Welding gloves	Sparklighter	pipe, 3″ long	Half-round file
Protective clothing	Welding tip		

Instructions

1. File each piece of pipe 1/4 inch back on both the outside and the inside of the joint to remove any mill scale, rust, or varnish.
2. Match the **welding tip** size to the wall thickness, and correctly adjust the pressure settings on the regulators.
3. Set the root opening up to one-half the wall thickness of the pipe.
4. Tack weld the pipe in three places. Remember that this will be a square-groove weld.
5. Begin at a tack weld. Remelt the tack welds, adding **filler metal.**
6. Complete the weld in one pass.
7. Test the weld by visual inspection. Examine it for incomplete fusion on the inside and **incomplete fill** on the outside.
8. Check with the instructor for evaluation, if necessary.
9. Practice this exercise until you meet the standards established in the course.

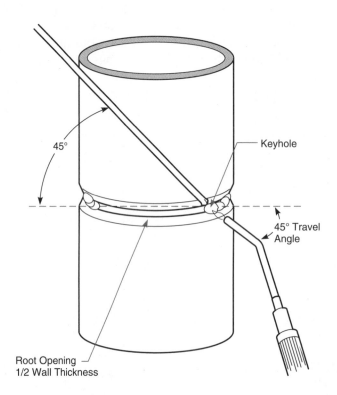

FIGURE 12–30   OAW butt joint in 2G position.

UNIT 12: EXERCISE 2

Oxyacetylene Welding (OAW): Butt Joint in 5G Position, 5G Pipe Welding

Welding begins in the **overhead position** and ends in the **flat position** after going through the **vertical position.** Once again, apply the same welding techniques used in welding sheet steel. The wall thickness of 2-inch Schedule 40 pipe is within the range of what would be considered **sheet metal.** Move the flame from side to side inside the joint while moving upward. The setup is illustrated in Figure 12–31.

Necessary Material and Equipment

Safety glasses	Goggles	2 pieces 2″ Schedule 40	Welding rod, 1/8″
Welding gloves	Sparklighter	pipe, 3″ long	Half-round file
Protective clothing	Welding tip		

Instructions

1. File each piece of pipe back 1/4 inch on both the outside and the inside of the joint to remove any mill scale, rust, or varnish.
2. Match the welding tip size to the wall thickness, and correctly adjust the pressure settings on the regulators.
3. Set the root opening up to one-half the wall thickness of the pipe.
4. Tack weld the pipe in three places.
5. Begin with a tack weld at the 6 o'clock position. Remelt each tack weld, stopping at the 12 o'clock position. Repeat on the other side of the pipe.
6. Complete the weld in one pass.
7. Test the weld by visual inspection. Examine it for incomplete fusion. Observe the quality of the weld bead.
8. Check with the instructor for evaluation, if necessary.
9. Practice this exercise until you meet the standards established in the course.

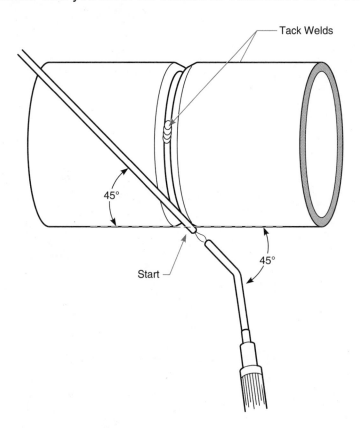

FIGURE 12–31   OAW butt joint in 5G position.

UNIT 12: EXERCISE 3

Shielded Metal Arc Welding (SMAW): Butt Joint in 2G Position, V-Groove Weld Designed to Meet AWS *Structural Welding Code*

The successful completion of these exercises begins with the preparations before welding. As you weld, keep the focus on the keyhole. The keyhole indicates complete root penetration on the important root bead. If a keyhole fails to develop, stop welding and prepare another joint. If the keyhole becomes too large, narrow the root opening, lower the amperage, or quicken the travel speed. If the keyhole does not develop, widen the root opening, raise the amperage, or slow down the travel speed. Figure 12–32 illustrates the setup for this exercise.

Necessary Material and Equipment

Safety glasses	Earplugs	2 pieces 4″ A36 Schedule	E6010 electrodes, 1/8″
Welding gloves	Chipping hammer	40 pipe,	Half-round file
Protective clothing	Wire brush	3″ long	Soapstone
Helmet	Grinder		

Instructions

1. Bevel each piece of pipe 30°.
2. File the pipe back 1/4 inch on both the outside and the inside to remove contaminants.
3. Grind the root face up to 1/8 inch.
4. Tack weld the pipe in four places with a root opening of up to 1/8 inch.
5. Feather the tack welds with the grinder for complete fusion.

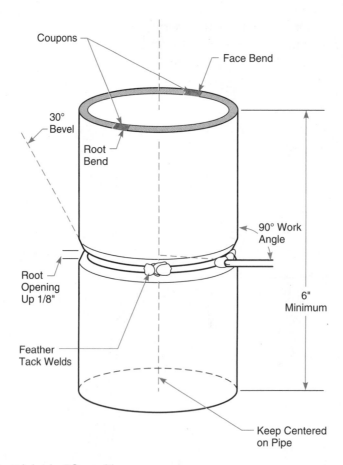

FIGURE 12–32 SMAW butt joint in 2G position.

6. Mark where the **coupons** will be taken for testing. Do not begin or end the arc where the coupons will be taken. Make all **arc strikes** within the groove angle.
7. Complete four passes, and make the second pass the hot pass. Run stringer beads for passes 3 and 4, which provide the cover passes.
8. Examine the root bead for complete penetration, and check the cover pass for complete fusion along the joint. The cover pass should extend no higher than 1/16 inch above the surface of the pipe.
9. Test the weld by passing one **face bend** and one **root bend.**
10. Check with the instructor for evaluation, if necessary.
11. Practice this exercise until you meet the standards established in the course.

UNIT 12: EXERCISE 4

Shielded Metal Arc Welding (SMAW): Butt Joint in 5G Position, V-Groove Weld Designed to Meet AWS *Structural Welding Code*

For this exercise, remember to keep the electrode moving and always pointed at the center of the pipe. Push the electrode into the weld pool quickly enough to keep the keyhole forming. Weave the second pass (hot pass) and the third pass. Clean completely after each pass. The setup for this exercise is shown in Figure 12–33.

Necessary Material and Equipment

Safety glasses	Earplugs	2 pieces 4″ A36 Schedule	E6010 electrodes, 1/8″
Welding gloves	Chipping hammer	40 pipe, 3″ long	Half-round file
Protective clothing	Wire brush		Soapstone
Helmet	Grinder		

Instructions

1. Bevel each piece of pipe 30°.
2. File the pipe back 1/4 inch on both the outside and the inside to remove contaminants.
3. Grind the root face up to 1/8 inch.
4. Tack weld the pipe in four places with a root opening of up to 1/8 inch.
5. Feather the tack welds with the grinder for complete fusion.
6. Mark where the coupons will be taken for testing. Do not begin or end the arc where the coupons will be taken. Make all arc strikes within the groove angle.
7. Complete three passes, and make the second pass the hot pass. Weave the second and third passes. Begin at the 6 o'clock position, and end at the 12 o'clock position.
8. Examine the root bead for complete penetration, and check the cover pass for complete fusion. The cover pass should extend no more than 1/16 inch above the surface of the pipe.
9. Test by passing two face bends and two root bends.
10. Check with the instructor for evaluation, if necessary.
11. Practice this exercise until you meet the standards established in the course.

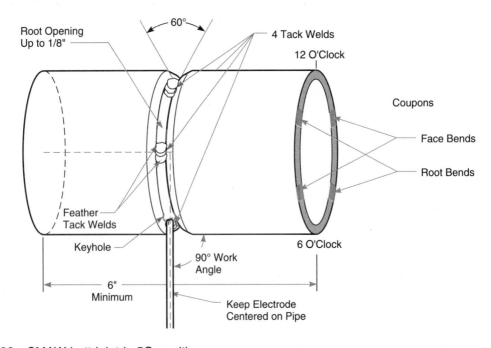

FIGURE 12–33 SMAW butt joint in 5G position.

UNIT 12: EXERCISE 5

Gas Tungsten Arc Welding (GTAW): Butt Joint in 2G Position, V-Groove Weld

This exercise and the next are designed for **low-carbon steel** pipe welding. The more-expensive **aluminum** and **stainless steel** pipe welding exercises can be added later. Remember that the welding of aluminum and stainless steel require a shielding gas on the inside of the pipe during the root bead. Choose a **nozzle** that fits comfortably into the groove angle forming the joint. With practice, you will learn to walk (move from side to side) the nozzle on the pipe without sticking the **tungsten electrode.** The success of the root bead depends on the formation and fill of the keyhole along the entire joint. Figure 12–34 gives the setup for this exercise.

Necessary Material and Equipment

Safety glasses	Grinder	2% thoriated tungsten,	Half-round file
Welding gloves	2 pieces 4" A36	3/32"	Soapstone
Protective clothing	Schedule 40 pipe,	Welding rod, 3/32"	Earplugs
Helmet	3" long	Shielding gas: argon	

Instructions

1. Bevel each piece of pipe 30°.
2. File the pipe back 1/4 inch on both the outside and the inside to remove contaminants.
3. Grind a root face of up to 1/16 inch.
4. Set the **output** power to **DCEN; high-frequency** is not required.
5. Tack weld the pipe in four places with a root opening of up to 3/32 inch.
6. Begin welding at one of the tack welds. Remelt the tack welds.
7. Complete four passes. The second pass is the hot pass. The third pass and the fourth pass are made as stringers, forming the cover passes.
8. Examine the root bead for complete penetration along the entire joint. The cover pass should

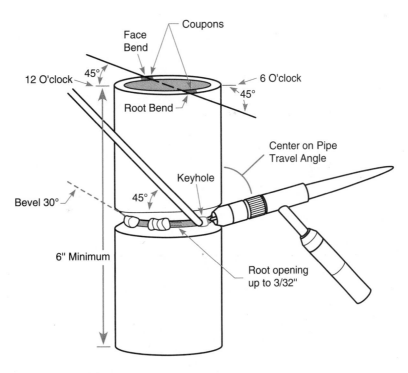

FIGURE 12–34   GTAW butt joint in 2G position.

have complete fusion with the **base metal,** with up to 1/16-inch **reinforcement** above the surface of the pipe.

9. Test by passing one face bend and one root bend.
10. Check with the instructor for evaluation, if necessary.
11. Practice this exercise until you meet the standards established in the course.

UNIT 12: EXERCISE 6

Gas Tungsten Arc Welding (GTAW): Butt Joint in 5G Position, V-Groove Weld

For this exercise, apply the techniques you developed during oxyacetylene welding. One difference, though, is that you must keep the torch aimed at the center of the pipe. Turn up the amperage if the **joint root** penetration is not complete. Turn down the amperage if the keyhole becomes too wide. Melt the filler metal in the weld pool. The setup for this exercise is shown in Figure 12–35.

Necessary Material and Equipment

Safety glasses	Grinder	2% thoriated tungsten,	Half-round file
Welding gloves	2 pieces 4″ A36 Schedule	3/32″	Soapstone
Protective clothing	40 pipe, 3″ long	Welding rod, 3/32″	Earplugs
Helmet		Shielding gas: argon	

Instructions

1. Bevel each piece of pipe 30°.
2. File the pipe back 1/4 inch on both the outside and the inside to remove contaminants.
3. Grind a root face of up to 1/16 inch.
4. Set the output power to DCEN; high-frequency is not required.
5. Tack weld the pipe in four places with a root opening of up to 3/32 inch.
6. Position tack welds at 3, 6, 9, and 12 o'clock, and remelt each tack weld. Begin welding at the 6 o'clock position, finishing at the 12 o'clock position. Repeat this procedure on the opposite side of the pipe.
7. Complete three passes. The second pass is the hot pass. The third pass, a weave, is the cover pass.

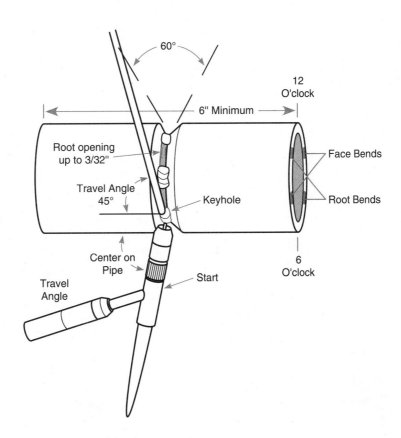

FIGURE 12–35 GTAW butt joint in 5G position.

8. Examine the root bead for complete penetration along the entire joint. The cover pass should have complete fusion with the base metal, with up to 1/16-inch reinforcement above the surface of the pipe.
9. Test the weld by passing two face bends and two root bends.
10. Check with the instructor for evaluation, if necessary.
11. Practice this exercise until you meet the standards established in the course.

UNIT 12: EXERCISE 7

Short Circuiting Arc Transfer (GMAW-S): Butt Joint in 2G Position, V-Groove Weld

The **short circuiting arc transfer** can be used for noncritical applications of pipe welding. The next two exercises are an opportunity for practice using a process not generally applied to pipe welding. Figure 12–36 illustrates the setup for this exercise.

Necessary Material and Equipment

Safety glasses	Earplugs	2 pieces 4" A36 Schedule	Half-round file
Welding gloves	Grinder	40 pipe, 3" long	Soapstone
Protective clothing	Sidecutter	Spool wire (0.030, 0.035)	Shielding gas: carbon
Helmet			dioxide

Instructions

1. Bevel each piece of pipe 30°.
2. File the pipe back on both the outside and the inside to remove contaminants.
3. Grind a root face of up to 1/8 inch.
4. Tack weld the pipe in four places with a root opening of up to 1/8 inch.
5. Feather the tack welds with the grinder for complete fusion.
6. Complete three passes. The second and third passes should be stringers. There is no hot pass.
7. Examine the root bead for complete penetration along the entire joint. The cover passes should fuse with the base metal and with each other.
8. Test the weld by passing one face bend and one root bend.
9. Check with the instructor for evaluation, if necessary.
10. Practice this exercise until you meet the standards established in the course.

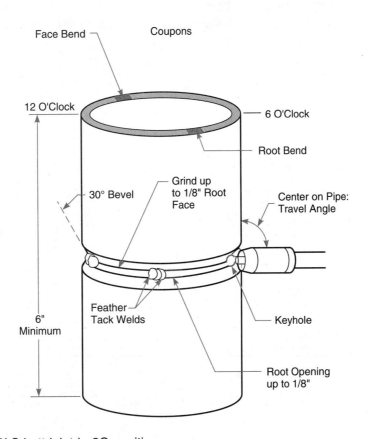

FIGURE 12–36 GMAW-S butt joint in 2G position.

UNIT 2: EXERCISE 8

Short Circuiting Arc Transfer (GMAW-S): Butt Joint in 5G Position, V-Groove Weld

Although this exercise is set up for downhill welding, uphill welding can be applied instead. Keep moving while holding the **wire** to a short **stickout.** Figure 12–37 illustrates the setup for this exercise.

Necessary Material and Equipment

Safety glasses	Grinder	Spool wire (0.030, 0.035)	Shielding gas: carbon
Welding gloves	Sidecutter	Half-round file	dioxide
Protective clothing	2 pieces 4″ A36	Soapstone	
Helmet	Schedule 40 pipe,		
Earplugs	3″ long		

Instructions

1. Bevel each piece of pipe 30°.
2. File back the pipe on both the outside and the inside to remove contaminants.
3. Grind a root face of up to 1/8 inch.
4. Tack weld the pipe in four places with a root opening of up to 1/8 inch.
5. Feather the tack welds with the grinder for complete fusion.
6. Position the tack welds at 3, 6, 9, and 12 o'clock. Begin welding at the 12 o'clock position, and finish at the 6 o'clock position. Repeat this procedure on the opposite side of the pipe.
7. Complete three passes.
8. Examine the root bead for complete penetration, and check the cover pass (weave) for complete fusion with the base metal.
9. Test the weld by passing two face bends and two root bends.
10. Check with the instructor for evaluation, if necessary.
11. Practice this exercise until you meet the standards established in the course.

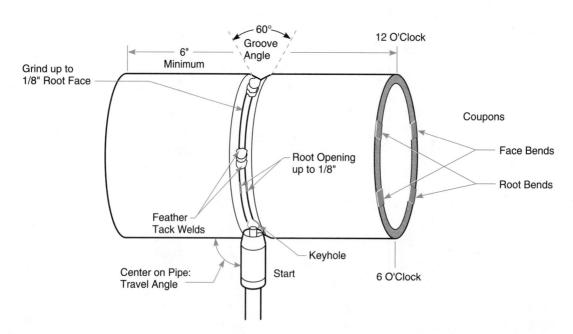

FIGURE 12–37 GMAW-S butt joint in 5G position.

UNIT 12: EXERCISE 9

Spray Arc Transfer (GMAW-SP): Butt Joint in 2G Position, V-Groove Weld Designed to Meet the AWS *Structural Welding Code*

Setting up the next two exercises for **spray arc transfer,** the root opening can be smaller, and perhaps no root opening is required. The greater heat generated by this process makes the difference. The setup for this exercise appears in Figure 12–38.

Necessary Material and Equipment

Safety glasses	Grinder	Spool wire (0.045)	Shielding gas: 98% argon,
Welding gloves	Sidecutter	Half-round file	2% oxygen
Protective clothing	2 pieces 4″ A36	Soapstone	
Helmet	Schedule 40 pipe,		
Earplugs	3″ long		

Instructions

1. Bevel each piece of pipe 30°.
2. File the pipe on both the outside and the inside to remove contaminants.
3. Grind a root face of up to 1/8 inch.
4. Tack weld the pipe in four places with a root opening of up to 1/8 inch.
5. Feather the tack welds with the grinder for complete fusion.
6. Complete three passes. The second and third passes should be stringers. There is no hot pass.
7. Examine the root bead for complete fusion along the entire joint. The cover passes should fuse with the base metal and with each other.
8. Test the weld by passing one face bend and one root bend.
9. Check with the instructor for evaluation, if necessary.
10. Practice this exercise until you meet the standards established in the course.

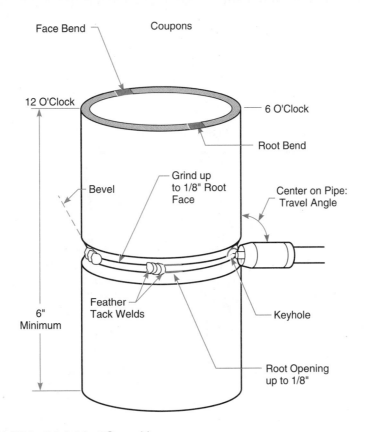

FIGURE 12–38   GMAW-SP butt joint in 2G position.

UNIT 12: EXERCISE 10

Spray Arc Transfer (GMAW-SP): Butt Joint in 5G Position, V-Groove Weld Designed to Meet the AWS *Structural Welding Code*

For successful completion of this exercise, apply the techniques you used when welding plate. Keep the gun moving to maintain the focus on the center of the pipe. Figure 12–39 shows the setup.

Necessary Material and Equipment

Safety glasses	Grinder	Spool wire (0.045)	Shielding gas: 98% argon,
Welding gloves	Sidecutter	Half-round file	2% oxygen
Protective clothing	2 pieces 4″ A36	Soapstone	
Helmet	Schedule 40 pipe,		
Earplugs	3″ long		

Instructions

1. Bevel each piece of pipe 30°.
2. File the pipe back on both the outside and the inside to remove contaminants.
3. Grind a root face of up to 1/8 inch.
4. Tack weld the pipe in four places with a root opening of up to 1/8 inch.
5. Feather the tack welds with the grinder for complete fusion.
6. Position the tack welds at 3, 6, 9, and 12 o'clock. Begin welding at the 12 o'clock position, finishing at the 6 o'clock position. Repeat this procedure on the opposite side of the pipe.
7. Complete three passes, weaving each one. Examine the root bead for complete penetration, and check the cover pass for complete fusion with the base metal.
8. Test the weld by passing two face bends and two root bends.
9. Check with the instructor for evaluation, if necessary.
10. Practice this exercise until you meet the standards established in the course.

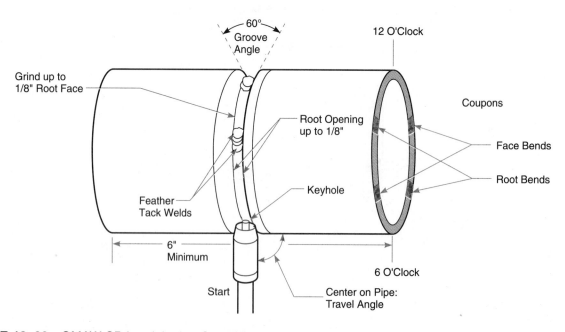

FIGURE 12–39 GMAW-SP butt joint in 5G position.

UNIT 12: EXERCISE 11

Flux Cored Arc Welding (FCAW): Butt Joint in 2G Position, V-Groove Weld Designed to Meet the AWS *Structural Welding Code*

With the size of the spool wire used for flux cored arc welding in this exercise, the travel speed may have to be faster. Clean between passes. Figure 12–40 shows the setup.

Necessary Material and Equipment

Safety glasses	Earplugs	Sidecutter	Spool wire, 1/16″
Welding gloves	Chipping hammer	2 pieces 4″ A36	Half-round file
Protective clothing	Wire brush	Schedule 40 pipe,	Soapstone
Helmet	Grinder	3″ long	

Instructions

1. Bevel each piece of pipe 30°.
2. File the pipe on both the outside and the inside to remove contaminants.
3. Grind a root face of up to 1/8 inch.
4. Tack weld the pipe in four places with a root opening of up to 1/8 inch.
5. Feather the tack welds with the grinder for complete fusion.
6. Complete three passes. The second and third passes should be stringers. There is no hot pass.
7. Examine the root bead for complete penetration along the entire joint. The cover passes should fuse with the base metal and with each other.
8. Test the weld by passing one face bend and one root bend.
9. Check with the instructor for evaluation, if necessary.
10. Practice this exercise until you meet the standards established in the course.

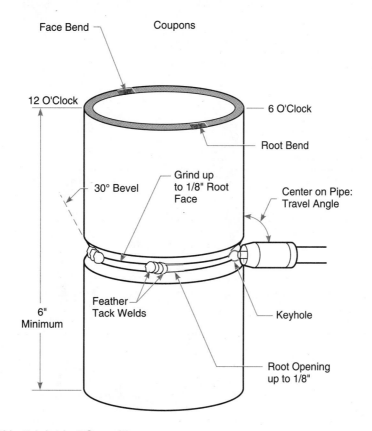

FIGURE 12–40   FCAW butt joint in 2G position.

UNIT 12: EXERCISE 12

Flux Cored Arc Welding (FCAW): Butt Joint in 5G Position, V-Groove Weld Designed to Meet the AWS *Structural Welding Code*

Leave the side edges of the joint until the cover pass. Weave the cover pass. The setup for this exercise is illustrated in Figure 12–41.

Necessary Material and Equipment

Safety glasses	Earplugs	Sidecutter	Spool wire, 1/16″
Welding gloves	Chipping hammer	2 pieces 4″ A36	Half-round file
Protective clothing	Wire brush	Schedule 40 pipe,	Soapstone
Helmet	Grinder	3″ long	

Instructions

1. Bevel each piece of pipe 30°.
2. File back the pipe on both the outside and the inside to remove contaminants.
3. Grind a root face of up to 1/8 inch.
4. Tack weld the pipe in four places with a root opening of up to 1/8 inch.
5. Feather the tack welds with the grinder for complete fusion.
6. Position the tack welds at 3, 6, 9, and 12 o'clock. Begin welding at the 12 o'clock position, finishing at the 6 o'clock position. Repeat this procedure on the opposite side of the pipe.
7. Complete three passes, weaving each one. Examine the root bead for complete penetration, and check the cover pass for complete fusion with the base metal.
8. Test the weld by passing two face bends and two root bends.
9. Check with the instructor for evaluation, if necessary.
10. Practice this exercise until you meet the standards established in the course.

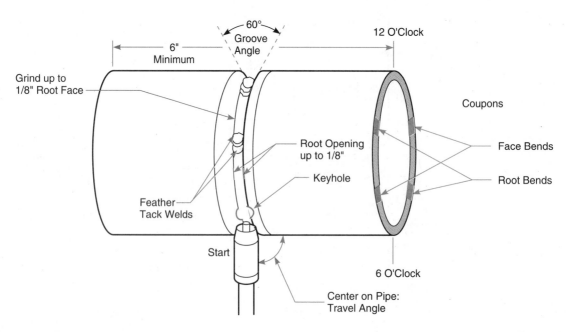

FIGURE 12–41 FCAW butt joint in 5G position.

Other Processes for Joining Metals

"Motorcycle maintenance gets frustrating. Angering. Infuriating. That's what makes it interesting."
—Robert M. Pirsig, *Zen and the Art of Motorcycle Maintenance*

GOAL

- Develop an awareness for some of the other welding processes

QUESTIONS

- What are some of the other welding processes?
- How do these processes work?
- When are some of these processes used?
- How do these processes differ from those learned in the shop?

WELDING TECHNOLOGY

There are nearly 100 welding and allied processes on the process wheel (Table 13–1). Some of these processes are becoming more popular, and the importance of others is diminishing. Every advance in technology brings change to the methods of production, and these changes can affect welding. Although change can be stalled, it cannot be stopped. Change can advance welding technology in what has to be an ongoing occupation in any productive enterprise.

This unit briefly describes only a few of nearly 100 welding processes that exist. Tomorrow, another process that no one has ever heard about will arrive on the scene, but no advancement in technology ever happens in isolation. Every advance in welding builds on present technology. Advances are made because someone is looking for a better way of doing things. Perhaps you will

be one of the innovators. Work hard, never stop asking questions, and never be satisfied with less than complete answers.

FORGE WELDING

Forge welding (FOW) must be considered one of the earliest processes for joining metal (Figure 13–1). In this welding process, the pieces of the **joint** are heated in a forge, or furnace, and are **fused** together with pressure produced by a hammer. Although forge welding is still being used to shape metal in manufacturing by electric-power-driven hammers, its popularity with the public rests on nostalgic images of the Old West and of early settlements on the North American continent.

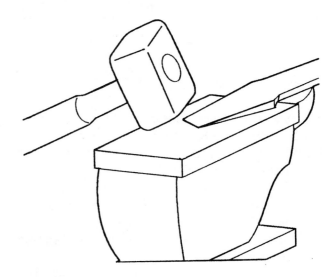

FIGURE 13–1 Blacksmith's anvil.

TABLE 13–1: PROCESS WHEEL

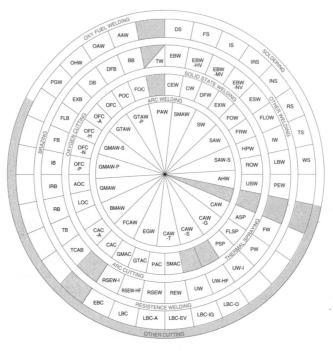

Arc welding (AW)		Plasma arc welding	PAW
Atomic hydrogen welding	AHW	Shielded metal arc welding	SMAW
Bare metal arc welding	BMAW	Stud arc welding	SW
Carbon arc welding	CAW	Submerged arc welding	SAW
Carbon arc welding—gas	CAW-G	Submerged arc welding—series	SAW-S
Carbon arc welding—shielded	CAW-S	**Solid state welding (SSW)**	
Carbon arc welding—twin	CAW-T	Coextrusion welding	CEW
Electrogas welding	EGW	Cold welding	CW
Flux cored arc welding	FCAW	Diffusion welding	DFW
Gas metal arc welding	GMAW	Explosion welding	EXW
Gas metal arc welding—pulsed arc	GMAW-P	Forge welding	FOW
Gas metal arc welding—short circuiting arc	GMAW-S	Friction welding	FRW
Gas tungsten arc welding	GTAW	Hot-pressure welding	HPW
Gas tungsten arc welding—pulsed arc	GTAW-P	Roll welding	ROW

Cont. on next page

SUBMERGED ARC WELDING

Submerged arc welding (SAW) was developed in the 1930s and is still popular today. In this welding process, a solid **wire** is fed continuously into the **weld pool.** Both the **arc** and the weld pool are protected by a granular **flux,** which can be added to the **weld** by air pressure. The flux keeps the weld pool and the arc from being seen. This eliminates both sparks and **spatter** (Figure 13–2).

Submerged arc welding can be used successfully for joining 3/8-inch to 1-inch **plate** in one **pass** in the **flat position** and **horizontal**

position. This **process** allows **base metal** up to 5/8 inch thick to be welded in a **butt joint** with a **square-groove weld.** It also allows the joining of even thicker steel with the preparation of only a single bevel. Submerged arc welding is commonly performed at up to 1,500 **amperes.** It provides deep **joint root** penetration without the problem of **distortion.**

ELECTROSLAG WELDING

Electroslag welding (ESW) is a modification of submerged arc welding. It is used to join together thick plate in the **vertical position.** This welding

TABLE 13–1: PROCESS WHEEL: (CONTINUED)

Ultrasonic welding	USW	Flow welding	FLOW
Soldering (S)		Induction welding	IW
Dip soldering	DS	Laser beam welding	LBW
Furnace soldering	FS	Percussion welding	PEW
Induction soldering	IS	Thermit welding	TW
Infrared soldering	IRS	**Oxyfuel gas welding (OFW)**	
Iron soldering	INS	Air acetylene welding	AAW
Resistance soldering	RS	Oxyacetylene welding	OAW
Torch soldering	TS	Oxyhydrogen welding	OHW
Wave soldering	WS	Pressure gas welding	PGW
Resistance welding (RW)		**Thermal spraying (THSP)**	
Flash welding	FW	Arc spraying	ASP
Projection welding	PW	Flame spraying	FLSP
Resistance seam welding	RSEW	Plasma spraying	PSP
Resistance seam welding—high frequency	RSEW-HF	**Oxygen cutting (OC)**	
Resistance seam welding—induction	RSEW-I	Flux cutting	FOC
Resistance spot welding	RSW	Metal powder cutting	POC
Upset welding	UW	Oxyacetylene gas cutting	OFC-A
Upset welding—high frequency	UW-HF	Oxyfuel gas cutting	OFC
Upset welding—induction	UW-I	Oxygen arc cutting	AOC
Brazing (B)		Oxygen lance cutting	LOC
Block brazing	BB	Oxyhydrogen cutting	OFC-H
Diffusion brazing	DFB	Oxynatural gas cutting	OFC-N
Dip brazing	DB	Oxypropane cutting	OFC-P
Exothermic brazing	EXB	**Arc cutting (AC)**	
Flow brazing	FLB	Air carbon arc cutting	CAC-A
Furnace brazing	FB	Carbon arc cutting	CAC
Induction brazing	IB	Gas metal arc cutting	GMAC
Infrared brazing	IRB	Gas tungsten arc cutting	GTAC
Resistance brazing	RB	Plasma arc cutting	PAC
Torch brazing	TB	Shielded metal arc cutting	SMAC
Twin carbon arc brazing	TCAB	**Other cutting**	
Other welding		Electron beam cutting	EBC
Electron beam welding	EBW	Laser beam cutting	LBC
Electron beam welding—high vacuum	EBW-HV	Laser beam cutting—air	LBC-A
Electron beam welding—medium vacuum	EBW-MV	Laser beam cutting—evaporative	LBC-EV
Electron beam welding—nonvacuum	EBW-NV	Laser beam cutting—inert gas	LBC-IG
Electroslag welding	ESW	Laser beam cutting—oxygen	LBC-O

process is similar to **electrogas welding** (EGW). With electroslag welding, the protective shielding **slag** causes resistance between the **filler metal** and the base metal, melting both. With electrogas welding, the heat generated by the arc melts the filler metal and the base metal, requiring a **shielding gas.**

Electroslag welding is really more like a casting process. Water-cool molded shoes restrict the flow of the weld pool, which is a combination of both the filler metal (two-thirds of the weld pool) and the base metal (one-third of the weld pool). The guide tube, which directs the flow of the filler metal,

moves up the joint as the weld is being made (Figure 13–3).

Electroslag welding is used on plate that is more than 1 inch thick and involves little preparation of the butt joint for the square-groove weld. This process is quite popular in the range of 400 to 700 amperes. The slag is easily removed with a chipping hammer when the joint is completed.

STUD ARC WELDING

Stud arc welding (SW) is a process for fusing studs (threaded or unthreaded) to a base metal.

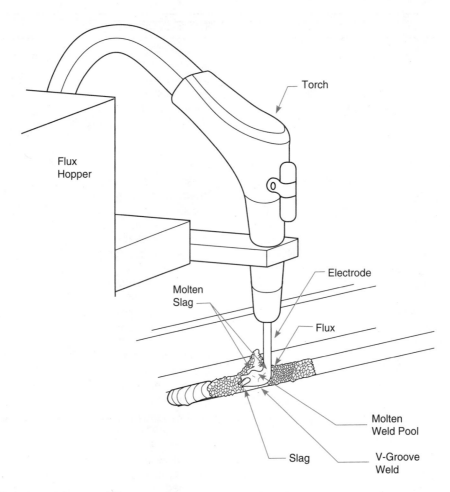

Torch

Flux
Hopper

Electrode

Molten
Slag

Flux

Molten
Weld Pool

Slag

V-Groove
Weld

FIGURE 13–2   Submerged arc welding process.

When the trigger of the gun is depressed, **output** power through the stud to the base metal creates an arc. The arc continues for a preset time, melting the stud into the base metal under pressure (Figure 13–4).

Stud arc welding operates on electric power supplied by a **motor generator** or a **rectifier/ transformer power source.** The specially designed gun uses various types of studs, including **low-carbon steel, stainless steel,** and **low-alloy steel.** Most studs contain a fluxing agent to protect the molten weld pool during the welding process. Stud arc welding can be used in some applications as a substitute for screwing in studs, saving the time required to tap the screw holes. This process is easy to learn and is popular for attaching floors and frames to metal structures.

RESISTANCE WELDING

There is a group of welding processes referred to as **resistance welding** (RW). Resistance welding was pioneered as a result of the development of electricity. The problems created by resistance of materials to the flow of electric **current** were turned into an opportunity with the development of resistance welding.

All of the welding processes that fall into the category of resistance welding are similar. Welding is caused by the resistance of the base metal to the flow of current, generating heat. Generally, neither filler metal nor fluxing material is ever used in these processes. Fusion results from melting together the sections of the joint with the addition of pressure.

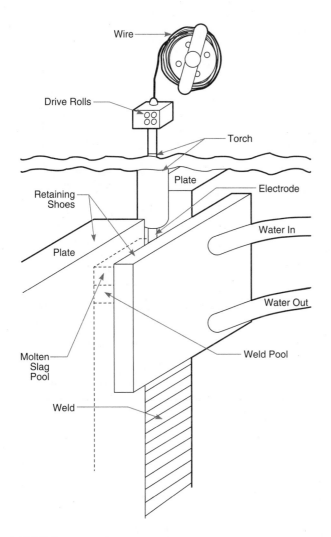

FIGURE 13–3 Electroslag welding process.

Resistance spot welding (RSW) is a process in which the **welding** area of a **lap joint** has been limited between two **electrodes.** It is a commonly used process found in many fabrication shops (Figure 13–5). One welding machine design consists of upper and lower electrodes (arms) made of **copper.** Water flows through the electrodes to keep them cool during the welding process. Pneumatic power, controlled by the operator's foot, provides the pressure necessary to force the sections of the lap joint together. As the upper electrode comes down, a timer can be adjusted to control the periods of both current and pressure. Resistance spot welding can easily join metals up to 1/8 inch thick (Figure 13–6).

Resistance seam welding (RSEW) is a process that can be a series of overlapping **spot** welds or continuous spot welds. With this process, the two copper electrodes are replaced by two rotating wheel electrodes, and water is applied either externally or internally to keep the rotating electrodes cool (Figure 13–7). The lap joint is the design used with this process. In some fabrication applications, gas- and liquid-tight seals can be achieved by resistance seam welding. Gasoline tanks on automobiles and trucks are joined together by this welding process.

High-frequency seam welding (RSEW-HF) and **induction seam welding** (RSEW-I) are forms of resistance seam welding. In high-frequency seam welding, a high-frequency welding current is conducted through electrodes that touch the base metal (Figure 13–8). In induction seam welding, the high-frequency welding current is magnetically passed from a copper coil without the electrodes touching the base metal (Figure 13–9). The high-frequency current used for these two seam welding processes limits the area of the **heat-affected zone,** resulting in stronger joints in a variety of metals. Both processes are popular in high-production operations for welding together **pipe** and **tubing** with wall thicknesses approaching 1 inch.

Projection welding (PW) is a resistance welding process in which the heat is limited to the projection (the raised surface). Pressure is used to force the pieces of the lap joint together in coordination with a preset time required to melt the projections (Figure 13–10). This process is used to attach machined parts to other parts. A common application involves joining table legs and fasteners (like nuts and bolts) to sheet metal.

Flash welding (FW) is a resistance welding process in which heat melts the surfaces along the edges of a butt joint. The pieces of the base metal forming the butt joint are brought together under pressure, forcing some of the metal out of the joint. The flashing (sparks) results in a momentary light show during the welding process (Figure 13–11). Window frames and propeller blades are among the many products joined by flash welding. A common application of flash welding is the fusion welding of band saw blades. Some band saws are sold with flash welding capabilities. The key to preparation is the removal of protective **mill scale** and grease in the joint area before welding.

Upset welding (UW) is another resistance

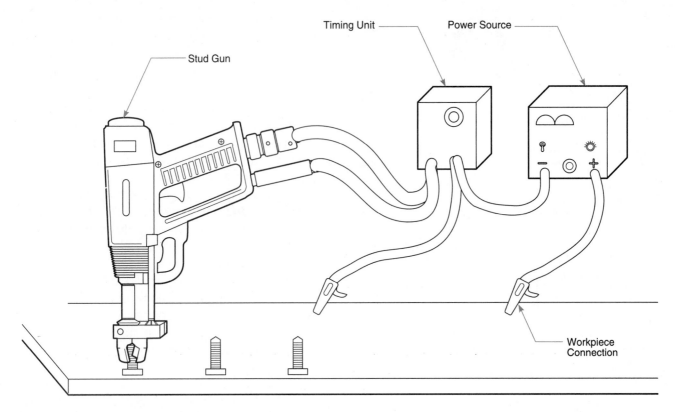

FIGURE 13–4 Stud arc welding process.

FIGURE 13–5 Resistance spot welding machine. *(Courtesy of Automation International, Inc., Danville, Illinois.)*

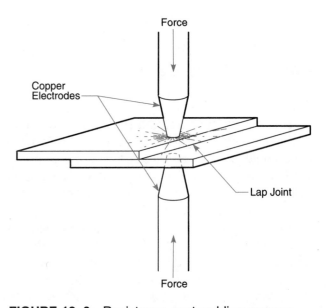

FIGURE 13–6 Resistance spot welding process.

FIGURE 13–7 Resistance seam welding machine. *(Courtesy of Automation International, Inc., Danville, Illinois.)*

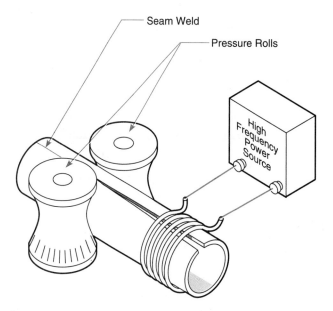

FIGURE 13–8 High-frequency seam welding process.

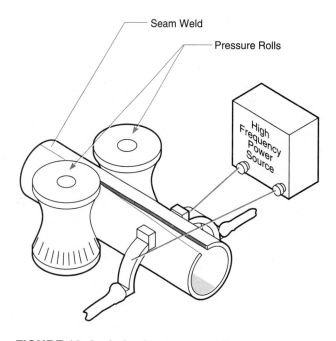

FIGURE 13–9 Induction seam welding process.

welding process, similar to flash welding. Heat is generated along the entire length of the butt joint, with pressure bringing the surfaces together. The pressure results in an **upset** (metal overflowing the joint) along the weld (Figure 13–12). No flash is produced during this welding process. Upset welding is used for joining tubes, rods, and cables of both **ferrous** and **nonferrous** metals. The recent development of **high-frequency upset welding** (UW-HF) and **induction upset welding** (UW-I) has increased the welding temperatures, resulting in greater production capabilities.

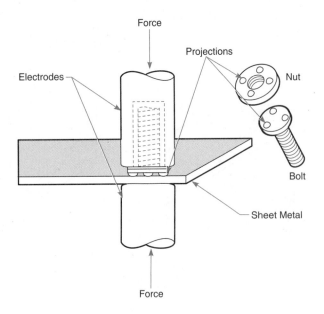

FIGURE 13–10   Projection welding process.

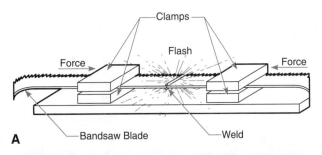

FIGURE 13–11   (A) Flash welding process. (B) Flash welder. *(Courtesy of Automation International, Inc., Danville, Illinois.)*

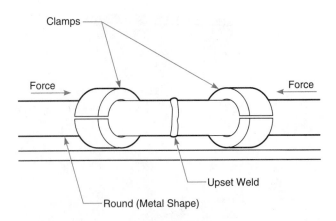

FIGURE 13–12   Upset welding process.

PLASMA ARC WELDING

Plasma arc welding (PAW) was developed in the 1950s. In this welding process, a narrow but focused arc is created between a **nonconsumable** electrode and a base metal. By constricting the arc and directing the flow of gas around the electrode, there is an increase in energy. An **ionized** gas results, producing a high-temperature plasma arc that can be in excess of 60,000° F (Figure 13–13).

Plasma arc welding can be compared to gas tungsten arc welding. It consists of a **torch** using a nonconsumable electrode for welding most metals. Under ordinary applications, filler metal is melted into the weld pool. The higher temperature of the arc results in a faster rate of travel and a higher deposition of filler metal than with gas tungsten arc welding. The higher welding temperature also allows the **welder** to hold a longer arc for better visibility in **manual** welding applications.

ULTRASONIC WELDING

Ultrasonic welding (USW) was also developed in the 1950s. In this welding process, pieces of metal are joined together by high-frequency vibrations under pressure. Ultrasonic welding is one of nine **solid-state welding** (SSW) processes. These are joining processes in which welding takes place below the melting temperature of the base metal.

With ultrasonic welding, the electrode is clamped to the pieces of the joint (usually a lap joint). The electrode is made to vibrate by high-frequency electric energy. The vibrations, together

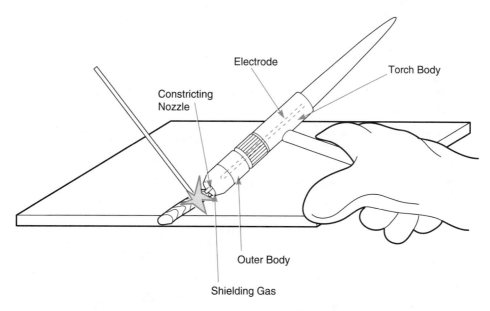

FIGURE 13–13 Plasma arc welding process.

with the additional pressure of clamping, cause surface tension and increased temperatures in the welding zone. A cleaning action takes place as the vibrations break down the surface **oxides.** The vibrations, increased temperatures, and pressure result in fusing the pieces together (Figure 13–14).

Ultrasonic welding is used to join metals such as **carbon steel,** copper, **nickel, gold,** and **alumi-**

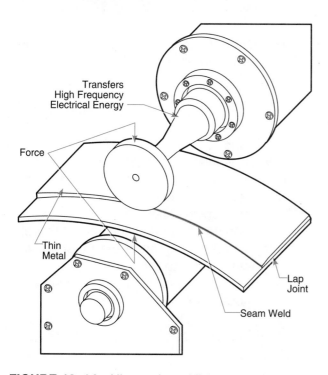

FIGURE 13–14 Ultrasonic welding process.

num. The absence of high temperatures makes this process advantageous for joining together dissimilar metals without the development of welding problems. **Spot ultrasonic welding** and **seam ultrasonic welding** are two applications for joining metals that fall under the heading of ultrasonic welding.

ELECTRON BEAM WELDING

Electron beam welding (EBW) is an expensive process developed in the 1950s to join metals that are difficult to weld by other means. In this process, high-velocity electrons strike the base metal and are converted to heat, melting the surfaces to be joined without the use of pressure. The welding can take place in a vacuum chamber or under a shielding gas.

Electron beam welding can achieve deep penetration with a high ratio of penetration to width, with low levels of distortion. By heating the emitter, electrons are generated that are focused by the beam-shaping electrode. The beam approaches the speed of light as it passes through a small hole. The electromagnetic focusing lens readjusts the beam before striking the joint (Figure 13–15).

Electron beam welding can be used with a variety of metals. **Titanium** alloy engine rings, the 9-foot-diameter aluminum ring on the world's largest telescope, and steel frame components on automobiles are examples of butt joints fused

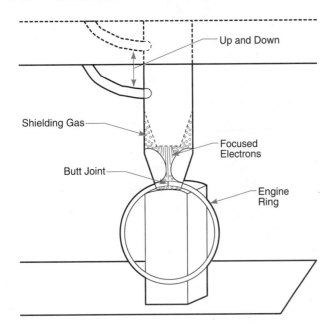

FIGURE 13–15   Electron beam welding process.

together by electron beam welding. This is the most promising of all welding processes for use in space. In the vacuum of space, lightweight guns of less than 10 pounds, powered by less than 1 kilowatt, are an advantage, not a disadvantage.

LASER BEAM WELDING AND LASER BEAM CUTTING

The laser beam was originally developed in the 1950s, although Buck Rogers's ray gun was in the minds of science fiction writers as early as the 1930s. *Laser* is derived from the term *light amplification by stimulated emission of radiation.* Laser beam welding (LBW) or **laser beam cutting** (LBC) occurs when the heat of a laser beam melts metal. In one form of this process, a gas mixture circulates in a tube raised to a high energy level. This energy is concentrated into a single wavelength of light, which is focused onto metal (Figure 13–16).

The use of laser beam technology is expensive but cost-effective for welding or cutting a variety of sheet metals. In some applications, laser beam welding provides better welding consistency with fewer weld failures than resistance spot welding.

REVIEW

1. What is forge welding?
2. What is submerged arc welding, and how is it different from **flux cored arc welding?**

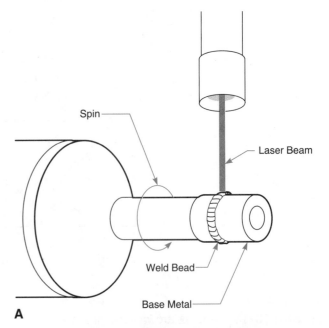

FIGURE 13–16   (A) Laser beam welding process. (B) Laser-welded gas filter. *(Courtesy of Larry Jeffus.)*

3. How is the base metal prepared for electroslag welding?
4. What is projection welding?
5. What is flash welding?
6. How is plasma arc welding similar to gas tungsten arc welding?
7. What is upset welding?
8. How are forge welding and ultrasonic welding different from all of the other welding processes discussed in this unit?
9. What are three advantages of using electron beam welding in space?

Create Three Questions

1.
2.
3.

Related English Questions

1. Write a paragraph explaining resistance welding.
2. Group together the welding processes listed below under one of these four classifications: arc welding, resistance welding, solid-state welding, other welding.

Electron beam welding	Induction seam welding
Ultrasonic welding	Submerged arc welding
Flash welding	
Electrogas welding	Electroslag welding
Forge welding	High-frequency seam welding
Plasma arc welding	
Projection welding	Stud welding
Laser beam welding	

For Further Thought

1. How do some of these welding processes differ from, say, shielded metal arc welding or gas metal arc welding?

2. What joint design is used for stud welding?
3. Why is a flux not usually required for any of the resistance welding processes?
4. What are some of the differences between resistance spot welding and spot welding with the gas metal arc welding process?
5. The laser beam can be used for both welding and cutting. What other processes mentioned in this unit for welding could be used for cutting?

SUGGESTED ACTIVITIES

1. Research one of the welding or allied processes listed on the process wheel.
2. Visit an industrial plant in which one or more of the processes mentioned in this unit is being used.
3. Think of a situation that requires a method for welding or cutting that has yet to be invented. Invent the method.

Glossary

AC *See* alternating current.
AC Vea corriente alterna.

acetone A liquid chemical used to stabilize acetylene gas.
acetona Un liquido químico que se usa para estabilizar al gas acetileno.

acetylene A compound of hydrogen and carbon produced from calcium carbide when released by water.
acetileno Un compuesto de hidrógeno y carbono remitido por el carburo calcio cuando éste combina con el agua.

ACHF Alternating current high-frequency. *See* high-frequency.
ACHF Corriente alterna de alta frequencia. Vea alta frequencia.

active gas A gas like CO_2 that forms compounds readily. With CO_2 as the shielding gas, some oxidation occurs on the surface of the weld.
gas activo Un gas como CO_2 que combina fácilmente. Con CO_2 como gas protector, resulta un poco de oxidación en la superficie.

actual throat The shortest distance from the weld root to the weld face.
garganta actual La distancia más corta de la raíz de una soldadura a la cara de la soldadura.

adhesion The attraction of molecules in a liquidus for the base metal.
adhesión La atracción de las moléculas en un liquidus del metal base.

air carbon arc cutting (CAC-A) An arc cutting process that will cut through any metal, using air to blow molten metal away.
corte con arco de aire y carbon (CAC-A) Un proceso de cortar con arco que puede cortar cualquier metal y que usa un soplo de aire para apartar el metal derritido.

alloy steel (1) A steel in which one of the following amounts is exceeded: 1.65% manganese, 0.60% silicon, or 0.60% copper. Or, (2) a steel to which a definite minimum quantity of aluminum, boron, chromium, or any other alloying element is added, to achieve a desired alloying effect. *See also* high-alloy steel; low-alloy steel.
acero mezclado (1) Un acero en el cual excede uno de los siguientes cantidades: 1.65% del manganeso, 0.60% del silicio, o 0.60% del cobre. O, (2) un acero al cual se ha añadido un cantidad definido minimo de aluminio, boro, cromo, o cualquier otro elemento de mezcla para producir el aleación deseado. Vea tambien acero de alta aleación y acero de baja aleación.

alternating current (AC) The back and forth motion of electric current, changing directions 120 times (60 cycles) per second.
corriente alterna (AC) El movimiento recíproco de un corriente eléctrico, que cambia dirección 120 veces (60 ciclos) por segundo.

aluminum A popular nonferrous metal that has one-third the weight of steel.
aluminio Un metal no-ferroso popular cuyo peso es la tercera parte del acero.

aluminum oxide A compound of aluminum and oxygen present on the surface of aluminum, having a higher melting temperature than aluminum, and which can cause trouble welding if not removed.
óxido de aluminio Un compuesto de aluminio y oxígeno que se encuentra en el superficie de aluminio capable de causar problemas al soldar si no se quite. Tiene una temperatura de derretimiento más alta que la del aluminio.

aluminum-silicon   A popular filler rod used for torch brazing aluminum.

aluminio-silicio   Una varilla de relleno popular que se usa para la soldadura fuerte con soplete.

American Iron and Steel Institute   An organization that classifies types of steel.

El Instituto Americano de Hierro y Acero   Una organización que clasifica los tipos de acero.

American Petroleum Institute (API)   An organization that issues codes and specifications for pipe welding.

Instituto de Petroleo Americano (API)   Una organización que emite los códigos y las especificaciones para la soldadura de tubo.

American Society of Mechanical Engineers (ASME)   An organization that issues codes and specifications for pipe welding.

Sociedad de Ingenieros Mecánicos Americana (ASME)   Una organización que emite los códigos y las especificaciones para la soldadura de tubo.

American Welding Society (AWS)   An organization that plays a major role in setting standards used throughout the welding industry.

Sociedad de Soldadura Americana (AWS)   Una organización que toca un gran parte en establecer las normas que se usan por toda la industria de soldadura.

amperage   (1) A setting on a welding machine. (2) A measurement in the rate of current flow.

amperaje   (1) Una colocación en una máquina de soldadura. (2)Una medida del razón del flujo del corriente.

ampere   A unit for measuring amperage.

amperio   Una unidad de medida de amperaje.

anti-spatter   A commercial spray or gel applied to keep spatter from sticking to the contact tube or nozzle of the torch used in gas metal arc welding.

anti-salpicadura   Un aerosol o gelatina comercial que se aplica para prevenir que la salpicadura se adhiere al tubo de contacto o a la boquilla del soplete que se usa en la soldadura de metal con arco protegido del gas.

arc   *See* welding arc.

arco   Vea arco de la soldadura.

arc blow   An erratic arc that refuses to go where the welder desires and results in sputter and spatter.

soplo del arco   Un arco errático que niega ir donde desea ponerlo el soldador y resulta en chisporroteo y salpicadura.

arc cutting (AC)   An arc welding process of cutting through metal by the heat generated between an electrode and a base metal.

corte con arco   Un proceso de soldadura de arco de cortar metal por medio del calor producido entre un electrodo y un metal base.

arc flash   A painful, but usually temporary, eye condition caused by the light of the welding arc on unprotected eyes.

deslumbramiento   Una condition de los ojos doloroso pero que suele durar poco tiempo causado por los efectos de la luz del arco de soldar en los ojos no protejidos.

arc length   The distance of the electrode from the weld pool.

largura del arco   La distancia del electrodo al charco de la soldadura.

arc seam weld   A series of spot welds or continuous welds between two overlapping pieces of metal.

soldadura de arco costura   Una serie de soldadura de puntos o soldadura continua entre dos pedazos de metal traslapados.

arc strike   The beginning of the arc for welding. An arc strike can lower the weld quality if made on the base metal outside of the welding area.

golpe del arco   El comienzo del arco para soldar.

arc time   The period of wetness in the weld pool.

tiempo del arco   El tiempo en que el charco de soldadura esté en forma líquido.

arc welding   Any of a group of welding processes that uses an arc generated by electricity.

soldadura de arco   Cualquier de un grupo de procesos de soldadura que utiliza un arco producido por la electricidad.

argon   An inert shielding gas used in some arc welding processes.

argon Un gas protector inerte que se usa en algunos procesos de soldadura.

arrow The part of a welding symbol that points to the joint to position the weld.

flecha La parte de un símbolo de soldadura que apunta a la junta para especificar la posición de la soldadura.

arrow side The side of the joint where the arrow is pointing.

lado de la flecha El lado de la junta al cual apunta la flecha.

automatic A term used to describe any equipment that can carry out welding, brazing, soldering, or cutting independently.

automático Un término que describe cualquier equipo que puede lleva acabo independientemente la soldadura, la soldadura fuerte o la soldadura blanda.

AWS *See* American Welding Society.
AWS Vea American Welding Society

AWS *Structural Welding Code* A book that lays out welding procedures for qualification testing of welders, welding operators, and tack welders.

AWS Código de Soladura Estructural Un libro que describe precisamente los procedimientos para calificación en los examenes de los soldadores, los operadores de soldadura y los soldadores de puntos aislados.

backfire A loud pop from the oxyacetylene torch.

contraexplosión Un chasquido fuerte que procede del soplete oxiacetileno.

backhand A welding technique in which the torch tip or gun nozzle is focused on the weld pool opposite the direction of welding.

soldadura en revés Una técnica de soldadura en que la boquilla del soplete o la boquilla de la pistola se enfoca en el charco de la soldadura en el sentido opuesto de soldar.

backing A material commonly used with groove welds on the back side of the joint.

respaldo Una materia que se usa comúnmente en la soldadura de ranura en el lado revés de la junta.

backing ring A strip of metal positioned inside the pipe at the joint root.

anillo de respaldo Una tira de metal localizado en el interior del tubo en la junta del raíz.

backing weld A weld bead laid on the back side of the joint before completing the weld.

soldadura de respaldo Un cordón de soldadura colocado en el lado revés de la junta antes de completar la soldadura.

back weld A weld bead laid on the back side of the joint after completing the weld.

soldadura de respaldo Un cordón de soldadura colocado en el lado revés de la junta despues de completar la soldadura.

bare electrode (wire) A welding electrode that does not contain a flux covering.

electrodo descubierto (alambre) Un electrodo de soldadura que no tiene revestimiento de flujo.

base material Any substance joined by welding, brazing, or soldering.

material base Cualquier substancia que se une por medio de la soldadura, la soldadura fuerte, o la soldadura blanda.

base metal The material being welded.
metal base El material en que se va soldar.

bauxite ore The principal raw material used in the manufacture of aluminum.

mineral bauxita La materia cruda principal que se usa en la fabricación del aluminio.

bevel An edge preparation of the base metal for welding in which any angle less than 90° is formed.

bisel (chaflán) Un corte preparado para la soldadura del metal base en el cual se forma un ángulo de menos de 90°.

bevel angle An edge preparation formed by a bevel of less than 90°.

ángulo del bisel Un corte preparado formado por un achaflán de menos de 90°.

bevel-groove weld The use of the butt joint with edge preparation given to one of two pieces forming the joint.

soldadura de bisel y ranura El uso de una junta de tope en que uno de las dos piezas formando la junta tiene un borde preparado.

bill of materials   The list of material needed to construct a given weldment.
lista de materiales   La lista de los materiales requeridos para construir un conjunto de partes soldadas específico.

blueprint   The instructions given for fabricating a part.
esquema   Las instrucciones específicas para la fabricación de una parte.

bonding   Holding together.
ligadura   Adheriendo el uno al otro.

boron   A nonmetallic element used as an alloy to increase hardness.
boron   Un elemento nonmetálico que se usa como mezcla para incrementar la dureza.

boundary device   A device used to mark off a robot's work area.
dispositivo de limites   Un dispositivo que se usa para definir la area de trabajo de un robot.

brass   An alloy consisting of copper and zinc. Overheating brass separates out the zinc.
latón   Un aleación que consiste del cobre y el zinc. Acalorando el latón hace separarse al zinc.

braze welding (BW)   A joining process in which the base metal is heated above 840° F but below the melting temperature of the base metal. Filler metal is deposited, but capillary action does not take place.
soldadura con bronze   Un proceso de unificación en el cual el metal base se calienta a una temperatura más alto que los 840° F (449° C) pero es una temperatura más bajo del derretimiento del metal base. El metal de relleno se deposita, pero no ocurre el acción capilar.

brazing (B)   A group of joining processes in which the filler metal is heated to liquidus above 840° F but below the melting temperature of the base metal, and capillary action takes place.
soldadura blanda   Un grupo de procesos de union en el cual el metal de relleno se calienta a un punto liquidus que es más alto que los 840° F (449° C) pero es una temperatura más bajo del derretimiento del metal base y en el cual ocurre el acción capilar.

breaking point   A measurement of when a material fails.

punto de quiebra   Una medida de cuándo falla un material.

Brinell hardness test   A test that calculates hardness by measuring the diameter of the impression left by a ball forced into the material. The test provides a standard of comparison for the hardness of different materials.
Examen Brinell de dureza   Una prueba que calcula la dureza por medio de tomar medidas del diámetro de una impresión de una bola que se oprime forzosamente en una materia. La prueba provee una norma de comparación de la dureza de varias materiales.

brittle   Very sensitive to cracking.
quebradizo   Muy disponible a grietarse.

broken arrow   The part of a welding symbol that points at the piece of the joint to receive the preparation for welding.
flecha incompleta   La parte del símbolo de la soldadura que apunta a la pieza de la junta que debe recibir la preparación para soldar.

buildup   A filler metal deposited so as to extend above the surface of the base metal.
recubrimiento   Un metal de relleno depositado para que se eleva sobre la superficie del metal base.

butt joint   A joint formed with pieces aligned in the same plane.
junta a tope   Una junta que se forma con las piezas alineadas en el mismo plano.

calcium silicate   A lightweight filler made of sand, lime, and asbestos used to stabilize acetylene within its cylinder.
calcio silicato   Un relleno ligero hecho de arena, cal, y asbesto que se usa para estabilizar al acetileno dentro de su cilindro.

cap   A part that holds the tungsten electrode and collet tube in position within the torch head.
tapa del collar   Una parte que sostiene al electrodo de tungsteno y al collar y los posicione en el soplete.

capillary action   The result when cohesion (attraction) between molecules in a liquidus for each other is overcome by adhesion (attraction) to the base metal.
acción capilar   El resultado cuando la cohe-

sión (atracción) entre las moleculas en un liquidus es predominado por la cohesión (atracción) al metal base.

carbon   A nonmetallic element added to steel and cast iron that affects hardness and, consequently, strength.

carbono   Un elemento nometálico que se añade al acero y al hierro vaciado que afecta la dureza y, por consequencia, la fuerza.

carbon dioxide (CO_2)   An active shielding gas used in some arc welding processes.

bióxido de carbono (CO_2)   Un gas protector activo que se usa en algunos procesos de soldadura con arco.

carbon steel   A general term covering a large range of steels. *See* low-, medium-, high-, and very high carbon steel.

acero al carbono   Un término general que se refiere a muchos tipos de acero. Vea acero de nivel bajo-, mediano-, alto- y muy alto carbono.

carburizing flame   An oxyacetylene flame in which acetylene dominates oxygen.

llama carburante   Una llama de gas combustible en el cual predomina el oxígeno.

cast iron   An alloy of iron that contains carbon (up to 4.5%) plus silicon.

hierro colado   Una aleación de hierro que contiene el carbono (hasta el 4.5%) y silicon.

cerium   A tungsten electrode, color-coded orange, that can be used as a substitute for a thoriated tungsten electrode.

cerio   Un electrodo de tungsten, distinguido por su color de código naranja, que se puede usar de substituto por un electrodo de tungsteno toriado.

cfh   Cubic feet per hour.
cfh   Pies cúbicos por hora.

chain   A type of intermittent fillet weld.
cadena   Un tipo de soldadura de filete intermitente.

check valve   A valve that allows the flow of pressure in one direction only.

valvula de seguridad   Una válvula que permite fluir la presión en un sólo sentido.

chipping (C)   One of five ways used to finish a weld.

desbastar   Uno de las cinco tratamientos para terminar una soldadura.

chromium   An element added to steel to increase hardness and corrosion resistance. Chromium is the principal alloying ingredient of stainless steel.

cromio   Un elemento añadido al acero para incrementar su dureza y su reistencia a la corrosión. El cromio es el ingrediente principal de aleación del acero inoxidable.

CO_2 *See* carbon dioxide.
CO_2 Vea bióxido de carbono

coefficient   A number that serves to measure a special characteristic of a material.

coeficiente   Un número que sirve para medir una característica especial de una materia.

coefficient of linear expansion   A number given to a material as a measurement of its expansion per inch per degree of rise in temperature.

coeficiente de expansión linear   Un número dado a una materia como una medida de su expansión por pulgada por cada grado de incremento en la temperatura.

cohesion   The attraction of molecules in a liquidus for one another. *Compare* adhesion.

cohesión   La atracción entre las moleculas en un liquidus. Compare con la adhesión.

cold laps   Incomplete fusion within a weld where a bridge of melted filler metal has been created without melting the joint root underneath.

solapes frías   Fusión incompleta dentro una soldadura donde un puente de relleno derritido se ha creado sin derritir la junta de raíz que se encuentra abajo.

cold-rolled steel   A steel that has been deformed into shape below crystallizing temperatures.

acero enrrollado a lo frío   Un acero ha tomado su forma en temperaturas más bajas de las de cristilización.

collet body   The part of an electric circuit that holds the collet tube.

conjunto del collar   La parte de un circuito eléctrico que sostiene el tubo del collar.

collet tube   The part of an electric circuit that holds the electrode in position.

collar   La parte de un circuito eléctrico que sostiene al electrodo en su posición.

color-coded   Color used as a means of identification.

código de colores   El uso de los colores para indentificar.

combination torch   A torch that can be used for welding, cutting, or heating by changing the tip.

soplete de combinación   Un soplete que se puede usar para soldar, cortar, o calentar con un cambio de la boquilla.

common sense   The application of knowledge that goes beyond books or instructions.

sentido común   La aplicación de los conocimientos que no provienen de ningún libro ni de los instrucciones.

complete root penetration   One hundred percent penetration along the entire joint root.

penetración completa de raíz   Un cien por ciento penetración por la entereza de la junta del raíz.

compressed air   Air held under pressure greater than atmospheric pressure. Air compressors supply air under pressure to operate tools, such as chipping hammers for removing slag.

aire comprimido   El aire que se sostiene bajo una presión más alta que la presión atmosférica. Los comprimidores del aire proveen el aire comprimido para operar las hierramientas neumáticas como por ejemplo, los desbastadores para quitar el escorial.

compression   The resistance to being crushed.

compresión   La resistencia a ser comprimido.

concave   A shape of the weld bead that is an inward or depressed surface. A crater provides a concave surface.

cóncava   La forma de un cordón de soldadura que tiene una superficie que va hacia abajo, de depresión. Un crater provee una superficie cóncava.

concave contour   (1) A weld face that curves inward. (2) A supplementary weld symbol.

perfíl cóncava   (1) Una cara de soldadura que encorva hacia abajo. (2) Un símbolo adicional de la soldadura.

conductor   A material with the ability to transfer heat. For example, aluminum is a better conductor than steel; therefore, more heat is needed to melt aluminum than steel of the same thickness.

conductor   Una materia con la abilidad de transferir el calor. Por ejemplo, el aluminio es un mejor conductor que el acero; por eso se requiere más calor para derritir el aluminio que un pedazo de acero del mismo espesura.

cone   A small flame visible inside a larger flame that indicates a neutral flame.

cono   Una pequeña llama visible dentro de una llama más grande que indica una llama neutra.

constant current   The electric output by which the welder can, to a limited degree, control voltage by raising or lowering the electrode.

corriente constante   La capacidad por el cual un soldador puede, a cierto grado, controlar el voltaje al subir o bajar el electrodo.

constant voltage   The electric output used for gas metal arc welding in which voltage remains constant with changes in arc length.

voltaje constante   La capacidad que se usa para la soldadura de metal con arco protejido por gas en el cual el voltaje se mantiene constante mientras que cambia la largura del arco.

consumable   The electrode or a filler metal that is melted into the weld pool, mixing with the base metal.

consumible   Un electrodo o metal de relleno que se derrite dentro del charco de soldadura, agregando con el metal base.

contacting sensor   A probe that touches the joint ahead of the weld pool.

sensor de contacto   Una tienta que toca la junta delante del charco de soldadura.

contact tube   The point of electric contact that energizes the wire (electrode).

tubo de contacto de electrodo   El punto de contacto eléctrico que da energía al alambre (o al electrodo).

contour   The shape of the weld face.

perfíl   El contorno de la cara de la soldadura.

controller *See* robot controller.
contralor Vea contralor robot.

convex A shape of the weld bead that is a raised, rounded surface.
convexa Un contorno del cordón de soldadura que tiene una superficie alzado y en curva.

convex contour (1) A weld face that curves outward. (2) A supplementary weld symbol.
perfíl convexa (1) Una superficie que curva hacia afuera. (2) Un símbolo de soldadura suplementario.

copper A metallic element that is readily soldered, brazed, or welded. Oxyacetylene welding is not recommended.
cobre Un elemento metálico que se dispone a la soldadura blanda, la soldadura fuerte, la soldadura de latón o la soldadura. No se recomiende la soldadura con oxiacetileno.

copper-zinc A popular filler rod used for torch brazing ferrous and nonferrous metals.
cobre con zinc Una varilla de relleno muy común que se usa para la soldadura de latón en los metales ferrosos y no-ferrosos.

corner-flange weld A weld made in a joint of which only one piece is flanged.
soldadura de reborde en esquina Una soldadura que se efectúa en una junta del cual sólo una pieza tiene reborde.

corner joint A joint formed with pieces aligned in different planes, involving the edge of at least one piece.
junta de esquina Una junta que se forma con las pieza alineadas en planos distinctos, involucrando al borde de al menos una de las piezas.

corrosion resistance The ability to withstand the stress of chemical reactions without breaking down.
resistencia a la corrosión La abilidad de resistir las tensiones de las reacciones químicas sin sostener daños.

coupon A dimensioned specimen from a weld or material for a destructive test to determine quality.
cupón Una probeta de dimensiones precisas que se toma de una soldadura o una materia y que se expone a una prueba destructiva para determinar su calidad.

covering Fluxing material that coats an electrode.
cubierto Material de revestimiento (flujo) que cubre un electrodo.

cover pass The final layer of a weld, consisting of from one to several weld beads.
pasada para cubrir La última capa de una soldadura que consiste de uno a varios cordones.

crack A fracture of the joint that is the result of welding.
grieta Una fractura en la junta que resulta de la soldadura.

crater An undesirable depression left when a weld bead is stopped or completed but not filled in.
crater Una depresión defectuoso que aparece cuando se interrumpa al cordón de soldadura o cuando se completa, pero no se rellena.

crater crack A crack originating in a crater.
grieta de crater Una grieta que origina en un crater.

critical pipe welding Welding performed on high-pressure pipe that is subject to specifications that govern welding procedures.
soldadura de tubo crítico La soldadura que se efectúa en la tubería de alta presión y que se somete a las especificaciones que reglan los procedimientos de la soldadura.

critical range The peak temperature at which a rapid rate of cooling will result in the formation of martensite.
extensión crítico La cima de la temperatura del cual un enfriamiento rápido resultará en una formación de la martensita.

critical welding Welding that must be done properly because failure of the welded joint will almost certainly lead to calamity.
la soldadura crítica La soldadura que requiere ser hecho correctamente porque la quebradura de la junta soldada seguramente causaría una tragedia.

crystallize To become solid.
cristalizar Solidificar.

current The flow or movement of electric input into the welding machine.

corriente   El flujo o movimiento de la electricidad invertida en la máquina de soldadura.

cutting tip   A torch tip designed for cutting.
punta para cortar   Una boquilla del soplete diseñado para cortar.

cylinder manifold   Several cylinders connected together to deliver a higher volume of gas.
conexción de cilindros múltiple   Varios cilindros enlazados que pueden entregar un volumen más alta de gas.

DC   *See* direct current.
DC   Vea corriente continua.

DCEN   *See* direct current electrode negative.
DCEN   Vea corriente continua electrodo negativo.

DCEP   *See* direct current electrode positive.
DCEP   Vea corriente continua electrodo positivo.

decimal equivalent   The use of decimal numbers as substitutes for common fractions.
decimal equivalente   El uso de los números decimales para substituir a los fracciones.

defect   A flaw in a weld as a result of welding that can affect weld quality.
defecto   Una falta de la soldadura resultando del proceso de soldar que puede afectar su calidad.

deformation   The change a material undergoes due to stress.
deformación   La transformación que padece una material debido a la tensión.

deoxidizer   An element, such as silicon or phosphorus, added to filler metal to bond with oxygen, to keep oxygen from having a negative effect on the weld.
desoxidante   Un elemento, como el silicio o el fósforo, que se añade al metal de relleno para combinar con el oxígeno para que el oxígeno no tenga efectos negativos en la soldadura.

destructive testing (DT)   A test of weld quality in which the joint is destroyed.
prueba destructiva   Una prueba de la calidad de la soldadura en la cual se destruye la junta.

direct current (DC)   Electric current that flows in one direction.

corriente continua   Una corriente eléctrica que fluye en una dirección.

direct current electrode negative (DCEN)   Welding with the electrode attached to the negative terminal and the workpiece connection attached to the positive terminal on the welding machine.
corriente continua electrón negativo   La soldadura en que el electrodo se conecta al terminal negativo y la pieza de trabajo se conecta al terminal positivo del equipo de la soldadura.

direct current electrode positive (DCEP)   Welding with the electrode attached to the positive terminal and the workpiece connection attached to the negative terminal on the welding machine.
corriente continua electrodo positivo   La soldadura en que el electrodo se conecta al terminal positivo y la pieza de trabajo se conecta al terminal negativo del equipo de soldadura.

discontinuity   A flaw that is not a defect unless it fails to meet the conditions established by the welding procedure.
discontinuidad   Una falta que no se considera un defecto a menos que no cumple con las condiciones establecidos por los procedimientos de la soldadura.

distortion   A usually uncontrolled change in the shape of metal, resulting from the heat of welding.
distorción   Un cambio en la forma del metal que suele ser fuera de control, que resulta del calor de la soldadura.

double-groove weld   A type of groove weld in which welding is done on opposite sides of the joint.
soldadura de ranura doble   Un tipo de soldadura de ranura que se hacen en lados opuestos de la junta.

downhill   Downward welding or brazing with gravity.
caída   La soldadura que se hace en sentido de decenso o la soldadura fuerte hecha con la gravedad.

drop transfer   A gas metal arc welding process in which drops of metal cross the arc to the weld pool. Two types are globular drop transfer and spray drop transfer.

transferencia de gotas Un proceso de la soldadura de metal con arco protejido por gas en el cual gotas de metal cruzan del arco al charco de la soldadura. Los dos tipos de gotas son gotas gobulares y transferencia de gotas en aerosol.

dross Oxidized metal formed on the underside of the kerf.

escoria Metal oxidado que forma en el lado inferior de una cortadura.

DT *See* destructive testing.

PD Vea prueba destructiva.

ductile cast iron *See* nodular cast iron.

hierro colado dúctil Vea hierro colado nodular.

ductility The resistance to deformation after stretching.

ductilidad La resistencia a la deformación despues de ser estirado.

duty cycle The recommended percentage of time that a given power source should be under the load of welding.

ciclo de trabajo El porcentaje recomendado en que una fuente de energía dada debe sujetarse a la carga de la soldadura.

dye penetrant test (DPT) A nondestructive test that uses a penetrative liquid with developer to locate surface defects in welds.

prueba de penetración de tintas Una prueba nodestructible que usa un líquido penetrativo con un desarrollador para localizar los defectos de la soldadura en la superficie.

edge-flange weld A weld in which two flanged pieces form the joint.

soldadura de reborde de la orilla Una soldadura en la cual dos piezas de reborde formen la junta.

edge joint A joint formed when the edges of two or more pieces are brought together parallel to one other.

junta de orilla Una junta que se forma cuando las orillas de dos piezas o más se unifican en manera paralela.

effective throat The shortest distance from the weld root to the weld face at the point of a flush contour.

garganta efectiva La distancia más corta del raíz a la cara de la soldadura que se calcula desde un punto nivel del perfíl.

elastic limit A measurement in the ability of a material to stretch under a load and return to its original shape.

limite elástico Una medida de la abilidad de una materia a estirarse bajo una carga y regresar a su forma original.

electrode A consumable or nonconsumable material through which electricity flows to create an arc.

electrodo Una materia consumible o no consumible que deja fluir la electricidad para crear un arco.

electrode angle *See* travel angle; work angle.

ángulo del electrodo Vea ángulo de avance; ángulo del trabajo.

electrode extension The length of a wire, measured from the end of the contact tube.

extensión del electrodo La longitud de un alambre que se mide de la extremidad del tubo de contacto.

electrode lead A cable that connects an electrode to a power source.

cable de electrodo Un cable que conecta a un electrodo al fuente de energía.

electrode negative *See* direct current electrode negative.

electrodo negativo Vea corriente continua electrodo negativo.

electrode positive *See* direct current electrode positive.

electrodo positivo Vea corriente continua electrodo positivo.

electrogas welding (EGW) An arc welding process in the vertical position in which the heat generated by the arc melts the filler metal and the base metal, requiring a shielding gas.

soldadura electrogas Un proceso de soldadura con arco en posición vertical en el cual el calor producido por el arco derrite al metal de relleno y al metal base, y requiere un gas protector.

electron beam welding (EBW) A welding process in which high-velocity electrons strike the base metal and are converted to heat, melting the surfaces to be joined without pressure.

soldadura a rayo de electrón   Un proceso de soldadura en el cual electrones de alta velocidad choquen contra el metal base y se convierten a calor, así derritiendo a los superficies y uniéndolos sin presión.

electronic warning device   A safety device which is activated when someone enters a robot's work area.

dispositivo de aviso electrónico   Un dispositivo de seguridad que se activa al entrar una persona en el area de trabajo de un robot.

electroslag welding (ESW)   An arc welding process from the vertical position in which the protective shielding slag causes a resistance between the continuous solid wire and the base metal.

soldadura de electroescoria   Un proceso de soldadura con arco en posición vertical en el cual la capa protejidora de escoria causa una resistencia entre el alambre sólido continuo y el metal base.

element   Any one of more than 100 substances made up of only one kind of atom.

elemento   Cualquier de los 100 o más substancias que se componen de un sólo tipo de atomo.

elongation   The stretching a metal undergoes as a result of some load or force.

extensión   El estiramiento que aguanta un metal como resultado de un cargo o fuerza.

emery cloth   An abrasive emery-coated sheet for light-duty cleaning of metal.

tela de esmeril para bruñir   Una hoja raspante cubierta de esmeril que se usa para la limpieza ligera del metal.

engine-driven generator   A welding machine that can produce alternating current, direct current, or both as output, supplied by an internal combustion engine.

generador de motor   Una máquina de soldadura que puede producir corriente alterna, corriente continua, o los dos, y se sostiene por un motor de combustión interna.

etch   A term used to describe a method to examine a weld for defects. A solution of one part nitric acid and two parts water can be used to etch iron or steel.

grabar   Un término que describe un método de examinar los defectos en una soldadura. Se usa una solución de ácido nítrico con dos partes de agua para grabar al hierro o al acero.

eutectic composition   The single lowest melting temperature of two metals combined, lower than either pure lead or pure tin.

composición de tipo eutéctico   La temperatura de derretimiento más baja de dos metales combinadas, más baja de la del plomo puro o del latón puro.

extrusion process   A process in which measured lengths of bare wire are forced through fluxing materials that cling to the wire and become baked on.

proceso de expulsión   Un proceso en el cual segmentos medidos de alambre descubierto se impulsan entre materiales de flujo o revestimiento que se adhieren al alambre y se endurecen al horno.

face bend   Stretching the weld face of a coupon to test the weld.

pliegue de cara   Una prueba de la soldadura que consiste en estirar la cara de una probeta.

fatigue strength   The resistance to changing forces or loads.

resistencia a la fatiga   La resistencia a los cambios que vienen con tensiones o cargas.

feather   To use a grinder to thin the ends of tack welds to aid complete fusion with the root bead.

achaflanar   Usar una molidor para adelgazar a las colillas de las soldaduras de puntos para completar la fusión con el cordón del raíz.

feed   The movement of filler material into the weld pool.

alimentar   El movimiento del material de relleno al charco de la soldadura.

feeder system   A part of the welding process that contains the drive rolls and controls for semi-automatic welding.

sistema de alimentación de alambre   Un parte del proceso de la soldadura que contiene los rodillos alimentadores y los controles para la soldadura semi-automatica.

ferrite   The formation of pure iron in the grain structure of steel and cast iron.
ferrita   La formación de hierro en la estructura de fibra del acero y el hierro colado.

ferrous   Containing iron.
ferroso   Lo que contiene hierro.

field welding   Welding performed away from the shop.
soldadura en sitio   Cualquier soldadura que se efectúa afuera del taller.

field weld symbol   A flag on the welding symbol used to indicate that welding will be completed away from the shop.
simbolo de soldadura en sitio   Una bandera en un símbolo de la soldadura que indica que la soldadura se efectuará fuera del taller.

filler metal   Metal added to a joint.
metal de relleno   Un metal que se añade a la junta.

filler pass   A weld bead laid between the root bead and the cover pass.
pasada para rellenar   Un cordón de soldadura que se coloca entre el cordón de raíz y la pasada para cubrir.

fillet weld   A weld that joins the surfaces of two pieces, usually at right angles to each other.
soldadura de filete   Una junta que une los superficies de dos piezas, por lo regular en un ángulo recto.

fillet weld gauge   A tool for the accurate measurement of fillet welds.
manómetro de soldadura filete   Un instrumento que mide precisamente las soldaduras de filete.

fillet weld legs   A part of the weld used to determine the size of a fillet weld, measured from the joint root to the toe.
patas de soldadura filete   Una parte de la soldadura que se usa para determinar el tamaño de una soldadura de filte, que se mide del raíz al pie de la junta.

film   A substance, placed intentionally or formed unintentionally on the base metal, that becomes a contaminant during welding. For example, paint helps preserve metal but becomes a contaminant if not removed before welding, brazing, or soldering.
membrano   Una substancia, aplicada intencionalmente o que aparece involuntariamente en el metal base, que contamina la soldadura. Por ejemplo, la pintura proteje al metal pero es un contaminante si no se quita antes de la soldadura, la soldadura blanda, o la soldadura fuerte.

finish   Any one of several methods used to achieve the required weld contour.
acabado   Cualquier de varios métodos que se usan para producir el perfíl requerido de la soldadura.

fixture   A device used to restrain movement during welding.
fijación   Un aparato que pone restricciones en el movimiento mientras que se efectúa una soldadura.

flammable   Easily ignited.
combustible   Lo que se enciende facilmente.

flange   Sheet metal bent to form a curve near the edge for the purpose of welding.
reborde   Una lámina de metal torcida de tal manera para formar una curva cerca del borde para efectuar una soldadura.

flare-bevel-groove weld   A weld in which one of the two pieces forming the joint is a curved surface (like the end of a pipe welded to plate).
soldadura de bisel ranura abierta   Una soldadura en la cual una de las dos piezas formando la junta es un superficie curvada (como la colilla de un tubo que ha sido soldado a una lámina).

flare-V-groove weld   A weld in which both members of the joint are curved surfaces.
soldadura abierta de ranura en V   Una soldadura en la cual las dos piezas de la junta son superficies curvadas.

flashback   When the flame burns back inside the torch, usually producing a squealing noise.
flama en retroceso   Cuando la llama sigue quemando dentro de la boquilla, comúnmente produciendo un chillido muy fuerte.

flash off   To take the torch away from the weld pool momentarily during oxyacetylene welding.
llamarada   Quitar al soplete del charco mo-

mentáneamente mientras que se efectua el soldeo de oxiacetileno.

flash welding (FW)   A resistance welding process in which heat melts the surfaces along the edges of a butt joint brought together by pressure with a flash of sparks during welding.
soldadura de flash   Un proceso de soldadura por resistencia en el cual el calor derrite las superficies bordando una junta de tope que se unen por la presión y produciendo un relampagueo de chispas en el proceso.

flat bar   Metal more than 3/16 inch thick and usually not exceeding 6 inches in width.
lamina plana   Un pedazo de metal midiendo más del 3/16 pulgadas (.48 cm) de grueso pero que no suele ser más largo que las 6 pulgadas (15.24 cm) de ancho.

flat position   A position in which the plane of welding is from 0° to 15°.
posición plana   Una posición en la cual el plano en donde se va soldar es de 0° a 15°.

flowmeter   A gauge that measures gas volume by its flow.
manómetro de fluidez   Un manómetro que mide el volumen del gas por su fluidez.

fluorescent penetrant test (FPT)   A nondestructive test that involves the examination of a weld under black light.
prueba de penetrante flourescente   Una prueba no destructivo que involucra una examinación bajo un luz ultravioleta.

flush contour   (1) A weld face that is a straight line from toe to toe. (2) A supplementary weld symbol.
contorno nivel   (1) Una cara de la soldadura que es una linea recto de pie a pie. (2) Un símbolo suplementario de la soldadura.

flux   A material that produces a gaseous shield to protect the weld pool and to stabilize the arc, making welding easier. Flux prevents the formation of oxides, helps filler metal to flow, and removes impurities that can affect weld quality.
flujo   Una materia que produce un escudo gaseoso para proteger al charco de la soldadura e estabilizar al arco, facilitando a la soldadura. El flujo previene que forman los óxidantes, hace más fluido al relleno, y remueve las impurezas que pueden afectar a la calidad de la soldadura.

flux cored arc welding (FCAW)   A semiautomatic arc welding process in which a tubular wire, filled with flux, is fed continuously into the weld pool.
soldadura de arco con nucleo de fundente   Un proceso semiautomático de soldadura de arco en el cual un alambre tubular, con relleno de flujo, se alimenta continuamente al charco de la soldadura.

forehand   A welding technique in which the torch tip or gun nozzle is focused on the weld pool in the direction of welding.
soldadura directa   Una técnica de soldar en la cual la punta del soplete o la boquilla de la pistola se enfoca en el charco de la soldadura en la dirección de la soldadura.

forge welding (FOW)   A welding process in which the pieces of the joint are heated in a forge or furnace and worked together with pressure produced by a hammer.
soldadura por forjado   Un proceso en el cual las piezas de la junta se calienten en una fragua o un horno y se junten con presión aplicado con un martillo.

freehand   A term used to describe welding or cutting performed without aid to steady the hand.
a mano alzada   Un término que describe la soldadura o los cortes que se efectuan sin algo que ayude en fijar a la mano.

frequency   The number of electrical oscillations (cycles) per second.
frequencia   El número de las oscilaciones eléctricas (cíclos) por segundo.

fuse   To weld a joint together with or without the use of filler material.
fundir   Soldar una junta con una materia de relleno o sin ella.

fusion   The melting and mixing together of the pieces that form the base metal with and without filler metal.
fusión   El derretimiento y mezcla de las piezas que componen el metal base con una materia de relleno o sin ella.

1G position   A position in which pipe is laid

out horizontally with 0° to 15° of slope. The pipe is not fixed and can be rolled.

posición de 1G Una posición en la cual el tubo inclina horizontalmente de 0° a 15° de descenso. El tubo no es fijo y se puede rodar.

2G position A position in which pipe is laid out vertically with 0° to 15° of slope.

posición de 2G Una posición en la cual el tubo inclina verticalmente de 0° a 15° grados.

5G position A position in which pipe is laid out horizontally with 0° to 15° of slope so that the pipe cannot be rolled.

posición de 5C Una posición en la cual el tubo inclina horizontalmente de 0° a 15° grados de descenso pero en la cual el tubo no se puede rodar.

6G position A position in which pipe is laid out on an incline of 45° (plus or minus 30°) of slope.

posición de 6C Una posición en la cual el tubo inclina un 45° (con más o menos 30°) de inclinación.

6GR position A position in which pipe is laid out on an incline of 45° (plus or minus 30°) of slope and a ring is positioned around it, causing a restriction making the welding more difficult.

posición de 6GR Una posición en la cual el tubo inclina un 45°(con más o menos 30°) y tiene al su alrededor un anillo causando una restricción y una dificultad en efectuar la soldadura.

galvanized A term used to describe metal that has been coated to prevent oxidation.

galvanizado Un término que describe un metal que ha recibido una capa protectiva para prevenir la oxidación.

gas metal arc cutting (GMAC) An arc cutting process that uses an electrode (wire) ordinarily used for gas metal arc welding.

cortes de metal con arco protegido gas Un proceso de corte con arco que usa un electrodo (alambre) que suele usarse principalmente para la soldadura de arco metálico con gas.

gas metal arc welding (GMAW) A semiautomatic arc welding process in which wire is fed continuously into the weld pool.

soldadura de metal con arco protegido por gas Un proceso de soldadura de arco semi-automático en el cual un alambre se alimenta continuamente al charco de la soldadura.

gas tungsten arc cutting (GTAC) An arc cutting process that uses an electrode (tungsten) ordinarily used for gas tungsten arc welding.

corte de arco de tungsteno con gas Un proceso de corte de arco que usa un electrodo (tungsteno) que suele usarse principalmente para la soldadura de arco de tungsteno con gas.

gas tungsten arc welding (GTAW) A gas shielding arc welding process in which an arc is produced at the end of a nonconsumable tungsten electrode.

soldadura de arco de tungsteno con gas Un proceso de la soldadura de arco protegido por un gas en el cual un arco se produce en el punto de un electrodo no consumible de tungsteno.

gauge number A measurement of metal thickness.

numero de manómetro Una medida de lo grueso del metal.

globular transfer A gas metal arc welding process in which a drop greater than the diameter of the wire is produced.

tranferencia globular Un proceso de la soldadura de arco metálico con gas en el cual se produce una gota de un diámetro más grande del que mide el alambre.

gold A metallic element that is a good conductor of heat, is resistant to corrosion, with a melting temperature of 1,945° F.

oro Un elemento metálico que es un conductor muy bueno para el calor, resiste la corrosión, y cuyo temperatura del derretimiento es el 1,945° F (1063°C).

gouging tip A torch tip designed to scoop molten metal out of a joint without cutting completely through it.

punto para escoplear en gubia Una boquilla del soplete diseñado para excavar el metal derritido de una junta pero sin cortarla completamente.

grain structure Recognizable shapes in the formation of metal that are readily visible under a microscope. Grain structure is affected by both heat treatment and the heat of welding.

estructura de grano Formas que se pueden reconocer en la formación del metal que se pueden percibir fácilmente con un microscopio. La estructura de grano se afecta por el tratado de calor y por el calor de la soldadura.

gray cast iron A weldable type of cast iron. It should be kept cool during welding or should be preheated and cooled slowly.
hierro colado gris Un tipo de hierro colado que se puede soldar. Se debe mantener fría al soldarlo o se debe precalentar y dejarlo enfriarse lentamente.

grinding (G) One of five ways used to prepare or finish a weld.
esmerilado Uno de los cinco tratamientos para acabar o preparar la soldadura.

groove angle The measurement in degrees for the bevel made on the edge of metal in preparation for welding.
ángulo de ranura Una medida en grados para el bisel en el borde de una materia en preparación para la soldadura.

groove face A surface after beveling has been completed.
cara de ranura Una superficie en la cual se ha completado un bisel.

groove weld Any one of eight joint designs for the butt joint.
soldadura de ranura Cualquier de los ocho diseños de la junta de tope.

guided bend test A test in which coupons are bent, stretching to examine for flaws in the welding.
prueba de pliegue guiada Una prueba en la cual probetas se doblan, estirando a los defectos de la soldadura para que se examinen.

gun A device that, in one application, can transfer welding output to the filler wire coming out of its nozzle with an on/off control switch.
pistola Un dispositivo que, con una aplicación, puede transferir la capacidad de la soldadura al alambre de relleno que sale de su boquilla por medio de un pulsador.

hammering (H) One of five methods used to finish a weld.

martillar Uno de los cinco métodos que su usan para acabar una soldadura.

hardening A heat treatment used to harden a material by changing its grain structure.
endurecimiento Un tratamiento de calor se usa para endurecer una material por medio de un cambio en su estructura de grano.

hardness The resistance to penetration or denting.
dureza La resistencia a la penetración o la abolladura.

HAZ *See* heat-affected zone.
HAZ Vea zona afectada por el calor.

heat-affected zone (HAZ) The area along a weld that is affected by the heat of welding.
zona afectada por el calor El area alrededor de una soldadura que se afecta por el calor del soldeo.

heating tip A torch tip designed strictly for heating.
boquilla de calentamiento Una boquilla que es diseñado específicamente para el calentamiento.

heat treatment The controlled heating of a metal.
tratamiento de calor El calentamiento controlado de un metal.

helium An inert shielding gas used in some arc welding processes.
helio Un gas inerte de protección que se usa en algunos procesos de soldadura con arco.

hertz (Hz) A unit of frequency equal to cycles per second.
hertz Una unedad de frecuencia que egala a ciclos por segundo.

high-alloy steel A steel in which chromium, manganese, or nickel content equals 12% or better.
acero de alta aleación Un acero en el cual el contenido del cromo, manganeso o el níquel egala o supera el 12%.

high-carbon steel A steel that contains 0.45% to 0.65% carbon.
acero alto carbono Un acero que contiene el 0.45% al 0.65% del carbono.

high-frequency   A method used to superimpose high voltage on the welding output at radio frequencies to maintain the arc through the entire cycle without freezing the electrode to the base metal.

alta frequencia   Un método de superponer un voltaje alto en la capacidad de soldar por medio de frecuencias de radio con el propósito de mantener un arco por un ciclo completo sin que el electrodo se pega al metal base.

high-frequency seam welding (RSEW-HF)   A resistance welding process in which high-frequency welding current is conducted through electrodes touching the base metal.

soldadura de costura de alta frequencia   Un proceso de soldadura de resistencia en el cual un corriente para el soldeo de alta frecuencia se conduce por los electrodos que tocan al metal base.

high frequency-upset welding (UW-HF)   An upset resistance welding process that uses high-frequency current.

soldadura de recalada de alta frequencia   Un proceso de soldadura de recalada de resistencia que usa una corriente de alta frequencia.

high-pressure gas cylinder   A cylinder designed to handle gas under pressure. Oxygen and argon cylinders are two examples that require special attention to avoid accidents.

cilindro de gas de alta presión   Un cilindro diseñado para bodegar al gas bajo presión. Los cilindros de oxígeno y de argón son ejemplares de cilindros que requieren atención especial para prevenir a los accidentes.

horizontal position   A position in which the plane of welding is from 15° to 100°.

posición horizontal   Una posición en el cual el plano de soldar puede ser de 15° a 100°.

hot pass   A pass that follows the root pass and involves a higher amperage.

pasada caliente   Una pasada que sigue la pasada del raíz y que involucra un amperaje más alto.

hot-rolled steel   A steel that has been deformed into shape at temperatures that allow course grains to recrystallize.

acero laminado en caliente   Un acero cuyo forma se ha tomado bajo temperaturas que permiten recristalizar los granos gruesos.

hydrogen   An odorless but flammable gas, the lightest of all elements. Hydrogen can cause underbead cracking in the arc welding of some steels.

hidrógeno   Un gas combustible sin olor, de todos los elementos el más ligero. El hidrógeno puede causar agrietura abajo de los cordones en la soldadura de arco de algunos aceros.

impact strength   The resistance to sudden force without fracture.

fuerza de impacto   La resistencia a un impacto repente sin causar fractura.

incomplete fill   A weld in which not enough filler metal has been deposited in the joint.

relleno incompleto   Una soldadura en la cual no se ha depositado bastante metal de relleno en la junta.

incomplete fusion   Lack of fusion between the weld bead and the base metal.

fusión incompleto   Una carencia de fusión entre el cordón de soldadura y el metal base.

inductance   A fine adjustment on some gas metal arc welding machines that can reduce spatter and, more important, regulate weld pool time by increasing or decreasing arc time.

inducción   Un reglaje fino disponible en algunas máquinas de soldadura de arco de protección de gas que puede reducir la cantidad de la salpicadura, y, más importante, puede regular el tiempo de que un charco de solea sea líquido por medio de incrementar o decrementar el tiempo del arco.

induction seam welding (RSEW-I)   A resistance welding process in which high-frequency welding current is magnetically passed from a copper coil without electrodes touching the base metal.

soldadura de costura por inducción   Un proceso de soldadura por resistencia en el cual un corriente de soldar de alta frecuencia se translada magnéticamente de una bobina de cobre sin que los electrodos tocan el metal base.

induction upset welding (UW-I)   An upset resistance welding process that uses high-frequency current passed by induction.

soldadura de calada por inducción   Un proceso de soldadura de calada de resistencia que usa un corriente de alta frequencia que se translada por inducción.

inert   A term used to describe noble gases, such as neon, argon, and helium, that neither react with oxygen nor form any other compounds readily.
inerte   Un término que describe a los gases nobles, como el neón, el argón, y el helio, que no reactivan fácilmente con el oxígeno ni otros tipos de compuestos.

infrared rays   Invisible light, absorbed by the body as heat, that can damage the eyes.
rayos infrarrojos   La luz invisible que el cuerpo percibe como calor que puede dañar a los ojos.

inorganic flux   A flux made of corrosive acids and salts that is used for soldering ferrous and nonferrous metals.
revestimiento inorgánico   Un revestimiento que se compone de los sales y de acidos corrosivos que se usa en el soldeo de metales ferrosos y no-ferrosos.

input   High-voltage, low-amperage electrical power that flows into a welding machine or power source.
potencia absorbida   El poder de alto voltage y bajo amperaje que fluye a la máquina de soldadura o un fuente de poder.

insulator   A device that does not conduct electricity and keeps the gun nozzle from becoming part of the arc and short-circuiting against the base metal.
insulador   Un aparato que no conduce la electricidad y previene que la boquilla de la pistola se devenga parte del arco y que causa un corto circuito con el metal base.

intermittent welding   Welding in which the welds are spaced and not continuous over the entire joint.
soldadura intermitente   La soldadura en la cual se deje espacio entre las soldaduras, y no son continuas por medio de la junta.

inverter   A lightweight welding machine that takes advantage of computer technology and is versatile enough to be connected to most electric input systems.

convertidor   Una máquina de soldadura ligera que se aprovecha de la tecnología de computadores y es bastante versátil para poder conectarse con la mayoría de fuentes de poder eléctricos.

ionized   A term used to describe atoms that carry a positive or a negative electrical charge.
ionisado   Un término que describe a los atomos que poseen una carga positiva o negativa.

iron   A magnetic metallic element not found naturally in usable quantities. *See* cast iron.
hierro   Un elemento metálico magnético que no se encuentra naturalmente en gran cantidades. Vea hierro colado.

iron ore   The raw material used in the manufacture of cast irons and steels.
mineral de hierro   El material crudo que se usa en fabricar el hierro colado y el acero.

iron soldering (INS)   A welding process that uses a tool for transferring heat to the base metal.
soldadura fuerte de hierro   Un proceso de la soldadura que utiliza una herramienta para transferir el calor al metal base.

J-groove weld   A weld in which the edge preparation of one piece in the joint has a curved surface.
soldadura de ranura en J   Una soldadura en la cual la preparación del borde de una pieza de la junta tiene una superficie curvada.

jig   A device for conducting a guided bend test.
montaje   Un dispositivo para efectuar una prueba de pliegue guiada.

joint   The junction of two or more pieces joined together by welding, brazing, or soldering.
junta   La unión de dos piezas o más por medio de la soldadura, la soldadura fuerte, o la soldadura blanda.

joint design   A configuration for joining materials together by welding.
diseño de la junta   Una configuración para unir materias por medio de la soldadura.

joint root   The closest part of the pieces forming a joint.

junta de raíz El sitio en donde las piezas que formarán una junta son más cercas.

joint root opening A separation at the joint root, usually for greater penetration.
apertura de la junta de raíz Una separación en la junta de raíz que se suele dejar para mejor penetración.

kerf An opening produced by cutting.
muesca Una apertura que se produce por medio de un corte.

keyhole A partial melting away of the edges of a joint root at the front of the weld pool.
ranura de llave Un derretimiento parcial de los bordes de una junta de raíz en la parte delantera del charco de la soldadura.

kindling temperature (1) A temperature between 1,400° and 1,600° Fahrenheit. (2) A temperature at which oxidized metal can be blown out of a cut.
temperatura de enciendo (1) Una temperatura entre 1,400° y 1,600°F (760°C y 871°C). (2) Una temperatura en la cual el metal oxidado se puede soplar de una corte.

lap joint A joint formed with two pieces extending over each other in parallel planes.
junta de solape Una junta que se forma con dos piezas que se extienden el uno sobra la otra en planos paralelos.

laser beam cutting (LBC) A cutting process in which the heat of a laser beam melts metal.
cortes con rayo laser Un proceso de cortar en el cual el calor de un rayo lazer derrite al metal.

laser beam welding (LBW) A welding process in which the heat of a laser beam melts metal.
soldadura con rayo laser Un proceso de soldar en el cual el calor de un rayo lazer derrite al metal.

laser sensor A sensing device mounted on a robot to search out the point for beginning the weld.
detector de laser Un dispositivo detector que se monta sobre un robot que le permite buscar el punto en que se debe comenzar una soldadura.

layer Overlapping weld beads on a metal surface.

capa Cordones de soldadura que se depositan el uno sobre la otra en una superficie metálica.

lead The cable that extends from the welding machine to the workpiece connection or the electrode holder.
cables de la soldadura Un cable que extienda de la máquina de soldadura al trabajo o a la mesa a soldar (cable de masa) o al portaelectrodo (cable de pinza).

lead A metallic element with a melting temperature of 620° F. Lead is a popular alloy for solder.
plomo Un elemento metálico cuyo temperatura de derretimiento es el 620° F (327°). El plomo es un metal de mezcla muy popular para la soldadura blanda.

left-handed threads Threads that require a fitting to be turned counterclockwise to tighten.
filete de paso a izquierda Los filetes que requieren un accesorio que debe dar la vuelta en sentido inverso para apretarse.

legs *See* fillet weld legs.
patas Vea patas de la soldadura de filete.

length The extent of a weld.
longitud La extensión de una soldadura.

limited thickness A qualification for welding steel up to a definite thickness.
espesor limitado Una calificación para el soldeo del acero de un espesor definido.

liner A device that carries the wire from the drive rolls to the contact tip.
forro Un dispositivo que lleva el alambre desde las ruedas de alimentación al tubo de contacto del electrodo.

liquidus The lowest temperature at which a metal is completely liquid.
liquidus La temperatura más baja en la que un metal se encuentra completamente en una forma líquida.

load A force, such as weight or movement of air in a wind.
carga Una fuerza, como el de un peso o el movimiento del aire de un viento.

longitudinal cracking A flaw that occurs in the weld face or the base metal parallel to the toes of the weld.

agrietura longitudinal Un defecto que ocurre en la cara de la soldadura o en el metal base que es paralelo a los pies de la soldadura.

low-alloy steel A steel in which chromium, manganese, or nickel content is less than 12%.

acero de bajo aleación Un acero en el cual la cantidad del cromo, manganeso, o níquel es menos del 12%.

low-carbon steel A steel that contains 0.1% (0.001) to 0.3% (0.003) carbon.

acero de bajo carbono Un acero que contiene de 0.1% (0.001) al 0.3% (0.003) del carbono.

low-hydrogen electrode An electrode used when underbead cracking as the result of hydrogen entrapment within the weld is a problem. These electrodes must be kept dry for quality welds.

electrodo de bajo hidrógeno Un electrodo que se puede usar para aliviar un problema de las grietas que se abren abajo de los cordones por consecuencia del hidrógeno que se atrapa durante el soldeo. Estos electrodos deben mantenerse secos para asegurar una soldadura de alta calidad.

machinability The tendency of a material to yield to mechanical shaping. Welding can affect the hardness of metal and thus its ability to be machined (shaped, cut, or pierced).

maquinabilidad La tendencia de una material de dejarse formar mecánicamente. La soldadura puede afectar la dureza del metal y así su abilidad a ser maquinado (sea formado, cortado, o perforado).

machining (M) One of five ways to finish a weld.

maquinado Una de las cinco maneras que se puede terminar una soldadura.

magnesium A silver-white metallic element that is one-third lighter than aluminum of equal dimensions but has the tensile strength of mild steel. It is commonly used in the aircraft industry.

magnesio Un elemento metálico de color blanco plateada que pesa una tercera parte menos del aluminio en dimensiones iguales, pero que tiene la resistencia al tensión comparable al del acero dulce. Se suele usar este materia en la industria aeronave.

magnetic particle test (MT) A nondestructive test used to locate surface defects.

prueba de partículas magnéticas Una prueba no destructivo que se usa para localizar a los defectos en la superficie.

malleable cast iron A type of cast iron produced by heat-treating white cast iron. Braze welding is a convenient method for repairing malleable cast iron.

hierro colado maleable Un tipo de hierro colado que se produce por medio de un tratamiento de calor del hierro colado blanco. La soldadura fuerte es un método apto para reparar al hierro colado maleable.

manganese A metallic element added to steel to increase its hardness.

manganeso Un elemento metálico que se añade al acero para aumentar su dureza.

manipulative Hand-to-eye coordination.

manipulativo La coordinación de mano a ojo.

manual A term used to describe welding, brazing, soldering, or cutting performed using hand-held equipment.

retención por la operadora Un término que describe la soldadura, la soldadura fuerte, la soldadura blanda y los cortes que se efectuan usando herramienta sujeto por la mano.

martensite A hard, pinlike grain structure that forms when steel cools too rapidly for the pearlite grain structure to form.

martensita Una estructura de grano dura y parecida a un alfiler que se forma cuando el acero se enfria demasiado rápido para formar la estructura de grano de perlita.

material The base metal subject to welding, brazing, or soldering.

material El material base en el cual se llevará acabo la soldadura, la soldadura fuerte, o la soldadura blanda.

material specification The information given to describe a material. Length, width, and weight per foot are three specifications.

especificaciones del material La información que describe al material. Longitud, espesor, y peso por pie son tres especificaciones.

medium-carbon steel   A steel that contains 0.3% (0.003) to 0.45% (0.0045) carbon.
acero de nivel mediana de carbono   Un acero que contiene del 0.3% (0.003) al 0.45% (0.0045) del carbono.

melt-thru   Uncontrolled overheating that results in an unexpected keyhole.
derritir de un lado al otro   Un calentamiento fuera de control que resulta en una ranura de llave inesperada.

melt-thru symbol   Instructions for complete root penetration from one side.
símbolo de un derrite de un lado al otro   Instrucciones para una penetración del raíz de un lado.

MIG   A common but inaccurate term for gas metal arc welding when an active shielding gas is used.
MIG   Un término comun si inexacto que se refiera a la soldadura de metal con arco protejido de gas cuando se utiliza un gas activo.

mild steel   *See* low-carbon steel.
acero dulce   Vea acero de bajo carbono.

mill scale   Oxides (blue-black in color) on the surface of steel that were formed during manufacturing and can affect weld quality.
escamas de laminado   Los óxidos (de color negro y azul) que se forman en la superficie en el proceso de la fabricación que pueden afectar la calidad de la soldadura.

molten pool   *See* weld pool.
charco derritido   Vea charco de la soldadura.

molybdenum   A metallic element added to steel to increase its hardness and tensile strength.
molibdeno   Un elemento metálico que se añade al acero para aumentar su dureza y resistencia al tensión.

motor generator   A welding machine that uses direct current as output for welding.
generador   Una máquina de la soldadura que utilice la corriente continua para la capacidad de la soldadura.

multipass   A term that describes welding with more than one weld bead.
soldadura de pasadas múltiples   Un término que describe la soldadura que consiste de más de un cordón.

NDT   *See* nondestructive testing.
NDT   Vea pruebas no destructivas.

neutral axis   An imaginary line in metal where there is no distortion.
fibra neutra   Una linea imaginaria en el metal en donde no hay distorción.

neutral flame   An oxyacetylene flame in which neither oxygen nor acetylene dominates.
llama neutra   Una llama de oxiacetileno en la cual no predomina el oxígeno ni el acetileno.

nick-break test   An impact test that uses visual inspection to uncover defects in a weld.
prueba de hendidura y rotura   Una prueba de impacto que utiliza a una inspección visual para descubrir a los defectos en una soldadura.

nickel   An alloying element that increases the hardness of steel. Nickel is often used in combination with chromium.
níquel   Un elemento de aleación que aumenta la dureza del acero. El níquel suele usarse en combinación con el cromo.

nitrogen   An active shielding gas used in some arc welding processes.
nitrógeno   Un gas activo que se usa en algunos procesos de soldadura de arco.

nodular cast iron   A high-tensile-strength type of cast iron in which the addition of magnesium causes graphite to form nodular (round) lumps in its microstructure. To weld, preheat and cool slowly.
hierro colado nodular   Un tipo de hierro colado de alta resistencia al tensión en el cual la agregación del manganeso causa una formación de cachos nodulares (redondos) de grafito en su estructura microscópica. Para soldar, hay que precalentarla y dejarla enfriar muy lentamente.

nonconsumable   A term used to describe an electrode that is not readily melted into the weld pool.
no consumible   Un término que se refiere a un electrodo que no se derrite fácilmente en el charco de la soldadura.

noncontacting sensor   A sensing device

that does not require physical contact with the weldment.

detector sin contacto Un dispositivo de detección que no requiere contacto físico con una soldadura.

noncritical pipe welding Welding on low-pressure pipe that may or may not require welder performance qualification.

soldadura de tubo no crítico La soldadura que se efectúa en la tubería de baja presión que puede ser que no require una calificación de ejecución del soldador.

nondestructive testing (NDT) A test of weld quality in which the joint is not destroyed.

prueba no destructiva Una prueba de la calidad de una soldadura en la cual no se destruye la junta.

nonferrous Containing no iron.

no-férreo Que no contiene hierro.

nozzle A device that fits on the end of a torch or gun to focus gas flow.

boquilla Un dispositivo que se ajusta en la extremidad de un soplete o una pistola para dirigir el flujo del gas.

operator's manual A pamphlet provided by the manufacturer that gives instructions for the care and operation of equipment.

manual del operador Un folleto proveido por el fabricante que da instrucciones para la operación y el cuidado de la maquinaria.

organic flux A flux made of not very corrosive acids and bases derived from living organisms. Organic flux burns easily if exposed to a flame.

revestimiento orgánico Un revestimiento hecho de ácidos no muy corrosivos y de bases derivado de organismos vivientes.

orifice A hole that has been machined into the tip used for oxyacetylene welding and oxyacetylene cutting.

orificio Un hoyo que se ha maquinado en la boquilla que se usa para las soldaduras de oxiacetileno y los cortes de oxiacetileno.

other side The side of the joint opposite to where the arrow is pointing.

lado opuesto El lado de la junta opuesto a donde apunta una flecha.

output Low-voltage, high-amperage electric power that flows into the welding arc from a welding machine or power source.

capacidad Fuerza eléctrica de bajo voltaje y alta amperaje que fluye al arco de la soldadura de una máquina de soldadura o de un fuente de poder.

overhead position A position in which the plane of welding is from 0° to 80°.

posición de sobrecabeza (entecho) Una posición cuyo plano de la soldadura es de 0° a 80°.

overheating When heat has a negative effect on a metal.

sobrecalentimiento Cuando el calor empeora al estado del metal.

overlap The amount that one weld bead extends over another.

traslape La cantidad que un cordón de la soldadura se extiende sobre otra.

oxidation The formation of oxides that can affect weld quality.

oxidación La formación de los oxidantes que pueden afectar a la calidad del soldeo.

oxide A compound of oxygen and iron that forms on steel and can affect the strength of the weld. Oxides can cause possible weld failure if not removed.

óxido Un compuesto del oxígeno y el hierro que forma en el acero y que puede afectar la fuerza de la soldadura.

oxidizing flame An oxyacetylene flame in which oxygen dominates acetylene.

llama oxidante Una llama de oxiacetileno en el la cual predomina el oxígeno.

oxyacetylene cutting (OFC-A) A cutting process that requires the use of oxygen and acetylene.

corte de oxiacetileno Un proceso de cortar que requiere el uso del oxígeno y acetileno.

oxyacetylene flame A flame used for welding, brazing, and cutting. The oxidizing flame approaches 6,300° F.

llama de oxiacetileno Una llama que se usa

para la soldadura, la soldadura fuerte y para cortar. La llama oxidante llega a la temperatura de 6,300°F (3482°C).

oxyacetylene welding (OAW) A welding process in which the necessary heat is supplied by an oxygen and acetylene flame.
soldadura de oxiacetileno Un proceso de la soldadura en el cual el calor necesario se provee por medio de una llama de oxígeno y acetileno.

oxygen An element used under pressure for torch welding, brazing, soldering, and cutting. Oxygen makes up 21% of the air and, when not controlled, can lower weld quality.
oxígeno Un elemento que se usa bajo presión para la soldadura, la soldadura fuerte, la soldadura blanda y los cortes con un soplete. El oxígeno compone el 21% del aire y, si no se controla, puede empeorar la calidad de la soldadura.

parameters Adjustments that affect the characteristics of the weld pool.
parámetros Los reglajes que pueden afectar las características de un charco de la soldadura.

pass (1) A length of weld bead. (2) Several weld beads overlapping each other.
pasada (1) Un longitud de un cordón de la soldadura. (2) Varios cordones que se traslapan.

pearlite The formation of ferrite and cementite (hard iron carbide) in the grain structure of steel and cast iron.
perlita La formación de ferrita y cementita (carburo de hierro duro) en la estructura del grano del acero y el hierro colado.

peening The use of force (applied with a ball peen) to relieve the stress generated in some welds.
martillazo El uso de la fuerza (aplicado por un martillo de bola) que puede aliviar las tensiones que se generan en algunas soldaduras.

performance test The application of welding under a given procedure to evaluate quality welding.
prueba de ejecución La aplicación del soldeo bajo procedimientos específicos para evaluar el soldeo de calidad.

phosphorus A nonmetallic element added as an alloy to increase the strength and corrosion resistance of some steels.
fósforo Un elemento no férreo que se agrega para aumentar la fuerza y la resistencia al corrosión de algunos aceros.

physical property A characteristic of a material, measured against mechanical force.
propriedad física La característica de un material, que se mide oponiéndose a una fuerza.

physical stress A condition caused by irritants like heat and smoke that can affect the human body and, consequently, weld quality.
tensión física Una condición causada por los irritantes como el calor y el humo que puede afectar al cuerpo humano y, por consequencia, a la soldadura.

pipe A structural, steam- and fluid-carrying medium measured from inside diameter up to and including 12-inch diameter. Pipe over 12 inches is measured by outside diameter.
tubo Un agente de estructura, vapor, y líquido que se mide por el interior hasta e incluyendo la medida de 12 pulgadas (30.48 cm). Los tubos mayores de 12 pulgadas (30.48 cm) de tamaño de miden desde el exterior.

pipe clamp A device used to hold pipe in position for welding.
abrazadera para tubo Un dispositivo que sujeta a un tubo en posición para soldar.

pitch The center-to-center spacing for intermittent fillet welds.
separación Los espacios que se dejan en la soldadura del filete intermitente que se miden del centro al centro de los punteados.

plasma arc cutting (PAC) An arc cutting process that uses ionized gas.
corte con arco de plasma Un proceso de cortar con arco que usa el gas ionizado.

plasma arc welding (PAW) An arc welding process in which a narrow but focused arc is created between a nonconsumable electrode and a base metal.
soldadura con arco de plasma Un proceso de la soldadura de arco en el cual un arco estrecho pero enfocado se crea entre un electrodo no consumible y un metal base.

plate   Metal that is more than 3/16 inch thick and usually over 6 inches wide.
placa   Un metal cuyo espesor mide más del 3/16 pulgada (.48 cm) y suele medir más de 6 pulgadas (15.24 cm) de anchura.

plug   A commercial or handmade device used to close off the ends of a pipe when gas shielding is used on the inside of a weld.
tapón   Un dispositivo comercial o hecho a mano para tapar las extremidades de un tubo cuando se emplea una atmósfera de gas en el interior del tubo.

plug weld   A weld in a circular hole of one piece joined to a second piece in a lap joint.
soldadura de tapón   Una soldadura de junta de solape que une a un hueco circular de una pieza a una segunda pieza.

poor root penetration   When the weld does not extend far enough into the base metal of the joint.
carencia de penetración de raíz   Cuando una soldadura no se extiende lo bastante dentro del metal base de una junta.

porosity   Pinholes in a weld as the result of gas pockets.
porosidad   Agujeritos en una soldadura que resultan por las cavidades de gas.

positioner   In robotic welding, a device that holds the weldment in place and moves in coordination with the movements of the robot.
montaje de posición   En la soldadura automática, un dispositivo que sostiene al conjunto de la soldadura y sincronisa sus movimientos con los de un robot.

postflow   The continual flow of gas after the arc has been stopped.
poscorriente   El flujo de gas que continua después de apagar el gas.

postheating   The application of heat to the base metal after soldering, brazing, or welding.
poscalentamiento   La aplicación del calor al metal base después de la soldadura, la soldadura fuerte o la soldadura blanda.

power adapter   A device that connects the power source and the shielding gas to the air-cooled torch used in gas tungsten arc welding.
adaptador de energía   Un aparato que conecta el fuente de energía y el gas de protección al soplete enfriado por aire que se emplea en la soldadura de arco con tungsteno de protección de gas.

power lead   A cable that connects the torch to electric current in gas metal arc welding and gas tungsten arc welding.
cable de energía   Un cable que conecta a la pistola con el corriente eléctrico en la soldadura de arco protejido por gas y en la soldadura de arco con tungsteno de protección de gas.

power source   See welding machine.
fuente de energía   Vea máquina de soldadura.

precoating   The first step of adding a thin layer of filler metal to a joint.
precubertura   El primer paso en añadir un metal de relleno de una capa.

preheating   The application of heat to the base metal before soldering, brazing, or welding.
precalentamiento   La aplicación del calor al metal base antes de la soldadura, la soldadura fuerte o la soldadura blanda.

prepositioning   Positioning pieces of the joint in anticipation of distortion.
preposicionar   Posicionar las piezas de una junta con la anticipación de que éstos se distorcionarán.

pressure regulator   See regulator.
regulador de presión   Vea regulador.

preweld   Before welding has begun.
presoldar   Antes de que el soldeo comienza.

procedure   See welding procedure.
procedimiento   Vea procedimiento de la soldadura.

procedure qualification   One of four AWS tests in which a welded joint of a specified base metal is tested.
calificación de procedimiento   Uno de cuatro de las pruebas de la AWS en la que una junta soldada de un metal base específico se prueba.

procedure specification   One of many instructions that must be followed when welding under a given procedure.
especificación de procedimiento   Una de muchas instrucciones a la que se debe prestar

atención al soldar bajo procedimientos específicos.

process A method recognized by the AWS used to join or cut metal.

proceso Un método empleado para unir o cortar el metal que es reconocido por la AWS.

projection weld A weld in which predetermined raised surfaces melt, resisting the flow of electric current.

soldadura de proyección Una soldadura en la cual se derriten las superficies elevadas predeterminadas, resistiendo el flujo del corriente eléctrico.

projection welding (PW) A resistance welding process in which the heat of welding is limited to the projections (raised surfaces).

soldadura de proyección Un proceso de soldadura de resistencia en el cual el calor del soldeo se limite a los proyecciones (las superficies elevadas).

propane A liquefied petroleum gas used with oxygen to produce a flame with a temperature of approximately 4,500° F.

propano Un gas petróleo en forma líquido que se usa con el oxígeno para producir una llama de una temperatura de aproximadamente 4,500°F (2482°C).

psi Pounds per square inch.

psi Libras por pulgada cuadrada.

pulsed gas metal arc welding (GMAW-P) An arc welding process that produces at least two current levels: one at the spray level and one at a level too low for spray to occur.

soldadura de metal con arco pulsado protegido de gas Un proceso de soldadura con arco que produce al menos dos niveles del corriente: uno en el nivel de atomización y uno de un nivel demasiado bajo para que ocurre la atomización.

pulsed gas tungsten arc welding (GTAW-P) A variation of gas tungsten arc welding in which high and low currents alternate within the arc.

soldadura de arco pulsado tungsteno protegido de gas Una variación de la soldadura de arco tungsten protegido de gas en la cual alternan corrientes altas y bajas en el arco.

pulsed spray *See* pulsed gas metal arc welding.

atomización pulsado Vea soldadura de metal con arco pulsado protegido de gas.

pure tungsten An electrode, color-coded green, with a melting temperature of 6,170° F.

tungsteno puro Un electrodo, código de color verde, cuyo temperatura de derritir es 6,170° F (3410°C).

quality control Inspection, testing, and qualification in manufacturing and maintenance.

control de calidad La inspección, examinación, y calificación en la fabricación y el mantenimiento.

quality welding/welds Welding that meets a credible standard of excellence.

soldaduras de calidad El soldeo que cumple con los modelos de la excelencia.

quenching The rapid cooling of metal.

amortiguamiento El enfriamiento rápido del metal.

radiographic test (RT) A nondestructive test that uses X rays to locate weld defects.

prueba radiográfica Una prueba no destructiva que usa los rayos equis para localizar las soldaduras defectivos.

real time As an event happens.

tiempo actual Mientras que ocurre un evento.

rectifier A device that converts alternating current to direct current.

rectificador Un dispositivo que convierte al corriente alterna al corriente continua.

reference line The part of a welding symbol that separates the arrow side from the other side.

lina de referencia La parte de un símbolo del soldeo que separa el lado con la flecha del lado contrario.

regulator An apparatus used to control the release of gas held under pressure.

regulador Un aparato que se usa para controlar a la evacuación del gas bajo presión.

reinforcement A weld with filler metal added beyond the flush contour.

reinfuerzo Una soldadura con material de relleno que se ha añadido más allá del perfíl nivel.

remote-panel switch A current adjustment on a torch, gun, or foot control that is separate from the control panel of the welding machine.

interruptor del cuadro remoto Un reglaje del corriente que se ubica en el soplete, la pistola, o un control a pie que es una pieza distincta del cuadro del control de la máquina de la soldadura.

resistance seam welding (RSEW) A resistance welding process in which a series of overlapping spot welds or continuous spot welds are made.

soldadura de costura de resistencia Un proceso de la soldadura de resistencia en el cual se efectuan una serie de puntos aislados traslapados o puntos aislado continuos.

resistance spot welding (RSW) A resistance welding process in which the welding area of a lap joint has to be confined between two electrodes.

soldadura de puntos aislados de resistencia Un proceso de la soldadura de resistencia en el cual una area de una junta de solape tiene que ser restrigido entre dos electrodos.

resistance welding (RW) A group of welding processes in which welding is caused by the resistance of the base metal to the flow of electric current generating heat.

soldadura de resistencia Un grupo de procesos de soldar en los cuales el soldea se causa por la resistencia del metal base al flujo eléctrico de un corriente la cual crea el calor.

reverse polarity *See* direct current electrode positive.

polaridad inversa Vea corriente continua electrodo positivo.

robot A machine that carries out the commands set by a computer program.

robot Una máquina que lleva acabo los ordenes delineados por una programa de computadores.

robot controller The communication center of robotics that runs programming.

controlador de robot El centro de comunicaciones de los procesos automáticos que maneja la programación.

robotic welding Automatic equipment that can be programmed to do welding.

soldeo automático Las máquinas que pueden ser programados para soldar.

Rockwell C hardness test A test that determines hardness by measuring the depth of penetration. The Rockwell C hardness test provides a standard of comparison for the hardness of different materials.

prueba de dureza Rockwell C Una prueba que determina la dureza por medio de una medida del profundidad de la penetración. La prueba de dureza Rockwell C provee un modelo para comparar la dureza de varios materiales.

rod *See* welding rod.
varilla Vea varilla del soldeo.

rolling (R) One of five ways to finish a weld.
laminado Una de las cinco maneras en que se puede acabar una soldadura.

root bead A weld that reaches into the joint root partially or up to 100%.
cordón del raíz Una soldadura que penetra la junta del raíz parcialmente o hasta el 100%.

root bend Stretching the root of a coupon to test the weld.
pliegue del raíz Estirar al raíz de una probeta para examinar a la soldadura.

root face An edge that has not been removed by a bevel, establishing the boundary of the root opening.
raíz de la cara Un borde que no se ha achaflanado, que establace el limite de la apertura del raíz.

root opening *See* joint root opening.
apertura del raíz Vea apertura de la junta de raíz.

root penetration The depth of penetration by the weld into the joint.
penetración del raíz La profundidad de la penetración de una soldadura en una junta.

rosin flux A nearly noncorrosive flux used in the electronics industry. Brasses and coppers are the metals most commonly soldered with rosin fluxes.

revestimiento de resina Un revestimiento casi no corrosivo que se usa en la industria electrónica. Se suele usar el revestimiento de resina en la soldadura blanda del cobre y del latón.

rust The usually corrosive iron oxides that form on iron and steel.

orín Los óxidos que ordinariamente son corrosivos que forman en el hierro y el acero.

safety glasses A critical piece of safety equipment with side shields to protect the eyes from damage relating to welding.

gafas de seguridad Un accesorio crítico de seguridad que tiene escudos en ambos lados para proteger a los ojos de los daños que se asocian con la soldadura.

scarf-groove weld A weld in which edge preparation is formed when one of the two bevels has been turned over in a butt joint.

soldadura de escharpe y ranura Una soldadura en la cual la preparación del borde se forma cuando uno de dos biseles se ha volteado en una junta de tope.

schedule The number assigned to pipe as an indication of the wall thickness.

tabla de espesor El número designado a la tubería para indicar el espesor del pared.

seam ultrasonic welding A solid-state welding process in which the welds result from vibration and pressure between a roller and an anvil.

soldadura de costura ultrasonico Un proceso de la soldadura de estado sólido en el cual las soldaduras resultan de la vibración y la presión formado entre una rueda y un yunque.

seam weld A series of spot welds or continuous welds between two overlapping pieces, forming a joint.

soldadura de costura Una serie de soldeos punteados o soldeos continuos entre dos piezas traslapados, formando una junta.

secure Restricts the movement of gas cylinders so they cannot fall.

seguro Restringe los movimientos de los cilindros de gas para que no se pueden caer.

self-shielding A term used to describe flux cored arc welding without a shielding gas.

autoprotectivo Un término que se refiere a la soldadura de arco con un electrodo con núcleo de revestimiento sin gas de protección.

semiautomatic A term used to describe welding, brazing, soldering, or cutting performed by equipment that is automatically controlled in some way. For example, gas metal arc welding is semiautomatic to the extent that wire is fed automatically into the weld pool.

semiautomático Un término que se refiere a la soldadura, la soldadura fuerte, la soldadura blanda, o los cortes que se efectuan por medio de un tipo de control automático. Por ejemplo, la soldadura de metal con arco protejido de gas es semiautomática en que el alambre proyecta automáticamente al charco de la soldadura.

shear force The force attempting to slide the pieces of a joint past each other. Whereas tensile is the force pulling a joint apart, shear is the opposite force, pushing against the joint from opposite directions.

fuerza cizallada La fuerza que intenta estropiar a una junta deslizando a una pieza alrededor de la otra. Mientras que la elasticidad es la fueza que separa a una junta, la fuerza cizallada funciona de manera opuesta, aplicando a la fuerza de direcciones opuestas.

sheet metal Metal that is 3/16 inch or less in thickness and over 6 inches wide.

lámina de metal El metal cuyo espesor es de 3/16 pulgada (.48 cm) o menos y mide más que 6 pulgadas (15.24 cm) de ancho.

shielded metal arc cutting (SMAC) An arc cutting process with electrodes ordinarily used for shielded metal arc welding.

corte de metal con arco protejido Un proceso de corte con arco que usa a un electrodo que suele usarse para la soldadura de metal con arco protegido por gas.

shielded metal arc welding (SMAW) An arc welding process that uses an electric arc, usually between a flux-covered consumable electrode and the weld pool.

soldadura de metal con arco protejido Un proceso de soldadura con arco que emplea a un arco eléctrico, comunmente entre un electrodo consumible revestido y el charco de la soldadura.

shielding gas A gas used to protect the weld pool from contaminants in the air.

gas protector   Un gas que se usa para proteger al charco de la soldadura de los contaminantes del aire.

short   To arc to remove the accumulation of filler metal on the end of a tungsten electrode.

rayado   El cebar un arco para quitar al acumulación del metal de aporte de la extremidad del electrodo de tungsteno.

short circuiting arc transfer (GMAW-S)   A gas metal arc welding process in which a solid wire is extinguished by the weld pool only to reignite and begin again, up to 200 times per second.

transferir por corto circuito   Un proceso de la soldadura de metal con arco protegido por gas en el cual un alambre sólido se extingue en el charco de la soldadura y luego se enciende de nuevo, un evento que ocurre hasta 200 veces por segundo.

side bend   Stretching the side of a coupon to test the weld.

pliegue de lado   El estirar a una probeta para examinar a la soldadura.

silicon   A nonmetallic element that can increase the hardness of steel and improve its mechanical properties.

silicio   Un elemento no metálico que puede aumentar la dureza del acero y mejorar sus propiedades mecánicas.

silver alloy   A popular filler rod used for torch brazing ferrous and nonferrous metals, except aluminum and magnesium.

aleación con plata   Una varilla de relleno muy popular que se usa para la soldadura blanda con soplete de metales férricos o no férricos, con la excepción del aluminio y el magnesio.

single-phase power   Input to the welding machine from one alternating voltage.

energía monofásico   La capacidad a una máquina de soldar de un sólo voltaje alterna.

single-stage regulator   A device that reduces pressure in one step.

regulador monoetápico   Un dispositivo que reduce a la presión por una etapa.

size   A dimension given on a welding symbol for any of several weld symbols.

tamaño   Una dimención que se expresa por medio de un símbolo de la soldadura entre los varios símbolos de soldar.

slag   Oxidized metal that can affect weld quality.

escoria   El metal oxidado que puede afectar la calidad de la soldadura.

slag inclusion   Nonmetallic material trapped within the weld.

inclusión de escoria   Material no metálica que queda preso en la soldadura.

slope   A fine adjustment on some gas metal arc welding machines that can reduce spatter.

aparejo de reglaje   Un reglaje fino de algunas máquinas de la soldadura de metal con arco protegido por gas que puede reducir la cantidad de salpicadura.

slot weld   A weld that is an elongated plug weld.

soldadura de ranura elongada   Una soldadura que parece una soldadura de un tapón alargado.

Society of Automotive Engineers (SAE)   An organization that classifies types of steel.

Sociedad de Ingenieros Automovilistas   Una organización que clasifica a los tipos de acero.

soldering (S)   A group of processes in which the filler metal is heated to liquidus below 840° F and below the melting temperature of the base metal.

soldadura blanda   Un grupo de procesos en los cuales un metal del relleno se calienta al punto liquidus el cual es menos del 840°F (449°C) y menos del temperatura de derrite del metal base.

soldering gun   A tool through which electricity furnishes the heat.

pistola de soldar   Una herramienta en la cual el calor se produce por la electricidad.

solid ring   A type of backing ring used in pipe welding for exact fitting.

anillo sólido   Un tipo de anillo de respalde que se usa para soldar la tubería que no queda exactamente.

solid-state welding (SSW)   A group of joining processes in which welding takes place by the use of pressure below the melting temperature of the base metal.

soldadura de estado sólido   Un grupo de procesos de juntamiento en los cuales el soldeo se efectúa por medio de la presión en temperaturas más bajas del derretimiento del metal base.

spacer material   Material used to set a space or separation in the joint of a groove weld.

material para separar   Un material que se usa para designar un espacio o una separación en la junta de una soldadura de ranura.

spatter   Molten metal thrown out of the weld pool.

salpicadura   El metal derritido que salte del charco de la soldadura.

split ring   A type of backing ring used in pipe welding for not-so-exact fitting.

anillo partido   Un tipo de anillo de respalde que se usa para soldar la tubería que no queda exactamente.

spool gun   A gun that carries wire (up to 2-pound spools) that is pulled by drive rolls through the gun.

pistola de bobina   Una pistola que lleva adentro el alambre (en bobinas de hasta 2 libras/ .9 k/) que se ensarte por la pistola por medio de las ruedas de alimentación.

spot ultrasonic welding   A solid-state welding process in which welds result from the vibration and pressure between an electrode tip and an anvil.

soldadura de puntos aislados ultrasonico   Un proceso de la soldadura de estado sólido en el cual las soldadura resulten del la vibración y la presión entre el punto del electrodo y un yunque.

spot weld   A weld made on or between two overlapping pieces of metal.

soldadura de puntos aislados   Una soldadura que se efectúa sobre o entre dos piezas de metal traslapados.

spray arc transfer (GMAW-SP)   A gas metal arc welding process in which the molten metal drops change to spray. A popular mixture to initiate spray drop transfer is 98% argon/2% oxygen.

transferencia de arco de rocío   Un proceso de la soldadura de metal con arco protegido por gas en el cual las gotas del metal derritido se convierten en un rocío. Una mezcla común para iniciar el transferencia de gota rociado es el 98% de argón y el 2% de oxígeno.

spray transfer   *See* spray arc transfer.

transferencia de rocío   Vea transferencia de arco de rocío.

square-groove weld   A weld that uses a butt joint without edge preparation.

soldadura de ranura escuadra   Una soldadura que use una junta de tope sin preparación.

staggered   A type of intermittent fillet weld.

en cadena   Un tipo de soldadura de filete intermitente.

stainless steel   A high-alloy steel with at least 12% chromium. Stainless steel is well suited for corrosive applications.

acero inoxidable   Un acero de alta aleación que contiene al menos el 12% de cromo. El acero inoxidable se recomienda para aplicaciones corrosivas.

steel   *See* low-, medium-, high-, and very high carbon steel.

acero   Vea acero de bajo, mediano, alto y muy alto nivel de carbono.

stickout   The length of the wire, measured from the end of the contact tube.

proyección   La longitud de un alambre que se mide desde la extremidad del punto de contacto.

straight polarity   *See* direct current electrode negative.

polaridad directa   Vea corriente continuea electrodo negativo.

strain   The deformation of a material due to stress. Strain can lead to weld failure.

alargamiento   La deformación de una material debido a la tensión. El alargamiento puede causar que las soldaduras fallen.

strength   The ability to resist strain or deformation.

fuerza   La abilidad de resistir al alargamiento o a la deformación.

stress   Forces acting on a material.

esfuerzo   Las fuerzas que actuan sobre una material.

striking   Beginning the arc for welding.
cebando (rayando) un arco   Comenzar un arco para soldar.

stringer   A weld bead deposited without weaving.
encordador (nervadura)   Un cordón depositado sin balanceo.

strip   Metal that is 3/16 inch or less in thickness not exceeding 12 inches in width.
banda   El metal que tiene el 3/16 de pulgada (.48cm) o menos de espesor y que no mide más de 12 pulgadas (30.48 cm) de ancho.

strong back   Metal welded to the back side of a test joint to prevent the weld from bending the joint over.
refuerzo   El metal que se suelda al lado revés de una junta de prueba para prevenir que se dobla la junta con la soldadura.

stubbing   Pushing the gun backward when the wire is driven through the weld pool.
troncar   Empujando a la pistola en sentido revés cuando el alambre es conducido por el charco de la soldadura.

stud arc welding (SW)   An arc welding process used for fusing studs (threaded and unthreaded) to a base metal.
espárrago (tachón) para soldadura de arco   Un proceso de la soldadura con arco que se usa para fundir las espigas (fileteado o sín filete) a un metal base.

submerged arc welding (SAW)   An arc welding process in which a continuous solid wire is fed into the weld pool and both the arc and the weld pool are protected by a granular flux.
soldadura de arco sumerjido   Un proceso de la soldadura con arco en el cual un alambre sólida se alimenta continuamente al charco de la soldadura y ambos el arco y el charco se protejan por medio de un flujo granular.

surfacing weld   A weld applied to build up the surface of base metal that is subject to wear.
soldadura de la superficie (cara)   Una soldadura que se aplica para elevar a una superficie del metal base que se sujeta a mucho desgaste.

sweat soldering   A soldering method by which precoating the joint with solder is followed by reheating the joint without adding more solder.

soldadura blanda por exudar   Un método de la soldadura blanda en el cual la junta recibe una capa del material de aporte antes de soldar y luego se recalienta la junta sín añadir más material de aporte.

tack weld   A short, temporary weld used to hold a weldment in position to complete welding.
soldadura de puntos   Una soldadura pequeña y temporario que se usa para sujetar las piezas para completar la soldadura.

tack welder qualification   A performance test under a given procedure for those who make tack welds for the welders.
calificación para soldador de puntos aislados   Una prueba de ejecución bajo procedimientos específicas para los que efectuan los puntos soldados para los soldadores.

tail   The part of the welding symbol that contains information for welding that cannot be provided another way.
cola   La parte del símbolo de la soldadura que contiene la información para soldar que no se puede proveer de otra manera.

tap method   A technique to initiate welding by striking the end of an electrode on the base metal, raising it quickly.
metodo del golpecito   Una técnica para comenzar el soldeo que consiste de golpear el extremo de un electrodo en el metal base, y levantarlo rápidamente.

teach pendant   A remote-control panel for programming robots.
instrumento para enseñar   Un cuadro de control remoto para programar a los robots.

tee joint   A joint formed with two or three pieces that are usually at right angles to one another.
junta en T   Una junta que se forme con dos o tres piezas que suelen posicionarse en ángulos rectos.

tempering   A heat treatment by which hardened steel is toughened.
templar   Un tratamiento del calor por el cual el acero se endurece.

tensile strength   The resistance to being pulled apart.

resistencia al tensión La resistencia de separarse.

tensile test A test used to measure the strength of a material subjected to a force trying to pull the material apart.
prueba de tensión Una prueba que se usa a medir la fuerza de una material que se expone a una fuerza que intenta a separar a los materiales.

theoretical throat The distance from the joint root to the hypotenuse of the largest triangle that would fit inside a fillet weld.
garganta teórico La distancia del raíz de la junta a la hipotenusa del triángulo más grande que quedaría adentro de la soldadura de filete.

thoriated A tungsten electrode that contains small amounts of thoria and is well suited for welding DCEN. One percent thoriated electrodes are color-coded yellow; 2%, red.
toriado Un electrodo de tungsteno que contiene cantidades pequeñas del torio y que se recomienda para la soldadura DCEN (corriente continua electrodo negativo). Los electrodos toriados de 1% reciben el color de código de amarillo; los de 2%, el rojo.

three-phase power Input to the welding machine from a combination of three circuits. Three-phase power is most cost-effective for industrial purposes.
energía trifásico La capacidad para una máquina de la soldadura que viene de una combinacón de tres circuitos. La energía trifásico es la más económica para los usos industriales.

through-the-arc sensor An oscillating welding gun or torch plus voltage- and amperage-sensing variables used to keep the welding on track within the joint.
detector por entre el arco Un soplete o pistola de soldadura oscilante que tiene tambien unos detectores variables de voltaje y amperaje que se usa para mantener a la soldadura en la posición correcta para la junta.

TIG Tungsten inert gas. A common but inaccurate term for gas tungsten arc welding when an active shielding gas is used.
TIG Tungsteno con gas inerte. Un término común si inexacta de la soldadura de arco con tungsteno protejido por gas cuando se usa una atmósfera protectiva de gas.

tin A metallic element with a melting temperature of 450° F. Tin is a popular alloy in solders.
estaño Un elemento metálico cuya temperatura de derretimiento es el 450°F (232°C). El estaño es una aleación común en el metal de aporte para la soldadura blanda.

tin-lead A most widely used solder for joining metals.
plomo-estaño El metal de aporte para la soldadura blanda que más se usa en unir los metales.

tip A part of the torch that concentrates heat for welding, brazing, cutting, and soldering. *See also* contact tube.
boquilla Una parte del soplete que enfoca al calor para la soldadura, la soldadura fuerte, los cortes y la soldadura blanda. Vea tambien tubo de contacto del electrodo.

titanium An alloy combined with other metals that has corrosion resistance and strength with uses in the aerospace industry.
titanio Una aleación que se combina con otros metales cuyos propriedades incluyen una alta resistencia a la corrosión y la fuerza y que se emplea en la industria aeronautica.

toe *See* weld toe.
pie Vea soldadura de pie.

tool steel A medium- to high-carbon steel used in making the hardest tools.
acero para herramientas Un acero de nivel mediana a nivel alta de carbono que se emplea en fabricar a las herramientas duras.

torch A device that transfers electricity or gases for welding, brazing, soldering, or cutting.
soplete Un dispositivo que traslada a la electricidad o a los gases para la soldadura, la soldadura fuerte, la soldadura blanda o los cortes.

torch brazing (TB) A process that uses the heat of a flame to join metal.
soldadura fuerte con soplete Un proceso que usa el calor de una llama par unir el metal.

torch soldering (TS) A process that uses the heat from a flame to join metal.
soldadura blanda con soplete Un proceso que usa el calor de una llama para unir al metal.

toughness   The resistance to fracture from a constant force.
tenacidad   La resistencia a la grietura debido a una fuerza continua.

transformer   A welding machine that uses alternating current as the output for welding.
transformador   Una máquina de la soldadura que usa el corriente alterna para la capacidad para soldar.

transverse cracking   A defect that occurs in the weld from toe to toe.
agrietura transversal   Un defecto que ocurre en la soldadura de pie a pie.

travel angle   The angle of the rod, electrode, gun, or torch in relation to the direction of travel along the joint.
ángulo de avance   El ángulo de la varilla, el electrodo, el soplete o la pistola en relación a la dirección de avance en la junta.

tubing   A structural shape measured from outside diameter.
tubería   Una forma estructural que se mide por su diámetro de afuera.

tubular wire   A filler metal that has flux inside the wire.
alambre tubular   Un metal de aporte o relleno que contiene el flujo adentro del alambre.

tungsten   A chemical element with the highest melting point of any known metal, 6,170° F, and a boiling point temperature of 10,700° F.
tungsteno   Un elemento químico cuyo temperatura de derritir es el más alto de todos los metales conocidos, el 6,170°F (3,410°C) y cuyo temperatura de hervir es el 10,700°F (5,970°C).

tungsten electrode   The nonconsumable end of the electrical circuit, creating a stable arc for welding, brazing, or cutting.
electrodo de tungsteno   La extremidad no consumible del circuito eléctrico, que crea un arco estable para la soldadura, la soldadura fuerte o los cortes.

two-stage regulator   A device that reduces pressure in two steps.
regulador bifásico   Un dispositivo que reduce a la presión por dos etapas.

type   A category of metal; part of the classification system used to separate one metal from another. For example, 1100, 6061, and 6063 are three different types of aluminum.
tipo   Una categoría de los metales; una parte del sistema de clasificación que se usa para distinguir un metal de otro. Por ejemplo, el 1100, el 6061, y el 6063 son tres tipos distinctos del aluminio.

U-groove weld   A weld in which both pieces of the joint have curved surfaces.
soldadura de ranura con U   Una soldadura en la cual ambos piezas de la junta tienen superficies curvadas.

ultimate tensile strength   A measurement of maximum load acting on a material without failure.
carga límite de tenacidad   Una medida de la carga máxima cuyos efectos en la material no causan una falta.

ultrasonic test (UT)   A nondestructive test that uses high-frequency sound waves to locate defects in welds.
prueba ultrasónico   Una prueba no destructiva que usa las ondas de sonido ultrasónicos para localizar los defectos en la soldadura.

ultrasonic welding (USW)   A solid-state welding process in which pieces of metal are joined together by high-frequency vibrations under pressure.
soldadura ultrasónico   Un proceso de la soldadura de estado sólido en el cual las piezas del metal se unen por medio de vibraciones de alta frequencia bajo presión.

ultraviolet rays   Invisible light that can damage the skin and is especially harmful to the eyes.
rayos ultravioletas   El luz invisible que puede dañar al piel y que es dañoso particularmente a los ojos.

underbead cracking   A defect that can occur in high-carbon and alloy steels in the heat-affected zone.
agrietura bajo el cordón   Un defecto que puede ocurrir en los aceros aleados de alto nivel de carbono que se encuentran en la zona afectada por el calor.

undercut   Underfill along the toes of the weld.
mordedura (sovocación)   Una falta de relleno en los pies de una soldadura.

uphill Upward welding or brazing against gravity.

soldando hacia arriba Efectuando una soldadura o la soldadura fuerte hacia arriba, contra la gravedad.

upset Metal that overflows the joint along the weld.

recalada El metal que se vierte de la junta en una soldadura.

upset welding (UW) A resistance welding process in which heat has been generated along the entire length of a butt joint brought together by pressure but without a flash.

soldeo de recalada Un proceso de la soldadura de resistencia en el cual el calor se ha generado en la totalidad del longitud de una junta de tope que se une por medio de la presión pero sin relampagueo.

valve An apparatus that controls the release of gas from a cylinder used in welding, brazing, soldering, or cutting.

llave Un aparato que contrala el escape de gas del cilindro que se usa en la soldadura, la soldadura fuerte, la soldadura blanda y los cortes.

vanadium A metallic element used as an alloy to increase hardness.

vanadio Un elemento metálico que se usa en la aleación para incrementar la dureza.

vertical position A position in which the plane of welding is from 15° to 100° with 10° of pivot.

posición vertical Una posición en la cual el plano de la soldadura es el 15° hasta los 100° con 10° grados de giro.

very high carbon steel A steel that contains 0.65% (0.0065) to 1.5% (0.015) carbon.

acero de nivel muy alto de carbono Un acero que contiene el 0.65% (0.0065) a 1.5% (0.01) de carbono.

V-groove weld A weld in which the edges of both pieces forming the joint are beveled.

soldadura de ranura en V Una soldadura en la cual los bordes de las dos piezas formando la junta son achaflanados.

Vickers hardness test A hardness test with its own scale for hardness.

prueba de dureza Vickers Una prueba de dureza con su propio escala de la dureza.

visual inspection An examination of weld quality using the eyes.

inspeccion visual Un examen de la calidad de la soldadura usando los ojos.

volatile Capable of exploding.

volátil Capaz de explotar.

voltage A measurement of the force causing the flow of electricity.

voltaje Una medida de la fuerza que causa el flujo de la electricidad.

water-circulating unit A device designed to cool the torch or gun during high-amperage welding.

unidad de circulación de agua Un dispositivo diseñado para enfriar a una soplete o una pistola durante el soldeo de alto voltaje.

water coupling A device that connects the water hose of the torch to the hose of the water-circulating unit.

acoplamiento de agua Un dispositivo que conecta la manguera de agua del soplete a la manguera de la unidad de circulación de agua.

weaving Any one of several patterns of movement from side to side during welding.

tisaje Cualquier de muchas modelos de movimiento de un lado al otro durante la soldadura.

weld The melting together of the joint with or without filler material.

soldadura La fusión de una junta con o sin un metal de relleno.

weldability The degree of welding difficulty.

soldabilidad El grado de dificultad de soldar.

weld all around A circle where the arrow meets the reference line of the welding symbol.

soldar alrededor Un círculo en donde la flecha toca la linea de referencia del símbolo de la soldadura.

weld bead A weld from a single pass.

cordón de soldadura Una soldadura de una sóla pasada.

welder   A person with the skill and knowledge to make quality welds.

soldador   Una persona con la destreza y la sabiduría para efectuar las soldadura de calidad.

welder qualification   A performance test under a given procedure.

calificación de ejecución del soldador   Una prueba de ejecución que se efectúa con procedimientos específicos.

weld face   The surface of the weld, extending from the weld toe on one side to the weld toe on the other side.

cara de la soldadura   La superficie de la soldadura, que se extiende del pie de la soldadura en un lado al pie de la soldadura del otro lado.

welding   Any process that joins materials together by heat.

soldadura   Cualquier proceso que une a los materiales por medio del calor.

welding arc   A controlled short circuit by which the electric current causes intense heat in a weld pool between an electrode and a base metal.

arco de soldadura   Un corte circuito controlado por el cual el corriente eléctrico cause un calor intenso en un charco de soldadura entre el electrodo y el metal base.

welding engineer   A person who is responsible for deciding on the best possible way to manufacture a product requiring welding. The decision might include selection of the base metal, welding process, joint design, and filler metal.

ingeniero de la soldadura   Una persona cuyo responsabilidad es el decidir cómo mejor fabricar un producto que requiere la soldadura. La decisión puede incluir la selección del metal base, el proceso de la soldadura, el diseño de la junta, y el metal de relleno.

welding gloves   Gloves designed specifically for welding—not ordinary work gloves.

guantes para soldar   Los guantes que se han diseñado específicamente para el soldeo—no son guantes de trabajo ordinarios.

welding inspector   An examiner of welds who has passed tests sponsored by the American Welding Society.

inspector de la soldadura   Un examinador de la soldadura quien ha aprobado los examenes bajo los auspicios de la Sociedad de la Soldadura Americana.

welding machine   A piece of equipment designed for welding, cutting, or both.

máquina de soldar   Una pieza de equipo que se ha diseñado para la soldadura, los cortes, o ambos.

welding operator   The person who operates the controls of automatic or robotic welding equipment.

operador para la soldadura   Una persona quien opera los mandos de los equipos automatizados, o de los robots.

welding operator qualification   A performance test under a given procedure for those who operate robotic welding equipment.

calificación de ejecución del operador de la soldadura   Una prueba de ejecución que se administra bajo procedimientos específicos para los que operan los equipos de soldadura de robot.

welding procedure   Detailed instructions that must be followed to achieve quality welding using a given process.

procedimiento de la soldadura   Instrucciones detalladas que deben seguirse para efectuar la soldadura de alta calidad usando un proceso específico.

welding rod   Usually, a bare wire filler metal in standard lengths of 36 inches, with diameters from 1/16 to 3/8 inch. Welding rods are used in welding or brazing, not for the purpose of conducting electricity.

varilla de soldar   Suele ser un alambre sin revestimiento de metal de aporte que tiene una medida normalizada de 36 pulgadas (91.44 cm) con un diámetro del 1/16 a 3/8 de una pulgada (.16-.95 cm). Las varillas de soldar que se usan en la soldadura o la soldadura fuerte no se usan para conducir la electricidad.

welding symbol   The foundation of welding instructions, comprising the arrow, the reference line, and the tail.

símbolo de la soldadura   Los fundamentos de los instrucciones para soldar, incluyendo a la flecha, la linea de referencia, y la cola.

welding tip A torch tip designed especially for oxyacetylene welding.
boquilla para soldar Una boquilla del soplete que se ha diseñado especialmente pra la soldadura de oxiacetileno.

weld interface The point at which the base metal meets the filler metal.
interface de la soldadura El punto en el cual el metal base toca el metal de relleno.

weldment An assembly of welded parts.
conjunto de partes soldadas Una asamblea de partes soldadas.

weldor A nonstandard term. *See* welder.
soldador Un término fuera del normal. Vea welder.

weld pool Liquid metal formed at the point of welding.
charco (aguas) de la soldadura El metal en forma de líquido que aparece en el punto de soldar.

weld root The point at which the weld penetrates into the joint root; the farthest point from the weld face.
raíz de la soldadura El punto en el cual la soldadura penetra en la junta del raíz; el punto más lejos de la cara de la soldadura.

weld symbol A symbol for one of 19 different welds.
símbolo de la soldadura Un símbolo para uno de los 19 tipos distinctos de soldaduras.

weld toe Runs the length of the weld to establish the boundary of the weld face with the base metal.
pie de la soldadura Corre la longitud completa de la soldadura para establecer el límite entre la cara de la soldadura y el metal base.

wetness *See* weld pool.
fusión Vea charco de la soldadura.

wetting The technique of applying heat to a solder to dissolve the surface of the base metal, forming a compound between the pieces of the joint.
fundir Una manera de aplicar el calor a una soldadura para disolver la superficie del metal base, formando un compuesto entre las piezas de una junta.

whipping Quickly moving the electrode out of the weld pool and back in again.
latigar Moviendo el electrodo rápidamente, sacándolo del charco de la soldadura y metiéndolo de nuevo.

white cast iron A type of cast iron that is not readily welded.
hierro colado blanco Un tipo de hierro colado que no se suelda fácilmente.

wire The electrode or filler metal fed into the weld pool.
alambre El electrodo o metal de relleno que se alimenta al charco de la soldadura.

wire feed Parameters, such as amperage, that affect the wire and, consequently, the weld itself.
alimentación del alambre Los parámetros, como el amperaje, que afectan al alambre, y, por consequencia, a la soldadura misma.

work angle The angle of the rod, electrode, gun, or torch in relation to the base metal.
ángulo de trabajo El ángulo de la varilla, el electrodo, la pistola, o el soplete en su relación con el trabajo.

work envelope The area in which a robot operates.
alcance de de operación El area en el cual trabaja un robot.

workpiece *See* base metal or weldment.
pieza de trabajo Vea metal base o conjunto de partes soldadas.

workpiece connection Connects the base metal to the power source. Has replaced nonstandard term *ground.*
conexion de pieza de trabajo Conecta el metal base al fuente de la energía. Es un término que ha reemplazado al término no normal de tierra (ground).

workpiece lead A cable connecting workpiece connection to the power source.
cable de pieza de trabajo Un cable que conecta a la pieza de trabajo al fuente de energía.

X-plane The movement of a robot from side to side.

plano X   El movimiento de lado a lado de un robot.

yield point   A measurement of when a material first continues to stretch more than the load being applied. Wind and weight are examples of load.

punto de ceder   Una medida del punto en que un material continua a estirase mas allá de la carga que se le impone. El viento y el peso son ejemplos de carga.

Y-plane   The movement of a robot back and forth.

plano Y   El movimiento adelante y atrás de un robot.

Y-valve   A device used to divide the gas flow coming from the regulator or regulator/flowmeter.

válvula Y   Un dispositivo que se usa a dividir el flujo del gas que viene del regulador o el regulador/manómetro de flujo.

Z-plane   The movement of a robot up and down.

plano Z   El movimiento de arriba y abajo de un robot.

zinc   A silver-white metallic element used as a protective coating for galvanizing metal. When combined with copper it forms brass.

zinc   Un elemento metálico de color blanco-plateado que se usa como capa protectiva del metal galvanisado. Cuando se combina con el cobre, forma el bronce.

zirconia   A tungsten electrode color-coded brown containing zirconium oxide. The zirconia electrode is well suited for welding by direct current electrode negative.

zirconía   Un electrodo de tungsteno de color código cafe que contiene el óxido de zirconio. El electrodo zirconía se recomienda para la soldadura con corriente continua electrodo negativo.

Index

seam, 236
  symbol for, 259, **263**
slot, 235
  symbol for, 256, **260**, 261
spot, 235–36
  symbol for, 256–58, **262**
strength of, calculating, 249–50
surface, 237
  symbol for, 258, **264**
tack, 243, **244**
Whipping, 108
White cast iron, 14
  braze welding and, 210
Wire
  AWS classification, **129**
  feed system, 126
Work angle, 176

Work envelope, robotic welding and, 295, **296**
Wrench, adjustable, 61

X
X-ray, 277

Y
Y-fitting, 200
Y-valve, pipe welding and, 310
Yield point, defined, 280

Z
Zinc
  brazing and, 213–14
  metals and, 7–8
Zirconia alloy tungsten electrode, 171